Contents

Preface: Mathematics means everything to me... 7

Wow Factor Mathematical Index Explained 11

Introduction: Russian Sums in an English Pub, Circa 1946 13

Chapter 1: The Most Ancient Mathematical Legend 19

Chapter 2: The First Two Great Mathematicians 36

Chapter 3: The Great Age of Grecian Geeks 57

Chapter 4: Archimedes – the Greatest Greek of Them All 92

Chapter 5: The Glory That Was Alexandria 118

Chapter 6: Total Eclipse of the Greeks 147

Chapter 7: Maths Origins, Far and Wide 180

Chapter 8: Mathematics Was Never a Religion 211

Chapter 9: Discovering the Unknown World 243

Chapter 10: The Huge Awakening and New Age of Learning 265

Chapter 11: The New Age of Mathematical Discovery 295

Chapter 12: How to Calculate Anything and Everything 332

Chapter 13: A Mathematician With Gravitas 363

Chapter 14: The Simple Mathematics That Underpins Science 384

Chapter 15: The Many Tentacles of Mathematics 404

Wow Factor Mathematical Index 435

Bibliography 469

Image credits 471

Index 472

WONDERS BEYOND NUMBERS

A BRIEF HISTORY OF ALL THINGS MATHEMATICAL

JOHNNY BALL

To my dear wife Di, who is by far the greatest wonder in my life.

Bloomsbury Sigma
An imprint of Bloomsbury Publishing Plc

50 Bedford Square
London
WC1B 3DP
UK

1385 Broadway
New York
NY 10018
USA

www.bloomsbury.com

First published 2017

British Library Cataloguing-in-Publication Data
A catalogue record for this book is available from the British Library.

Library of Congress Cataloguing-in-Publication data has been applied for.

ISBN (hardback) 978-1-4729-3999-9
ISBN (trade paperback) 978-1-4729-3998-2
ISBN (ebook) 978-1-4729-3996-8

2 4 6 8 10 9 7 5 3 1

Bloomsbury Sigma, Book Twenty-eight

Typeset by Deanta Global Publishing Services, Chennai, India
Printed and bound by CPI Group (UK) Ltd, Croydon, CR0 4YY

Preface: Mathematics means everything to me...

Among my earliest recollections was playing 'fives and threes' with my parents when I was about six, using double-nine dominoes (Dad always said, 'Double-six dominoes are for wimps.'). Not yet able to hold more than two dominoes in my hands, I set them on their edges, and soon worked out that double 7 and double 8 were useless on their own, because 14 and 16 aren't multiples of five or three – but placed at each end they added up to 30, six fives and ten threes, making 16, the maximum score.

At Kingswood School, Bristol, aged seven, we used to bang on our desks shouting, 'We want homework!' So I can only assume our teacher was doing something right! He would give us 100 simple sums and say, 'If you do 10, I'll be happy.' While listening to the radio with my folks in the evening, I always relished the challenge of completing the entire 100.

When I was eight, Dad acquired and refurbished a 6ft snooker table, and let me win a game when I was 11. Snooker is nothing more than applied mathematics, as you learn to assess angles and the effects of controlling the cue ball. I played right through my teens and once, in a match, I played a shot off the rounded angle of the middle pocket on a full-sized table, and struck the distant yellow ball perfectly, laying it safely on the end cushion.

At school I was always top in maths, but just after sitting my 11+ exam, my parents moved back to their home town of Bolton, Lancashire. I went with them, and the local grammar school placed me in form 2B. I won a maths and a chess prize in my first year, but through an accident and an illness, I lost a good half of each of the next two autumn terms and regressed through forms 3C, 4D, lower 5E and finally 5E, as they didn't have a 5F. I had come off the rails and only achieved two O levels. However, my GCSE maths score was still 100 per cent, even though I hadn't taken a note for two years. On almost my last day at that school, our senior maths mistress saw my work for the first time and told me I had a strong mathematical brain.

A major aircraft manufacturer gave me a job on the condition I got three more O levels, which I did. More than that, however, I took to the engineering world like a duck to water, and my energy and maths ability helped me streak ahead of better qualified lads on my course, as I quickly taught myself to multiply double figures in my head. I personally worked out the individual costs of the 1,400 parts of the Blackburn Beverley propeller unit, including thousands of machine time costs. I was still only 16.

Aged 18, I signed up for the RAF and came top in my training course. I soon found myself among the boffins testing guided missiles and the latest radar technology in Wales. Later, working with Air Traffic Control in Hanover, Germany, I grasped all that was needed to identify aircraft on our radar screens, often just from the blip they produced, and their speed.

By now I had started to search out books on the subject I loved above all others, mathematics. The Penguin books of Eugene P. Northrop and W. W. Sawyer were soon joined by the puzzles of Ernest Henry Dudeney and Boris Kordemsky. It was Martin Gardner, however, who really broadened my maths horizons, with his republished articles on recreational mathematics, which had first appeared in *Scientific American*.

You could say the RAF had been my university and young man's playground; straight afterwards the fun continued, as I joined Butlin's as a Redcoat Host for three glorious years at 'smile school'. Stand-up comedy followed, which I loved, and soon TV and radio beckoned, and I was recommended to BBC Children's TV. Their integrity and care impressed me and soon I was writing comedy sketches for adult TV and a children's show called *Playaway.*

In 1977 I was asked what I would do if I had my own TV show, and without hesitation I said, 'Probably a show on maths!' All around me jaws dropped, but *Think of a Number* was born, winning a BAFTA in its first series. *Think Again* won more awards; in all I wrote and presented solo 20 series of TV shows based on maths and science. Now it seemed my sketchy education and lack of single-subject degree were perhaps my salvation. By researching every single topic from scratch, my across-the-board knowledge and understanding just grew and grew.

Mathematics has been my travelling companion throughout my entire life. Very often today people stop me to say that my TV shows helped them on their paths to become scientists, teachers, statisticians, model jet-engine makers, *Big Issue* sellers, bookmakers, nuclear physicists, authors and more. I am truly a very lucky man.

So now I have written *Wonders Beyond Numbers*, to celebrate my love of this subject, and to show how maths, science, technology, art, music, architecture and engineering all developed through a huge relay race of achievements. Running this race have been many brilliant minds, born in ignorance and innocence, but who progressed by wanting to know more and to see further, as Newton once said, 'By standing on the shoulders of the giants that have gone before.'

More importantly, I want to remove much of the fear in so many people who, when maths is even mentioned, actually shrink into their shells rather than stand tall. For so many, the hated concepts of addition, subtraction, multiplication and division are all there is to maths. It's rather like a tourist saying, 'Isn't the sea beautiful?' to which a guide might say, 'Well, yes, but you're only looking at the top of it.'

Mathematics is like an ocean, with our number system counting for little more than the surface. In fact, the true depth and wonder of mathematics makes it by turns a fun, exciting, exhilarating, empowering and often truly amazing playground.

I have always been a terrible swimmer, but where maths is concerned I have just dived in. Now I want you to strip off your inhibitions and come and join me – for the ocean of mathematics is full of amazing things, and extraordinary stories of the brave and heroic people who opened up this world of wonders for us.

If I can paraphrase Galileo (one of the greatest): 'Everything in the Universe is written in a language, through which we can understand absolutely anything and everything. That language is mathematics and the symbols are triangles, squares, circles and other geometric figures.' So the wonders of the mathematical world have always been well beyond just numbers.

Above all, I hope that this book is enjoyable for all levels of reader – young, old, learned or otherwise. My narrative will not be interrupted by mathematical ideas, but illustrated and clarified by them. Some maths topics will be in their own 'maths blocks' along the way, while others will be found in the Wow Factor Mathematical Index, where often quite complex concepts will hopefully be simplified and clarified so that their true Wow Factor can be understood. As Einstein once said, 'If you can't explain it to a seven year old, then you don't really understand it yourself.'

I sincerely hope that my book might become a companion to help you and others view mathematics, and indeed the whole world, in a clearer, more understandable – and hopefully more pleasurable – light. Maths has enriched my life every step of the way, and I know it can enrich your life too, if you are happy to let it. Enjoy.

Wow Factor Mathematical Index Explained

This short history of all things mathematical roughly follows a chronological order covered over 15 chapters. Some historic periods were heavy in maths developments and others less so.

So that the maths does not become too heavy going at certain points, selected mathematical items are placed in boxes separate from the text, so that the flow of the story isn't too fragmented.

However, some items have been removed from the text and replaced with references to the Wow Factor Mathematical Index, where they can be better explored in isolation.

The title Wow Factor refers to the fact that many concepts featured are quite a revelation, even to people who are already devoted fans of mathematics. Some Wow Factor items may feature more modern examples, which in the text would affect the chronological flow. Some are reference concepts, like charts of trigonomic ratios. Some are included just for the fun of it.

As an example, many people know that Archimedes once said, 'Give me a lever long enough and a firm place to stand, and I could move the Earth.' Well, how long do you think that lever would have to be? To the Sun? To the stars? In the Wow Factor Index, I made an estimated guess and the result, when I found it, made me think, 'Wow!'

INTRODUCTION

Russian Sums in an English Pub, Circa 1946

The history of mathematics is filled with legendary stories that have always fascinated me, many of which I have absorbed over the years. For me, it all started aged about eight, when an old chap in the 'Children's Room' of a country pub near Bristol asked if I had trouble with multiplication at school, and then said...

'Let me show you how the Russians do multiplication.' Then, with the stub of a pencil on some scrap paper, he began. 'Say you want to multiply 13 times 9? Well just write the numbers side by side:

13 and 9

'Now, keep splitting the left-hand number in half till you get to 1. So, half of 13 is...'

'Six and a half!' I exclaimed.

'Ah, yes, quite right,' he said. 'But in Russia they have "purges", where they get rid of things they don't like. And they don't like fractions. So we'll forget the half – let's call it 6. Good maths this, isn't it?! Now, half of 6 is 3 and half of 3 is 1½ – but we'll forget the half again and call it 1.

That gives us four numbers: 13, 6, 3 and 1. Now, for the 9 (on the right), let's double that till we get four numbers on each side. So 2 x 9 = 18, then 36, then 72. Very good.'

13	9
6 (½)	18
3	36
1 (½)	72

'Another thing Russians don't like,' he continued, 'is even numbers on the left-hand side. If they find one, they scrub out the whole row. So let's see … 13 – that's odd; 6 – that's even, so out goes the whole row … 3 is odd, and so is 1, so they're okay.'

13	9
3	36
1	72

'Then we add up the remaining numbers on the right-hand side:

$$9 + 36 + 72 = 117$$

'And 13 times 9 is 117. Clever, those Russians!'

It works every time, for any pair of numbers. Try it.

I don't know if a Russian ever taught anyone this method, but we know that self-educated people across Europe used it, and in fact it's a very old system indeed. One form of it was used by the Ancient Egyptians, who are the subject of the first chapter in our story. The earliest mathematician, as far as we know, was also an Egyptian, and his name was **A'h-mose**.

Papyrus, Papyrus, Read All About It!

Actually, A'h-mose may just have been a scribe who wrote down mathematical ideas. In any case, he was responsible for creating the Rhind Mathematical Papyrus, a quite remarkable relic of history (see plate section).

The Rhind Papyrus is an ancient scroll discovered in Luxor in Egypt in 1858, and named after Scottish antiquarian Henry Rhind. After Rhind's death the papyrus came into the possession of the British Museum, where it can still be seen today. When it was found it was in several pieces, but originally it would have been about 5.5m long.

A rough translation of the papyrus's title is the rather mysterious 'Accurate reckoning for inquiring into the knowledge of all dark things.' This seems to be implying that maths was a dark art in those days, which may have been the case – or perhaps those in the know realised that having an understanding of maths gave them an advantage over others. Nothing much has changed: as always, knowledge is power, and that includes mathematical knowledge.

A'h-mose, who lived around 1650 BC, claimed to have copied his scroll from an earlier work written, according to estimates, between 1849 and 1801 BC, although the ideas it contains could be much older. The Pyramids at Giza were built between 2561 and 2450 BC, some 700 years before the Rhind was created, yet it contains explanations for mathematics that would have been useful in those pyramid-building times. In one section, for example, there is text about dividing nine loaves between 10 workers, and 100 loaves between two groups of workers, one of which was entitled to a bigger share than the other. All useful stuff for those who had to cater to the needs of a vast army of pyramid builders.

Perhaps the most surprising thing about the Rhind Papyrus is the simple multiplication system it uses for most of its calculations – it's very similar to the Russian method I showed you earlier. Let's use it to multiply 9 x 23.

First, write down the numbers to be multiplied, side by side.

9	23

To the left of the 9, write a sequence of numbers starting with 1 and then doubling – so 1, 2, 4, 8, 16 and so on – but stop before you get to a number that's higher than your smallest multiplier. In this case you would stop at 8, as 16, the next number in the sequence, is larger than 9.

Next, double 23 three times to create a column on the right that's the same length as the one on the left.

1/	9	23/
2		46
4		92
8/		184/

By adding combinations of numbers in the left-hand column, you can get every number from 1 to 15 (1, 2, [1 + 2], 4, [4 + 1],

[4 + 2], [4 + 2 + 1] and so on). All you need to do for this puzzle is mark the numbers that add up to 9. Simple: 1 and 8.

Now mark the corresponding numbers in the right-hand column and add them.

$$23 + 184 = 207$$

And 9 x 23 = 207. Correct, and so simple (there is more on Egyptian maths in the Wow Factor Mathematical Index).

All the Sevens

One puzzle in the Rhind Papyrus might be familiar – problem number 79:

'Seven houses each have seven cats, which each kill seven mice, which would each eat seven *spelt* of wheat, which would each produce seven *hekat* of grain [spelt and hekat are measurements]. How much grain is saved?'

It's similar to a rhyming puzzle I learnt as a child:

As I was going to St Ives, I met a man with seven wives,
Every wife had seven sacks, every sack had seven cats,
Every cat had seven kits [kittens],
So – kits, cats, sacks, wives? How many were going to St Ives?

This is a trick question, of course: only one person – the 'I' in the rhyme – is going to St Ives; all the others are coming the other way. But there's no trick in the Rhind Papyrus example, which the Egyptians seem to have included in the scroll purely because they were fascinated by maths, and in this case the number 7. They also took the basic idea reflected in the St Ives version one stage further. The papyrus listed:

Houses	7
Cats	49
Mice	343
Spelt of wheat	2,401
Hekat of grain	16,807
Total	19,607

or $7 + 7^2 + 7^3 + 7^4 + 7^5 = 7 + (7 \times 7) + (7 \times 7 \times 7) + (7 \times 7 \times 7 \times 7) + (7 \times 7 \times 7 \times 7 \times 7) = 19{,}607$.

More importantly than all this, however, the Rhind Papyrus offers an explanation for the remarkable expansion in modern computer and communications technology. Just like the Rhind (and Russian) multiplication, the maths behind all our communications technology is based on nothing more complex than repeatedly halving or doubling.

The Binary Number System

So what's the connection? Going back to our earlier puzzle, make a new column on the left, and place a 1 next to the marked numbers, and a 0 next to the others.

1	1/	9	23/
0	2		46
0	4		92
1	8/		184/

This new column, reading from the bottom up, gives us 1,001, which is the *binary number* for 9.

Binary numbers are made up of just two digits: 1 and 0. To clarify, the binary number for 10 in our example is 1,010, because the two ones represent 8 + 2 = 10.

So, if we wanted to multiply 10 x 23, we would mark the numbers in our sequence that add up to 10, in this case 2 and 8. The corresponding numbers on the right are 46 and 184, which, added together, give 230. Right again!

The advantage of using binary numbers is that all of them are represented by just ones and zeroes, which can be translated into any number of things in the real world: on a computer screen, a 'dot' for 1 or 'no dot' for 0; on a CD, a tiny pit for 1 or no pit for 0; on a hard disc, an incredibly small splinter of material that points one way for '0', and, when magnetised, points another way for '1'; or the tiniest pulse of a radio wave for 1 and no pulse for 0.

And once you have your binary system, you can scale it. In digital terms, you can then form a huge broadband network, containing many billions of digits. By transmitting them at the speed of light to satellites, and then back to aerials on

Earth, we have the option of watching hundreds of films and TV shows at any given moment – all of them perfectly coded into ones and zeroes.

Without binary numbers there would be no satnavs, mobile phones, tablets or laptops. Binary numbers have made our lives infinitely richer, and empowered us in a whole panoply of ways. But so far we are only touching the surface. Thanks to modern technology, millions of new ideas will emerge in the not so distant future to further amaze and empower us, in ways we have not yet even thought of.

You could picture modern communications technology as a vast inverted pyramid of ideas and systems that are growing exponentially, becoming ever wider and more astounding. So, in the face of all this bewildering modernity, it might seem strange that the wonder of numbers all started at least 4,500 years ago, with the people who gave us the pyramids.

Maybe the way to understand our modern lives and what makes them tick is to look back at the history of mathematics: how we got to where we are, the influence of mathematics and mathematical thought through the ages, and the often weird, wonderful and downright amazing people who got us there. I do hope so, because that's what I want to achieve with this book. Wish me luck. Come on – let's get started.

CHAPTER 1

The Most Ancient Mathematical Legend

Wonder of Wonders

My story of mathematics begins with the oldest and only surviving member of the Seven Wonders of the Ancient World – the Pyramids at Giza, and specifically the Great Pyramid of Khufu (sometimes given the Greek name *Cheops*). It is the largest of the Giza pyramids, and was the first to be built there some 4,500 years ago. The Pharoah Khufu reigned for about 23 years, and because the Pyramid could be seen from many miles in all directions, its original name was 'Khufu belongs to the horizon.'

The pyramid contained about 2.3 million blocks of limestone mostly harvested from local quarries. Napoleon Bonaparte once said that the Great Pyramid contained enough stone to build a low wall around France. I have worked out that the original pyramid had enough stone to build a wall 2m high and 19cm wide that would stretch from the Pyramid site to the North Pole (see Figure 1.1). You might like to check my maths, so have a calculator handy. Otherwise just glance over it and skip the actual calculations.

Figure 1.1

A Wall to the North Pole

Giza is near enough at 30 degrees of latitude north. So the wall needs to stretch the remaining 60 degrees of latitude to the North Pole, or ⅙th of the way around the Earth. That is 40,000km which divided by 6 = 6,666.666km.

The original pyramid had a base side of 230.37m and a height of 146.6m. The volume of a pyramid is ⅓ the volume of the cube or cuboid shape it would fit snuggly into (a cuboid is a cube with one side shorter or longer than the other two).

So 230.37 x 230.37 (the base) x 146.6 (the height) divided by 3, equals the amount of stone.

We now need to divide the stone by our wall dimensions, which I said would be 2m high and 19cm wide. So the full equation is (don't be nervous, they are only numbers):

$$\frac{230.37 \times 230.37 \times 146.6}{3 \times 2 \times 0.19} = 6{,}824{,}658\text{m or } 6{,}824.658\text{km.}$$

To check this, should you have a mind to, feed the numbers into your calculator alternately: one from the top row divided by one from the bottom row, times another from the top row now, etc. Otherwise your calculator might overload.

So a wall of 2m high by 19cm wide would stretch 6,824.658km and would reach the North Pole with 158km to spare. If we made the wall 19.5cm thick, it would reach 6,649.67km and end upto 17km short of the North Pole.

Today, the pyramid is sadly incomplete as the outer casing stones have been removed – many to build the city of Cairo. But originally the pyramid had a polished finish with a gold pyramid cap at the top – it must have looked utterly amazing. The sun's rays bounced off the sides to cast a light shadow on the desert floor as well as a dark shadow to the north. The pyramid was very accurately aligned, with the entrance side and the east and west edges point exactly north, and the corners were perfect right angles. Achieving this would have been easy for the Egyptians. They would have marked the Sun's shadow at noon to find north, and then made two circular arcs with ropes and wooden stakes. A line through the two points where the arcs crossed would have given them their accurate west/east line. The base was also remarkably level and it is believed they cut channels in the ground and filled them with water to form a giant spirit level.

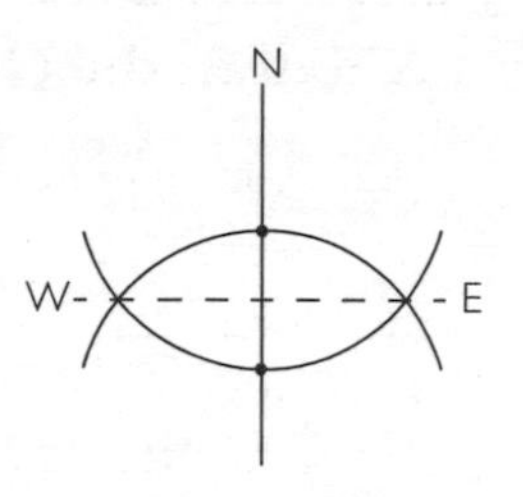

Figure 1.2

It is clear the Egyptians were already accomplished water users and land measurers. Each year the Nile burst its banks, flooding the land on either side. Rather than being a disaster, this was an annual miracle that made Egypt the most fertile country in the known world, and – thanks to the prowess of its farmers – the wealthiest. However, the waters also erased any field markings. So as soon as the waters receded, a team of surveyors armed with lengths of linen rope – and known as 'rope stretchers' (the phrase 'stretched linen' and 'straight line' both have the same ancient root) – measured and divided the land so that each farmer got what he was entitled to. They used rope knotted at regular intervals and there is evidence that they understood that a rope with 3, 4 and 5 equal unit lengths would form a perfect right-angled triangle.

To measure the land they would form a base line and then measure off a triangle. Then, using one of this triangle's new sides as a base, they would construct further triangles and rectangles until the whole area was divided. The triangles may have been haphazard or they could have been more formal with two triangles together forming rectangular strips of farmland. To keep the distribution of land fair, and the farmers happy, each farmer received a mixture of good land and not so good land. Now farming could begin in earnest until the next flood, when the whole farming community would suddenly be redundant and make a huge and handy workforce for more Pyramid building.

A Moving Account

Back at the pyramid, arranging nearly 2.3 million blocks of stone into a Pyramid 146m high was no mean feat, especially at a time when not even the wheel was known in Egypt. Most of the blocks were about 150cm tall, and the Egyptians were clearly adept at shifting huge weights up and down great heights, either using ramps, or possibly using a beefed up version of the *shadoof*, which is still used in Egypt to raise water.

To create the inner chambers of the pyramid, the Egyptians used giant blocks of granite (each weighing about 15 tons) to bear the weight of the stones above them. This granite had to be brought some 600 miles down the Nile. To move the huge blocks of granite from the ship to the pyramid site, the Egyptians used sloping causeways, some of which can still be seen to this day, and wooden sledges to drag the blocks to their destination. With enough manpower and strong enough ropes they could move pretty much any weight, using fine sand, or lime and water, to reduce friction.

The Magical Pyramid Dimensions

There's no doubt that Khufu's pyramid was a supreme feat of engineering. But for me, the most interesting aspect is not how it was built, but the mystery that surrounds its mathematical dimensions. The length of each base side was 230.37m (it still is today, because many of the base stones are still in place). The

Figure 1.3

accurate angle of the slope was 51 degrees, 50 minutes and 40 seconds, and this achieved the original height of 146.6m, making it the tallest building in the world for 4,000 years until the building of Lincoln Cathedral in England, which reached a height of 160m from 1311 to 1549, when its spire collapsed.

So, were these amazing dimensions chosen on purpose and so carefully worked out? Or could it all have just been a coincidence? I've taken these measurements from *The Pyramids,* by prolific author and respected Egyptologist Alberto Siliotti, and used them to explore the maths, which I think is fundamental to how the Great Pyramid was constructed.

What is Pi, and why?

As we progress through this history of all things mathematical, we will come across some areas of mathematics that need explanation for the general reader, but not too many. As an example, we will meet the term 'Pi' quite regularly, so here is a short explanation:

If you look at a ball from any direction what you see is a circle, and when you think about it, there are only two things you can measure on a circle – the distance around it and the distance across it. But in every case, for every circle, if you divide the distance around (the circumference) by the distance across (the diameter) you will always get the same answer, which is what we call Pi and this is represented by the Greek letter π. The Welsh mathematician William Jones suggested the symbol in 1706 because it is the first letter of the Greek word for 'periphery', or outer edge. A generation later the top mathematician of the day, Leonard Euler, approved the idea and the symbol stuck.

Now Pi is *irrational* and the decimals, starting 3.14159 ... go on forever. However, for most mathematical purposes the figure of 3.14 or 22/7 is usually accurate enough.

Now let's try some magic manipulation. If you form a circle with a radius of 146.6m (the height of the Great Pyramid), it would pass inside the corners of the base, but outside the sides for the most part. The distance around this circle would be $2\pi r$. Assuming π is $^{22}/_{7}$, this gives $2 \times {}^{22}/_{7} \times 146.6 = 921.48$m.

The distance around the four base edges of the pyramid is 230.37 x 4 = 921.48m – it's exactly the same figure (see Figure 1.3). Wow! But that's only the start.

The area of the sloping side of a pyramid, as with any triangle, is half the base times the height of the triangle, or the distance up the sloping side, originally 186.58m. So half the base is 115.185m x 186.58m equals 21,491.56m^2. But if you take the height at 146.6m and square it, you get 21,491.56m^2. It's exactly the same again. Amazing or what?

While there is no surviving evidence that the Egyptians were aware of the mathematical significance of the dimensions of the great Pyramid, if Silotti's dimensions are correct then it indicates the Ancient Egyptians did have a very strong understanding of mathematics and the value of Pi. However, Richard Gillings, in his book *Mathematics at the Time of the Pharoahs*, assessed this magical maths and concludes that it was a total myth, and the mathematical world tends to agree with him. So who is right, and what do we know of Egyptian mathematical knowledge 4,500 years ago?

It's clear by looking at the progress the Egyptians made with the six pyramids built before those at Giza, starting with the Step Pyramid of Djoser at Saqqara (see plate section), that their building skills became more and more sophisticated. And because architecture and mathematics go hand in hand, it's a fair bet that their maths also became more accomplished. Personally, I think what sold Khufu on the idea of building the largest pyramid ever was that, besides being his future tomb, it would also be a mammoth celebration of Ancient Egyptian mathematics.

Aside from the Great Pyramid, what other evidence is there that the Ancient Egyptians were adept at maths?

A Slice of Egyptian Pi

We know that the Greek historian Herodotus talked about the Pyramid maths and knew that the height used as a radius would form a circle equal to the distance around the base, but we don't know where he got his information from – only that he wrote it 2,000 years after the pyramid was built. But there are a few other sources.

According to the Rhind papyrus, which I mentioned in the Introduction, the Egyptians decided on a sophisticated value for Pi: 16^2 divided by 9^2, or 256 divided by 81, which equals 3.16. The Rhind also shows us the Egyptian method for calculating the area of a circle with a 9 unit diameter. They calculated it as equivalent to an 8 x 8 unit square, or 64 units (see Figure 1.4). A circle of 9 units diameter gives a radius of 4.5 units and πr^2 (using the Egyptian value for Pi at 3.16) equals 4.5 x 4.5 x 3.16 = 63.99. Wow. How close to 64 square units can you get?

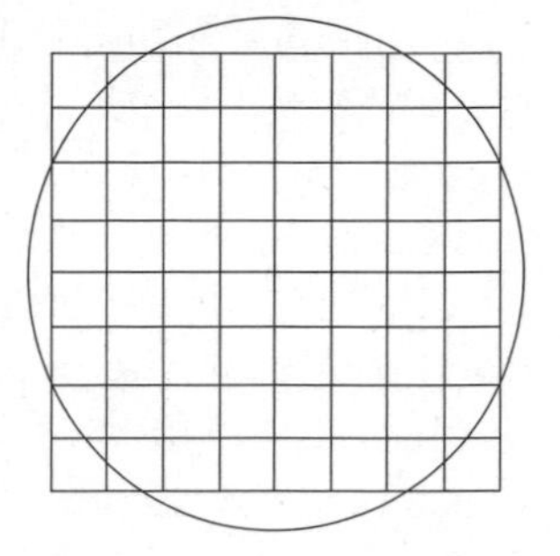

Figure 1.4

Now if we explore the pyramid's dimensions using the Ancient Egyptian Pi value of 3.16, the distance around the base would be 146.6 x 2 x 3.16 = 926.51m, and not the 921.5m figure claimed for it. The difference is about 5m or about ½ of 1 per cent. It's still very close, but it doesn't quite fit. So you have to ask, is the mathematics of the Great Pyramid 'good maths or old myths'? Whatever the answer, you have to admit it's a fascinating puzzle.

The Moscow Papyrus

Another important source of Ancient Egyptian Maths is another papyrus. The Moscow Mathematical Papyrus was discovered in 1892 in Thebes in Lower Egypt by Egyptologist Vladimir Golenishchev, and now resides in the Pushkin State Museum in Moscow. The 25 problems in the papyrus are mathematical in nature, and could all relate to the maths required to build a pyramid. There are also calculations for meeting the needs of the workforce, including providing enough food and beer for huge numbers of builders. According to estimates, the document might have been written around 1850 BC, about 550 years after the Great Pyramid was built.

Seven of the Moscow Papyrus problems are geometric and show the depth of their mathematical understanding.

We have already shown how to calculate the volume of a pyramid – ⅓ the height times the base. There are also several ways to prove this without the maths, by cutting a cube in pieces.

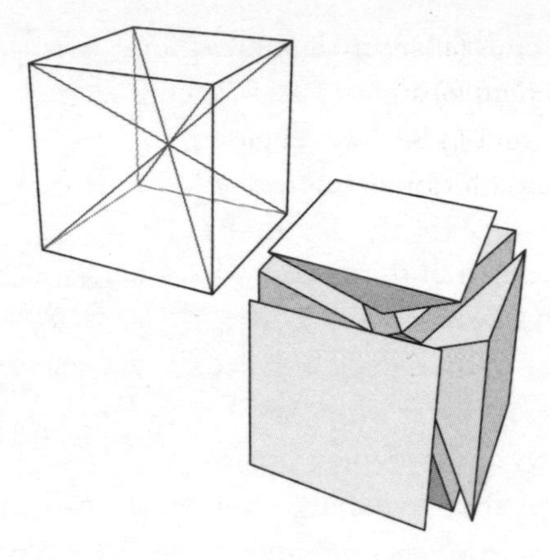

Figure 1.5

If you divide a cube (of cheese, say?) with three cuts that pass through two opposite edges, you produce six small pyramids, all meeting at their tips, or apexes (see Figure 1.5). Each has a side of the cube as a base and the centre of the cube as the height it reaches. Double the height of any of these six smaller pyramids and you double its volume. So three of them would be equal in volume to the original cube, although you wouldn't be able to assemble them back into a cube.

But what if only the bottom half of this pyramid had been built? What would its volume be, and how much stone would be needed to complete it? This involves calculating the volume of a pyramid with its top cut off, which is called a truncated pyramid or a *frustrum*.

The Moscow Papyrus – Problem 14

A pyramid is 12 cubits high and 4 cubits along its base edge (a cubit is 18 inches or 45cm). So its volume is:

4 x 4 x 12/3 = 64 cubed cubits.

The same pyramid is truncated halfway up, at a height of 6 cubits. Now what is its volume?

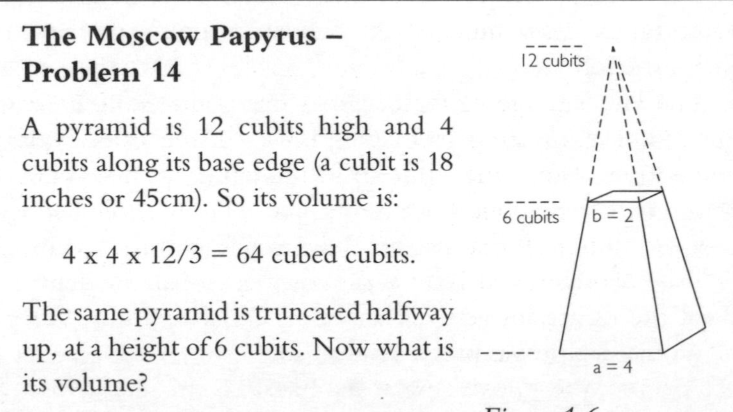

Figure 1.6

Its base edges (*a*) are each 4 cubits long.
The top edges (*b*) are now each 2 cubits long.
Its new height (*c*) is now 6 cubits
This is the solution given in the Moscow Papyrus:

First find the area of the square base with side *a*:	4 x 4 = 16 square cubits
Find the area of the square top with side *b*:	2 x 2 = 4 square cubits
Now multiply the two edges:	$a \times b$ or 4 x 2 = 8
Now add the three together:	$a^2 + ab + b^2 = 16 + 8 + 4 = 28$
Multiply by height *c* and divide by 3:	28 x 6/3 = 56 cubed cubits

Thou has found rightly (which is a translation of the ancient mathematicians' final statement). It's very neat.

We can prove this is correct by calculating the volume of the missing top:

Base x height divided by 3: 2 x 2 x 6/3 = 8 cubed cubits

Let's do a final check:

Truncated pyramid + lost top = whole pyramid:
56 + 8 = 64 cubed cubits

Using this clever bit of maths, pyramid builders could always calculate how much stone they had used and, more importantly, how much extra they were going to need to finish the job.

The Grecian age of mathematics lasted almost 1000 years and produced amazing discoveries, as we shall see. But Egyptian civilisation, from the time of Imhotep, the first known architect, lasted about 3,000 years, and quite possibly laid the foundations for all that maths. Sadly, other than the pyramids and the Moscow and Rhind papyri, there's little evidence of what the Egyptians actually achieved mathematically, but it's safe to say that it was probably awesome.

For me the most amazing revelation in the Moscow Papyrus is problem number 10, a geometric conundrum that starts by asking the reader to calculate the surface area of a semi-spherical basket with a diameter of 4.5 units. The answer the papyrus gives is 32 units (the complete workings are shown in the Wow Factor Maths Index).

The area of the opening of the basket, which would have been covered by a lid, is the area of a circle with radius 2.25 units. The formula to calculate this is πr^2. So, using the Egyptians' value for π, we get 2.25 x 2.25 x 3.16 = 15.99 or 16. According to the papyrus, then, the surface area of a semi-spherical basket is twice the area of its lid.

The maths (as shown in the Index) is pretty convoluted, but it is also quite incredible: for most historians of mathematics, we had to wait until **Archimedes** (287–212 BC) before calculations of curved surface areas became established (see Chapter 4).

It's possible that the Egyptians used this maths to describe something that was already familiar to them. Basket making was a vital occupation for the Ancient Egyptians, and craftsmen or women would have discovered from experience how much material was needed to make a half-sphere basket, and that the lid would require half as much. This is just the kind of knowledge that craftsmen and women would pick up and pass through generations.

The Sumerians and Their Sums

Egypt at the time of the Pyramids wasn't the only mathematical powerhouse around. To the north-east, nestled between the rivers Tigris and Euphrates, were the Sumerians, who, as it turned out, lived up to their name: they were rather good at sums.

The Sumerian civilization was pretty advanced by anyone's standards: it developed formal agriculture, built canals and dams, and invented the arch and even the wheel, which Sumerians were probably using before the Great Pyramid was built. Remarkably, news of this ground-breaking idea doesn't seem to have travelled the 800 or so miles to Egypt for about 200 years.

We know a great deal about the Sumerians' maths, because, unlike the Egyptians, they didn't use papyrus to record it (papyrus slowly rots away as the moisture in the air gets to it,

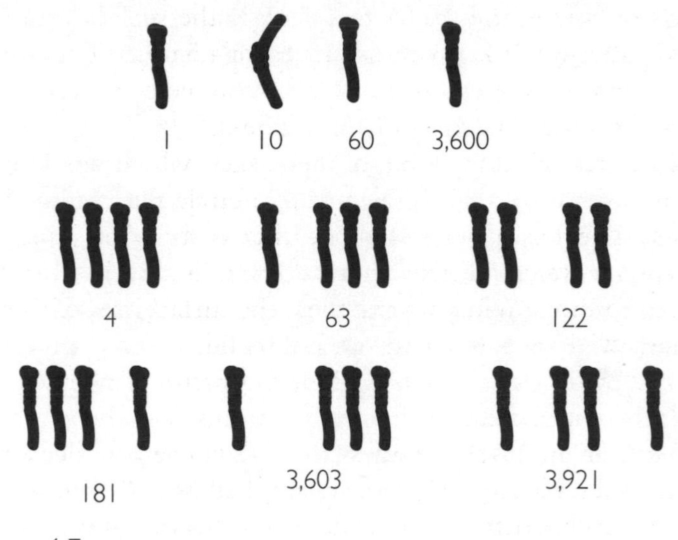

Figure 1.7

so other than a few extant examples, most of the documents the Egyptians produced have perished). To record both their language and their mathematics, the Sumerians made marks in a piece of clay (using a wedge-shaped stick called a *stylus*), which then hardened in the sun. Fortunately, thousands of examples of their writing and mathematics have survived for us to study today, including shopping lists, business accounts, schoolwork, times tables and even mathematical research. Before the Iraq war, when tourism was still possible, you could buy ancient tablets inscribed with calculations and lists. All tablets, regardless of their size, could be bought for roughly the same price (about $5), so the sellers would break large samples into smaller pieces. The overall loss for historians is hard to calculate, but tragically sad.

The Sumerians seem to have chosen a system of mathematics that was simple yet highly ambitious. The system varied over time, but they could represent any number using just two symbols – an upright unit symbol or two marks set at an angle for 10. But the same unit symbol could represent 1, 60 and 3,600.

The number 1 was a natural starting point, and 10 made sense because humans have 10 digits on their hands. But why did they them jump to 60 and then take a massive leap to 3600 (60 x 60)? Well it had a lot to do with the passage of time and the calendar.

For the Sumerians, as soon as the sun had set and the evening meal had been eaten, there was little to do at night except sleep or indulge in the odd social function. Stargazing was as attractive a pastime as almost anything else. The stars have no apparent pattern – unless, that is, you invent one. So, by effectively joining the dots, the Sumerians imagined shapes, and invented tales of heavenly animals, mythical beasts and heroes to go with them, stories that were taught to each new generation.

Even today, the most important star pattern is the Great Bear, or the Plough – or, if you're in the USA, the Big Dipper, so called because it looks like the little saucepan-like implements often chained to water fountains so passers-by can dip them in for a drink (see Figure 1.8). Looking at the Big Dipper, follow the star pattern along the bent handle, into the pan and along the bottom. Now pass up the other side, and continue in a straight line, and after a short while you get to the Pole Star. This star has the unique quality of being the only star that never changes its position. While the

Figure 1.8

heavens revolve, the Pole Star is always directly north – a fact that has benefitted hunters and travellers the world over for centuries.

By watching and recording the movements of the stars, and the angle of the Sun as it moved north, south and back again, the Sumerians calculated the length of a year at about 365 days. Although this number is accurate, mathematically speaking it's complicated and awkward. 365 has only two divisors – 5 and 73 – two prime numbers. So, in a 365-day year the Sumerians could have had 73 weeks of 5 days, or 5 months of 73 days, neither of which was an especially attractive idea.

Quite reasonably, the Sumerians figured that the maths would be easier if there were 360 days in a year. This is a beautiful number as 360 can be divided by 23 other numbers: 1, 2, 3, 4, 5, 6, 8, 9, 10, 12, 15, 18, 20, 24, 30, 36, 40, 45, 60, 72, 90, 120 and 180. The downside for the Sumerians was that it gave them an enormous number of day/month combinations to choose from – but they seem to have settled on 12 months of 30 days.

The Egyptians, and the Mayans of South America (who came much later; see Chapter 7), used the same system, and all added 5 or 6 days at the end of the year to round things off. However, the Sumerians might not have stuck rigidly to their system. From watching the heavens, they discovered how to predict when the very first sliver of new moon was likely to appear, and the night when it actually did, signalling the start of a new month. In fact, they became so good at understanding the movements of the heavens that before long they could also predict solar eclipses, many years into the future.

Finger-counting Good

For the maths of everyday life, Sumerian children may have learnt to count to 30 on their fingers – well they were handy. Try this – count the fingers (and thumb) on your left hand: 1, 2, 3, 4, 5. Now use the thumb on your right hand to represent 6 and go through the left hand again: 7, 8, 9, 10, 11. The first finger of the right hand then becomes 12. Carry on until you've used all the fingers on your right hand, which represent 6, 12, 18, 24 and 30. If you swap hands, you can count up to 60.

The Sumerians may also have used another way to count to 60 on their fingers. Try this method…

Each finger has three segments. On your left hand, use your thumb as a pointer to count the segments of the fingers, which gives you 12. Then raise thumb of your right hand to represent 12, then repeat the process on your left hand to count the segments until your left hand has recorded 12, 24, 36, 48 and 60. Both systems are really neat.

Once the Sumerians had given their year 360 days, and as a year is a cycle – or circular – it seemed like a natural step to divide a circle into 360 degrees, aiding tremendously their expanse of geometric maths. By contrast, the Chinese calculated the length of the year accurately at about 365.25 days, but then tried to divide a circle into 365 portions. They found it so difficult, they gave up developing circular maths for around 1,000 years, by which time the Sumerian method had reached them.

Working with 6s and 60s seems to have suited the Sumerians down to the ground and made for very elegant geometry.

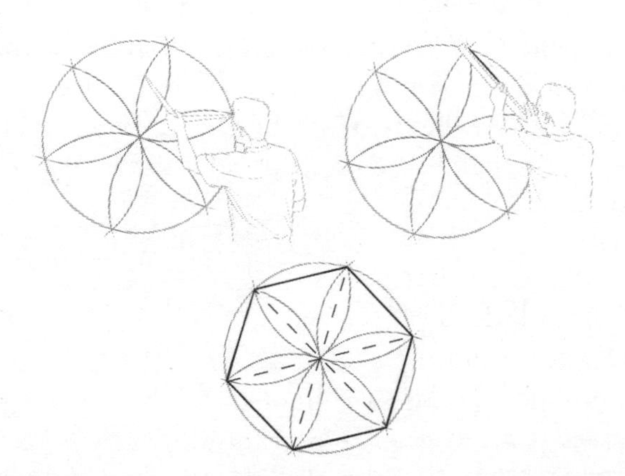

Figure 1.9

Using a pair of rusty compasses (where a change in their angle of separation is not required), you can form a circle and then mark off six equal points around it and inscribe a regular hexagon inside the circle. Include the diameter lines between opposite points and you have a figure where all of the 12 short lines are exactly the same length and the triangles formed are equilateral and equiangular, with every angle measuring 60 degrees. Even today we still use 60 a great deal in our everyday lives – one hour is divided into 60 minutes, and one minute into 60 seconds – largely thanks to the Sumerians' valuable mathematical legacy.

A Tablet That Takes Some Swallowing

In around 1800 BC the Sumerians were replaced by the Babylonians, a civilization centred around the city of Babylon (whose hanging gardens were another of the seven ancient wonders). Around that time, mathematics, which had remained relatively straightforward till then, was becoming more ambitious and more complex. One tablet from the period that gives examples of Babylonian maths was dug up and offered to the philanthropist G. A. Plimpton. It now bears the title Plimpton 322, and resides in the Plimpton Collection at Columbia University in New York (see Figure 1.10).

The tablet seems to show that the Babylonians might have understood what we now known as Pythagorean triples where $a^2 + b^2 = c^2$ and which links to Pythagoras' famous theorem about right angled triangles. But the Babylonian tablet has been dated to about 1,200 years before Pythagoras was born.

The tablet reads from right to left, with the numbers 1 to 15 down the right-hand side to denote the position of the different calculations. There is then a patterned dividing line, followed by a column of 15 numbers that turn out to be whole-number sizes for the *hypotenuse* (longest side) of a right-angled triangle. Written in cuneiform and the Sumerians sexagesimal system, the top number is 2, 49, which stands for $2 \times 60 + 49 = 169$. The next number to the left is 1, 59, which

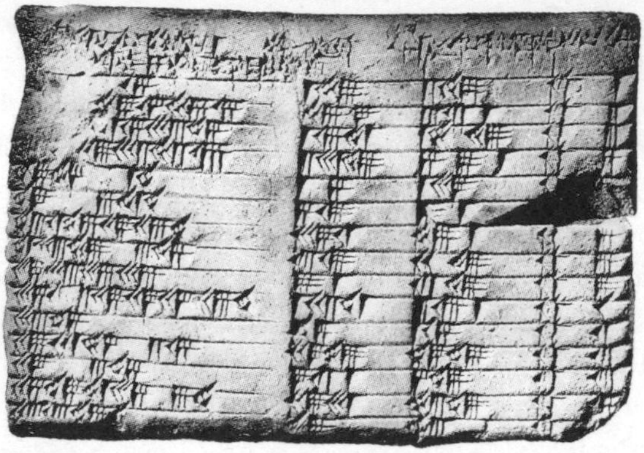

Figure 1.10

is 60 + 59 = 119. Sure enough, these are two sides of a right triangle in which the third side is 120.

Note that 119 and 120 are *almost* the same number. Were they exactly the same, the triangle they formed would be isosceles with two angles of 45 degrees, but the third side would not be a whole number. As it is, the three sides of length 119, 120 and 169 form a right triangle with the small angle at 44.8 degrees, so very close to 45.

The item at position 4 on the tablet is a remarkable monster. It gives a hypotenuse of 18,541 units and a second side of 12,709. That would make the third side 13,500 units, and the three sides would form a right triangle with an angle of 43.3 degrees.

The simplest example is at position 11, where the hypotenuse is 5 and the other side is 3 units. This gives us the 3, 4, 5 triangle, for which $3^2 + 4^2 = 5^2$, or 9 + 16 = 25. The smallest angle would be 36.9 degrees. The fifteenth position, meanwhile, gives 53 for the hypotenuse and 28 for the second side. The third side would be 45 units, and the right triangle formed would have a small angle of 31.9 degrees.

The bottom of the tablet seems to have been deliberately broken off below the fifteenth line, probably while the clay was still moist. Which makes sense, because there are just 15 right triangles with whole-number sides and angles between 30 and 45 degrees. Once the angle gets below 30 degrees, there are literally hundreds of them.

On the far left of the tablet there is some rather advanced trigonometry – a discipline that would not be developed and fully understood until modern times.

On this evidence, the Babylonians were pretty sophisticated mathematicians. But it seems they also had an exemplary idea of progress: keeping things simple for the general population, but enlisting accomplished mathematical scientists to push forward the boundaries of knowledge. One question remains: if the theory of right-angled triangles was around at the time of the Babylonians, why was it named after Pythagoras, who lived another 1,200 years into the future? To answer it we need to jump forward in time a bit, and about 600km west.

CHAPTER 2

The First Two Great Mathematicians

The Crafty Father of Greek Mathematics

The first person we can be certain was a 'mathematician' as we understand the term today was not Babylonian, but a Greek chap called **Thales** (624–546 BC). Born in the city of Miletus (on Turkey's Bodrum peninsula), in his lifetime his achievements ranged from trigonometric measurements to celestial predictions. Although Thales is held by many to be something of a genius, it was a term he himself was sceptical of, remarking that the secret of appearing to be a genius is largely about making the extra effort to learn and understand better.

The Sumerians and Babylonians had been predicting solar eclipses for many years, and Thales seems to have learnt how to predict them too – with remarkable accuracy: on one particular day he pinpointed, sure enough, there was an eclipse. It so happened that two armies, of the Medes and the Lydians, were about to go into battle that day. But as the skies darkened and the Sun disappeared, they decided to call a truce, reckoning that the gods were clearly more powerful than they were anyway. We have since been able to work out the precise date of that event: 28 May 585 BC. Amazingly, we don't know the exact date of any other battle of that era, but we do know when the one battle that didn't happen, didn't happen.

Thales was a colourful character by all accounts, and there are plenty of stories about him. One day, while walking along, contemplating the stars, he fell down a hole and an old lady passing said, 'What's the good exploring the sky when you don't know what's under your feet?' Thales himself often recounted this story, suggesting he was aware that the practicalities of life on Earth are perhaps more important than heavenly conjecture over which we have no control. It's a sentiment I agree wholeheartedly with, which is why I have always championed practical engineering over purely theoretical physics.

Thales once heard someone say of him 'If he is so clever, why isn't he rich?', so he decided to rectify the situation. He had previously warned the people of Miletus of unscrupulous traders cornering the market in one type of goods, then holding people to ransom by charging inflated prices. No one paid any attention, however, so in a crafty move Thales bought up all the olive presses in his area. For one season he controlled the entire crop of olives, and all the olive oil, making a fortune. But as people gradually started to heed his earlier warnings, Thales sold on the presses and returned to a more frugal life.

Thales and the Crafty Donkey

Thales was also once out-thought by a crafty donkey – or almost. A train of donkeys was employed to carry salt in roughly woven sacks strapped to their backs. One day, while wading across a stream, one donkey stumbled and dipped under the water, before scrambling to its feet and continuing with the others. But the water had washed away a good portion of the salt from the coarse sacks on its back, considerably lightening its load. From that day on, every time they crossed the stream, the donkey threw a wobbly, stumbled into the river, and continued with a lighter load.

A crafty donkey for sure, but not as crafty as Thales, who spotted what he was doing and why. Next time they took the journey, he filled the wily donkey's sacks with sponges. Arriving at the stream, right on cue, the donkey stumbled. Righting itself, however, it discovered that its load was more than twice as heavy as it had been before, due to the water the sponges had absorbed. Outsmarted by Thales, the donkey had stumbled its last.

Thales also travelled to Egypt, and was enamoured of the pyramids – so much so that he devised a system of measurement to work out how tall they were without having to climb them. On a particularly cloudless day, after choosing a pyramid he wanted to measure, he put a stick in the ground, then measured the height of the stick and the length of its shadow. Picking a point on the ground, level with the centre of the base of the pyramid, he then measured the distance from that point to the tip of the pyramid's shadow. By dividing that distance by the length of the stick's shadow, then multiplying it by the height of the stick, he worked out the height of the pyramid (see Figure 2.1). Using this simple method, he could then measure the height of anything that casts a shadow.

For convenience, Thales found it was better to do the measuring when the height and the shadow length of his stick were in a simple proportion. If the two lengths were equal, then the height of the object he was measuring would equal the length of its shadow, and the triangle formed would be a right-angled isosceles triangle with two angles of 45 degrees. This maths, although remarkable in its simplicity, wasn't new – in fact, the Babylonians had done most of it long before. But Thales' nifty measurement method was the beginning of *trigonometry*: the science of measuring things by knowing the relationships between the angles and the sides of triangles.

While he was in Miletus, Thales had a pupil called **Anaximander** (610–546 BC), who was just 14 years his junior. Anaximander was a traveller, who spent some time in the East, from where he brought back several scientific ideas, including the sundial. Again, both the Egyptians and the Babylonians had used sundials for centuries, but for the Greeks, timekeeping became a far more reliable art.

Anaximander saw that the heavens revolved around the Pole Star, which he explained by saying that the heavens were

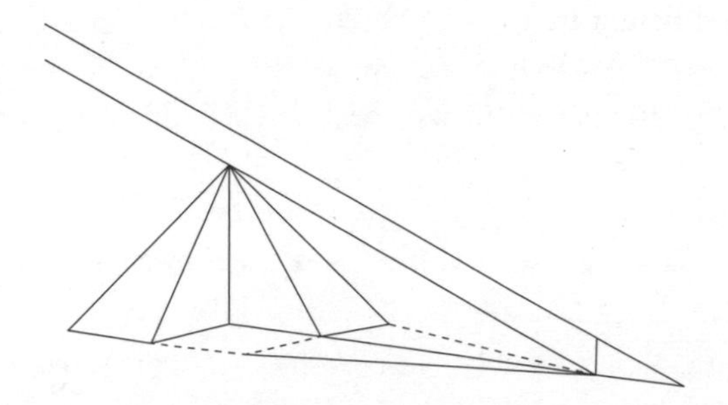

Figure 2.1

spherically shaped, rather than being a huge dome over a flat Earth. (This appears to have been the first time that spheres had been considered in Greek maths and science.) Anaximander also saw that the Earth itself must be curved, although at first he thought it was a cylinder, standing on its end (like a tin of beans), with a height of about one-third its diameter. As the Sun always moves across the Earth from east to west, this explained why more was known of the lands under its path than those lying further north or south. This viewpoint remained unchanged for some 700 years, right up to Ptolemy's map of the Earth (produced in about AD 150), which shows the world more as a cylinder than a sphere (for more on Ptolemy see Chapter 5).

Thales must have been influenced by Anaximander's ideas, and they probably inspired some simple maths he devised that proved useful to sailors: he worked out how to determine the distance from the shore to the horizon. Try it next time you're by the sea – lying on the shore, place your head on the sand right at the edge of the water. You can't see any distance at all. But if you walk up the beach a little so your head is about 6ft above the waterline, the horizon will now be about 3 miles away.

Thales realised he could work out the distance to the horizon according to how high above the water level his head was, even if he was standing on a cliff. Have a look at the table below – the distances are good approximations.

Distance to horizon in miles	0	3	6	9	12	15
Observation height in feet	0	6	24	54	96	150
Observation height in metres (approximate)	0	2	8	18	32	50
Height in units of 6ft/2m	0	1	+3=4	+5=9	+7=16	+9=25

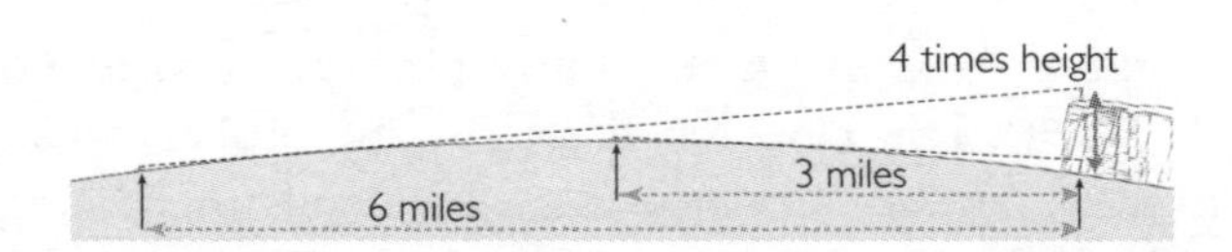

Figure 2.2

Thales showed that to see twice as far, your head needs to be four times higher; to see three times as far, you need to be nine times higher, and so on. In other words, he saw that to find progressive squares you simply add successive odd numbers. This amazing sequence crops up time and again in the history of mathematics, but this was the first known example of using it to solve a problem.

Of course, the opposite also applies. If you're 12 miles out to sea in a small boat, you can see a cliff only if it is 96ft (about 30m) high. And you have to be within 3 miles from shore to see people on a beach. That's why crow's nests on later ships were raised as high as possible. Today, a ship's bridge is pretty high, and because sailors know how far they are above the water, they can accurately assess distances to mountains, cliffs and lighthouses.

Thales' Theorem

Thales produced this theorem, which was subsequently named after him:

For any circle with diameter AB, any two lines drawn from A and B that meet at a point on the edge of the circle will always form a right angle (see Figure 2.3 – left).

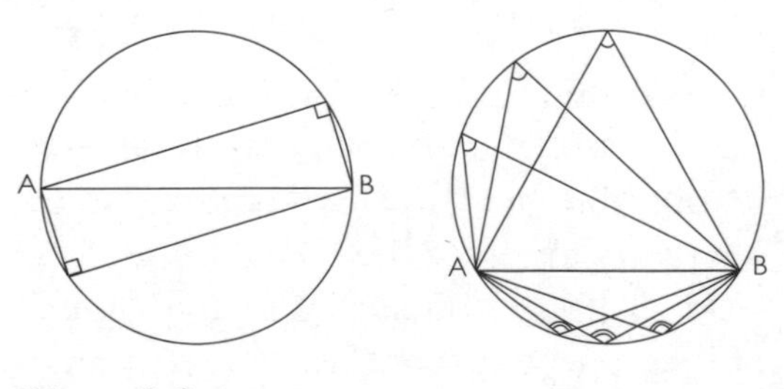

Figure 2.3

This was the first example of what's now known as an *invariance theorem*, in which all possible variations give the same result. The clue for the theorem probably came from the observation that both a square and any rectangle will fit perfectly into a circle for which their diagonal is also the diameter of the circle. And exactly half the rectangle or square, with its right-angled corner, will fit into half the circle.

This concept can be extended: if you take any line AB across a circle that is not the diameter, then all angles formed by triangles set within the circle on the larger side of AB will be the same but less than 90 degrees. All triangles within the circle on the smaller side of AB will have equal angles that are larger than 90 degrees, and any two opposite angles will always add up to 180 degrees (see Figure 2.3 – right).

Centuries after Thales' death, the Greeks drew up a list of the 'Seven Wise Men' of ancient times, and put him at the top. But one man, who might just possibly have been Thales' pupil, was destined for greater and wider fame. His name? **Pythagoras**.

So, Pythagoras ... Who's Theorem Was It, Really?

It's not Pythagoras' fault that we name the famous theorem after him. It states that:

In a right-angled triangle, the square on the hypotenuse (the longest side) equals the sum of the squares on the other two sides.

Although the theorem may have originated long before his day, we're often told that Pythagoras produced the first proof of it. But did he? In fact – brace yourself – did Pythagoras actually exist at all...?

There is no written evidence that Pythagoras actually lived – according to legend he refused to record his secrets or to allow anyone else to write them down. Everything we know about him has come to us from other scholars who called themselves 'Pythagoreans', and who were members of the 'cult' of Pythagoras. 'Pythagoreanism' was not just about mathematics: it was a semi-religious way of life, steeped in mysticism. To add to the confusion, those who first wrote about Pythagoras all lived shortly after he had supposedly died (or at least disappeared), so they could not have known him personally. Pythagoras was also, supposedly, the first person to use the term 'philosopher'. What we do know for certain is that the time in which Pythagoras lived marked the beginning of the glorious age of philosophical thought that made ancient Greece so influential in the history of intellectual development.

Before we get into the maths, however, here are some of the alleged facts about Pythagoras, his life and his work. He was born on the island of Samos and lived from around 560 to 480 BC, though no one is sure of the actual dates. When he was perhaps about 20, it seems he met the aged Thales, who admitted that his own mind was now past its best, but recommended that Pythagoras' education would not be complete unless he visited Babylon and Egypt. Heeding Thales' advice, Pythagoras took to the road, adorned as always in flowing white robes and unkempt hair. Pythagoras may well have been the original Greek geek...

It seems that while Pythagoras loved Babylon, he found little to admire in Egyptian maths or architecture, although he may have studied it in detail. It all seemed too heavy and rigid – for Pythagoras, mathematics had a simple but infinitely varied beauty: to him, a simple straight line was much more than a simple straight line. It could be a stretched rope marking the edge of a field, a shadow cutting a pavement of squares at an angle that changed over time, the shortest distance a bird might fly between two points or a string to be plucked.

Linking Music to Mathematics

Perhaps Pythagoras' most valuable and far-reaching discovery was the link between maths and music. The story goes that one day he heard a blacksmith hitting two anvils, which emitted two distinctly different notes. Intrigued, Pythagoras started to explore noises and the things that make them. By hitting, blowing and plucking different objects, he eventually discovered that all the sounds they make are governed by the same mathematical rules (see plate section).

Pick up a guitar and pluck a string. Now find the halfway mark on the string, press the string to the neck of the guitar at that point, and pluck again. The second note is the same note as the first, but an octave higher. Let's assume the string is tuned to C: open, it emits a low C and, at the halfway point, C one octave higher. Now, from the halfway point, add another third of the string. The new note is G. Add another half and you get an F, and so on (see Figure 2.4). Those are natural spacings, but then it gets a little more complex. Here is a full octave from C down to lower C. To C (half the string) add $\frac{1}{15}$ for B, $\frac{1}{5}$ for A, $\frac{1}{3}$ for G, $\frac{1}{2}$ for F, $\frac{3}{5}$ for E, $\frac{7}{9}$ for D and $\frac{1}{1}$, or the same distance again, for the lower C.

Pythagoras discovered that the same rules govern the size of bells, and the volume of air above a liquid in a row of identical flasks. Just a few years BT (before television), when I was a lad, people often played the 'bottlephone' – a xylophone made of bottles or glasses on the kitchen table. Try it, by filling glasses with water measured out using the fractions mentioned above.

For all musical instruments – whether strung, blown or bashed – simple maths defines the notes. Pythagoras even suggested that notes rise by an octave if you square the tension applied to them. So a string attached to a 1kg weight will produce a particular note, but to raise the note by an octave would take not 2kg but 4kg of weight. The maths of the modern music we know today can be very complex, with sharps and flats and subtle tone variances. But because computers can handle complex number crunching with ease, synthesisers can now produce any sound, pitch or timbre electronically, simply by applying the maths.

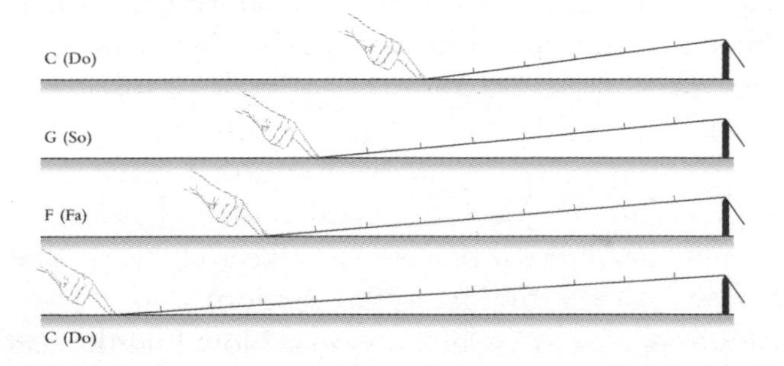

Figure 2.4

For Pythagoras it was much more straightforward. The sound emitted by bells or chiming glasses depends on the volume of the reverberating material; with stringed instruments it is the tension and length of the vibrating string that counts; with wind instruments (such as a simple reed pipe, or the opening of a bottle when you blow over it), it is the volume of the vibrating column of air that creates each note. But all are governed by the same simple mathematical proportions.

Pythagoras even suggested that musical sounds might govern the stars and planets, with each celestial body emitting a particular note – a concept known as the Music of the Spheres. Some 2,000 years later, a chap called Johann Bode discovered that the distances of the various planets from the Sun do have a mathematical pattern perhaps as beautiful as the maths that produces music (for more on Bode, and the mathematical law named after him, see Chapter 14).

The Order of Pythagoreans

When Pythagoras returned home from his travels, now aged about 30, Samos was ruled by Polycrates, a tyrannical bully. Pythagoras decided that the place simply wasn't big enough for both of them. By then, huge numbers of Greeks, often plagued by petty wars and quarrels, had upped sticks and moved west, setting up city states along the shores of the Mediterranean wherever the local inhabitants allowed them to gain a foothold. A great many settled in Sicily and southern Italy, and in Croton Pythagoras established a base and set up a centre of learning. Like-minded followers flocked to his side, forming a group that came to be known as the Order of Pythagoreans, or simply the Brotherhood.

Pythagoras and Number Theory

Pythagoras' first love was geometry, but he also played around with basic number theory, using geometry to explore and explain number relationships in an almost childlike way. He may have been the first to notice that numbers alternate between odd and even, and that concepts like 'threeness', 'fourness' and 'fiveness' might have their own particular magic. He associated the number 5 with marriage, for example, because it combines 2 (which Pythagoras regarded as female) and 3 (which he saw as male). No person was as simple as the number 1, it seems.

Pythagoras also demonstrated that groups of pebbles, as they expand, can form a series of identical mathematical shapes. Try placing one pebble on a surface, then add two more to make a triangle of three. Now add three more to form a larger triangle of six pebbles. Then add four more to make another triangle of 10, like a pack of pool balls at the start of a game (see Figure 2.5a). By adding five, six and seven more pebbles, you can form a progression of triangles, each one larger than its predecessor. This progression – 1, 3, 6, 10, 15, 21 and 28 – gives us the continuing series of numbers known as *triangular numbers*.

Pythagoras additionally saw that any two consecutive triangular numbers in an 'L' formation can be placed next to each other to form a perfect square, by turning one upside down (see Figure 2.5b).

But it was square numbers that Pythagoras found most fascinating. Try this: take one pebble, then add three more to create a backwards 'L' shape (which the Greeks called the *gnomon*, or carpenter's rule). The four pebbles make a perfect square of two rows of two.

Now add five more pebbles in a backwards 'L' shape, and you have a square of nine pebbles. Add another seven pebbles to give 16, then another nine to form 25 (see Figure 2.6). Can you spot the pattern? Each successive odd number, when added to

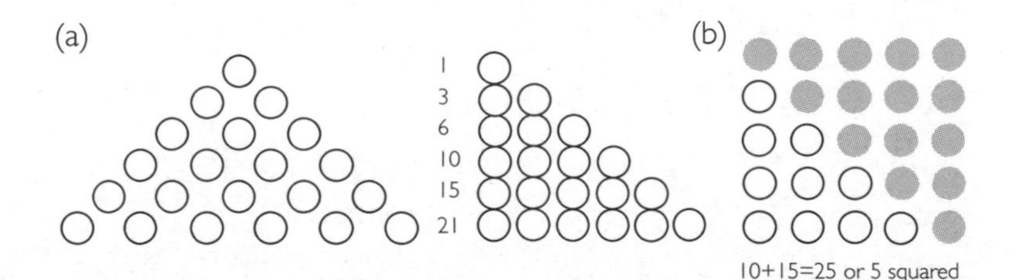

Figures 2.5 a and b

the group, produces the next highest square number – and this sequence goes on forever. As we saw earlier, Thales spotted this progression when judging distances at sea; this particular discovery would also prove very useful to Galileo and Newton some 2000 years later in working out how cannon balls fly and how gravity weakens as objects get further apart.

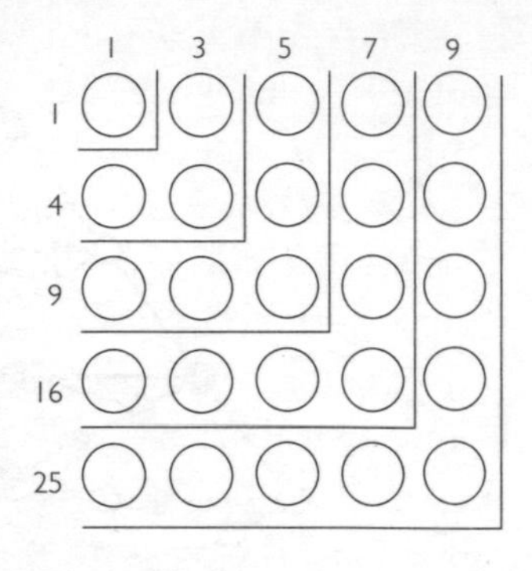

Figure 2.6

	1	**3**	**5**	**7**	**9**	**11**	**...**
Adding the above as you go:	**1**	+3 = **4**	+5 = **9**	+7 = **16**	+9 = **25**	+11 = **36**	...

Pythagoras also saw that two identical numbers, when multiplied together, always form the *square* of that number, so 2 x 2 = 4, 3 x 3 = 9 and so on. But when you multiply two different numbers, say 3 and 4, they form a rectangle, and if you multiply the two sides together you get the area of the rectangle (12 in our example). This concept makes multiplication much easier to understand. A rectangle of 3 x 7 pebbles, for example, contains three rows of seven or seven columns of three, making 21 pebbles in total. Simple!

Pythagoras continued to build on his ideas about number theory. If you place triangular numbers (1, 3, 6, 10, 15, 21 and so on) in a tower, they form *tetrahedral* layers and the tetrahedral sequence 1, 4, 10, 20, 35, 56... You can do the same with the square numbers 1, 4, 9, 16, 25, 36 – which form a three-dimensional square pyramid with the sequence 1, 5, 14, 30, 55 and 91 (see Figure 2.7).

From Plane to Solid Shapes

Pythagoras loved geometric shapes drawn with dots, and creating families of numbers that form polygons with a variety of sides (see Figure 2.8).

He worked with regular shapes that cover a plane – or flat – surface perfectly. Squares clearly do this, but so do other regular

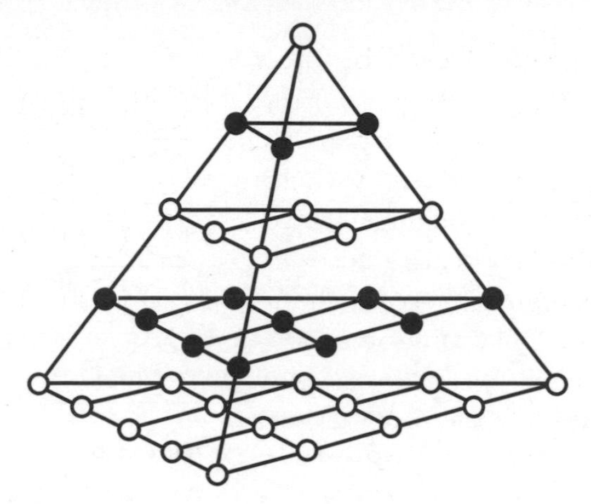

Figure 2.7

shapes, like equilateral triangles. From there, for Pythagoras, it was only a short step to three-dimensional shapes. And once he started looking for them, they kept on coming...

1. Six squares form the faces of a cube, or a *hexahedron* – a regular solid with six identical faces.
2. A group of six equilateral triangles, when laid together on a flat surface, produces a hexagon. But a group of four equilateral triangles will also form another regular, four-sided solid called a *tetrahedron*.
3. Four isosceles triangles form a pyramid with a square base (something the Egyptians were well acquainted with). By using equilateral triangles instead, you can join two pyramids at their bases to create an eight-sided *octahedron*, the third regular solid.

It may have been at this point that Pythagoras began to see something magical in these shapes. Consider this, for example:

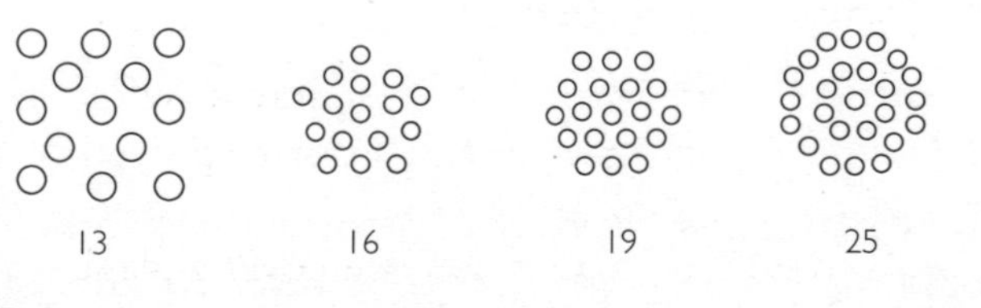

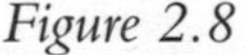

Figure 2.8

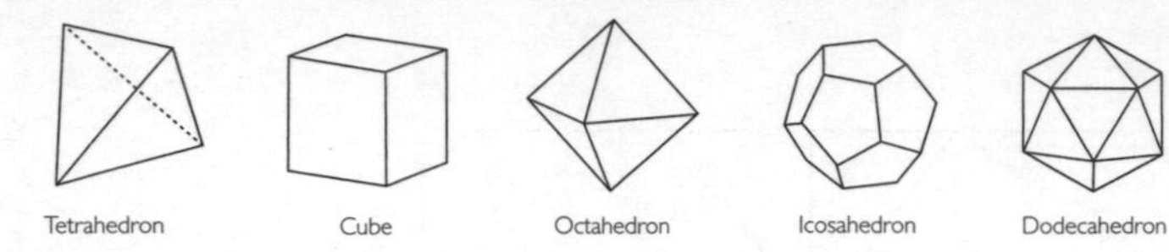

Figure 2.9

a *pentagon* (five-sided shape) won't cover a plane, but five equilateral triangles joined to the sides of a pentagon will form a shallow five-sided or pentagonal pyramid. By connecting more and more triangles in circular groups of five, you arrive at a spherical-shaped combination of 20 equilateral triangles – in doing this Pythagoras had discovered the fourth regular solid: the *icosahedron*.

There's more. Pentagons won't fit together to perfectly cover a flat plane, but if you join three at one corner, the shallow pyramid-like shape created will allow other pentagons to join along two free edges, and – lo and behold – just 12 together will form another regular spherical-like solid: the *dodecahedron*. Pythagoras had now discovered five regular solid shapes – the Five Platonic Solids, named after Plato.

Proving Pythagoras' Theorem

At some time during his geometric explorations, Pythagoras started thinking about right-angled triangles. At this point he happened upon perhaps the most famous of all mathematical discoveries, often described as the bane of every secondary school student's life. Today we call it Pythagoras' Theorem, but it wasn't his initially. The Chinese and the Babylonians knew of it well before Pythagoras, and the Egyptians had almost certainly used right-angled triangles in building pyramids and dividing up land.

But Pythagoras deserves the credit for 'proving' that in absolutely any right-angled triangle,

The square on the hypotenuse (the long side) is always equal to the sum of the squares on the other two sides.

Once again he seems to have arrived at this proof largely from playing around with shapes. Let's do the same. Look at Figure 2.10.

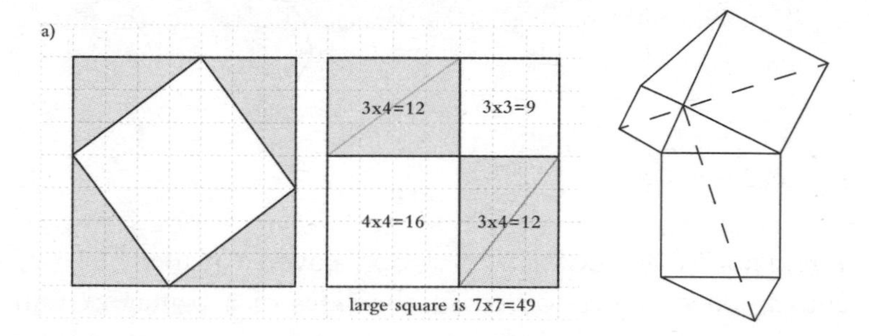

Figure 2.10a and b

We have a large square and, inside that, a slightly smaller square, tilted so that it touches the sides of the large square in four places to form four right-angled triangles. It's clear that the inner square sits on the hypotenuse of each of these triangles.

Now let's group the four triangles into pairs to form two rectangles. We can place them in the large square so that the space they leave forms two smaller squares, one on the shortest side of the triangle and the other on the second-longest side. Both large squares now contain four identical triangles. The rest of the space in one is taken up by a large square and in the other by two smaller squares. So the area of the large square in one diagram must equal the area of the two smaller squares in the other diagram. QED.

Here is another proof, discovered by Leonardo da Vinci. By adding two extra triangles and two long diagonal lines, he produced two very similar shapes and you can see that if you remove the triangles, the large square must equal the two smaller squares in area (see Figure 2.10b).

Legend has it that when Pythagoras devised his proof, he celebrated by sacrificing an ox. According to some reports it was 100 oxen, but that's probably fanciful – certainly it doesn't sit easily with what we know about the Pythagoreans, who believed in the transmigration of souls (that a dead person might be reincarnated as an animal). Besides, they were reputedly vegetarians. And while they probably had oxen to help plough their fields, it's unlikely they'd have had 100 spare to slaughter as a 'thank you' to the Muses or the Gods.

What's more likely is that Pythagoras threw a party for all his friends to celebrate his discovery. But why did he have so many friends? Surely they weren't just hanging around on the off-chance that he might make a major maths breakthrough and decide to celebrate in style...? Pythagoras' host of followers were devoted to him for many reasons. To begin with, the ideas he put forward were not being taught by anyone else. Secondly, he advocated peace, harmony and well-being, and was always trying to seek out natural solutions to ailments or medical problems, as well as the way to a healthy life. Thirdly, he encouraged his followers to seek out new ideas and solutions for themselves. In doing so, however, because he created so many disciples, we can't be sure which of his ideas were actually his, and which were proposed by his followers.

For a long time, it seems, Pythagoras and his clan were very happy immersing themselves in the beauty of maths. But the happiness and calm didn't last. Nevertheless, the cloud that emerged over Pythagoras and his idyllic community wasn't the result of death or other misfortune; rather it was due to the discovery that maths is not always neat, tidy and totally dependable. One fateful day, Pythagoras discovered what for him turned out to be a splinter in his banister of truth. He discovered *irrationals*.

It's an Irrational World

We can't be sure exactly what event shattered the Pythagorean mathematical utopia, but it may have been triggered by something as seemingly innocuous as dividing a simple line. Pythagoras saw that a line could be divided in many ways. Suppose, for example, you want to split up a line that's 17 units long. You could divide it into two units of 8 and one single unit (8 + 8 + 1 = 17), or you could split it into 7 + 7 + 3, or 6 + 6 + 5, or 5 + 5 + 5 + 2, 4 + 4 + 4 + 4 + 1, and so on.

As Pythagoras explored line-cutting further, a question occurred to him. Could you cut a line so that the two parts were in the same relationship to each other as the longer section was to the whole line?

Cut a line into two third and one third parts. One part is twice the length of the other – ratio 2:1. But compare the whole line to the largest part and the ratio is 1 to ⅔rds or $\frac{3}{2}$. If you compare the whole line to the smaller part, the ratio is 1 to ⅓rd or $\frac{3}{1}$. Neither of these is in the same proportions as the two parts – $\frac{2}{1}$. So is it possible to create two lines for which the ratios match? The answer is yes, and Pythagoras may have discovered it while exploring one particular shape: the pentagon – or perhaps the pentagram, which, after all, was the badge of the Pythagoreans (see Figure 2.11).

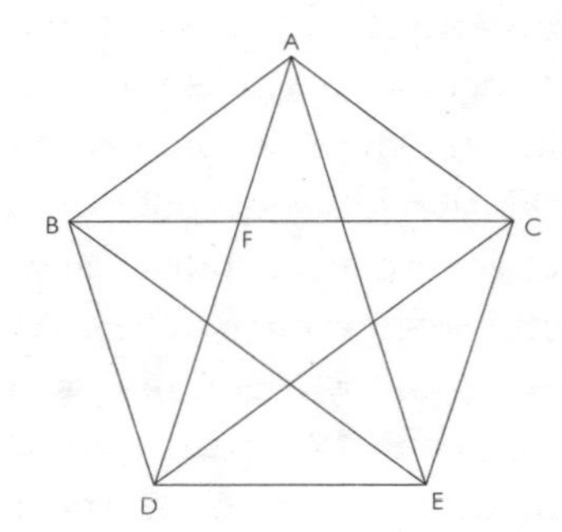

Figure 2.11

There are only two things that you can easily measure on a pentagon – the distance between two far points (B and C in the diagram, say) and the distance between two nearer points (D and E). Notice that the internal lines of the pentagram are always parallel to one side of the pentagon. This means that any diagonal line between two distant points is divided by any other diagonal line into a smaller length (BF) and a longer length (FC). By the strange and wonderful magic that is geometry, the ratio between BF and FC is the same as that between DE and BC. This ratio is known as *Phi* (say 'Fi') or – because of its beautiful mathematical properties – the *Golden Ratio* or *Golden Mean*.

We have no evidence that Pythagoras discovered or thought about dividing lines in this way, or about the Golden Ratio. However, within a few years of his death, around 480 BC, the sculptor and painter Phidias began supervising the construction of the Parthenon in Athens, and he is believed to have incorporated the Golden Ratio into many aspects of the main building. He may even have incorporated Phi into the sculptures in the Parthenon – although this is disputed by some scholars.

How to Produce a Golden Rectangle

Take a square and find the midpoint along the base. With a pair of compasses placed on that point and set to a radius that reaches to the top-right corner, draw an arc to the extended base line of the square. Complete the rectangle. The ratio between the sides of this rectangle is Phi (see Figure 2.12).

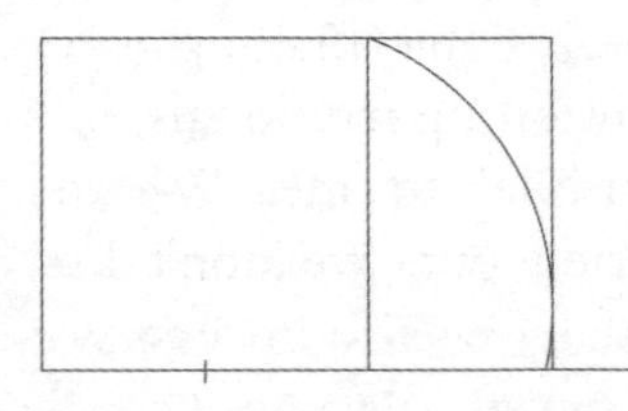

Figure 2.12

If Pythagoras did discover the Golden Ratio, either by observing pentagons or exploring golden rectangles, it must have come as a tremendous shock to him. If you divide the length of the long diagonal of a pentagon by the length of one of its sides, the result is not a whole number – to six decimal places it is 1.618034, although the decimal places don't stop there: they go on forever.

But the Pythagoreans knew nothing of decimals. Up to this point, the fractions they would have been familiar with were those known by the Egyptians, in which the number 1 is always the numerator (the top number). So Pythagoras and his followers may have known about ½, ⅓, ¼, ⅕ and so on, but they definitely did not understand more complex fractions or decimals, which came much later – the Chinese were using them around 300 BC, and the Hindus about 1,000 years afterwards.

What the Pythagoreans had discovered were numbers that didn't behave neatly, but were *irrational*. Let's look again at our pentagon. If you compare the lengths of a pentagon's sides and its diagonals, variations of the same number crop up time and time again.

> The Golden Ratio:
>
> If BC is 1 unit, then FC is 0.618034... units long.
> If FC is 1 unit, then BC is 1.618034... units long.
> If FC is 1 unit, then BF is 0.618034... units long.
>
> In reverse:
>
> If FC is 1 unit, BC is 1.618034... units long.
> If FC is 1.618034... units long, then BC is 2.618034...
>
> In all cases, the .618034... is only an approximation, because the decimal places go on forever, and never repeat.

These irrational numbers shattered Pythagoras' beautifully perfect mathematical world. So shocked was he by the discovery, we are told, that he took drastic measures to stop other people making it too. He banned his followers from mentioning irrationals, on pain of death – which seems a bit irrational to me. Whether such a punishment was ever carried out, we don't know. But when Hippasus, one of Pythagoras's followers, was drowned in a shipwreck, many remarked that he had been punished by the Gods for disclosing the irrational relationship between the sides and diagonals of a pentagon.

Although the Golden Ratio seems to have turned Pythagoras' world upside-down, it may not have been the first irrational number he had encountered. They crop up everywhere. One he must surely have known about relates to the diagonal of a simple square with sides of 1 unit in length.

From Pythagoras' Theorem, the squares on the short sides are both 1 x 1 = 1, and adding them gives 2. So, because this number must equal the square of the long side – the diagonal of the shape – the diagonal's length must be the square root of 2. To six decimal places this number is 1.414213, but the decimal places go on forever, making it irrational.

In fact, the Babylonians had already explored the concept of the square root of 2, as Figure 2.13 shows, and discovering wayward numbers that go on forever didn't seem to worry them. But it certainly troubled Pythagoras, and was a problem he never came to terms with.

Figure 2.13

Theodorus (465–398 BC), however, had less of an issue with irrationals. He was born and died in Cyrene in present-day Libya, and he followed the Pythagorean beliefs. But he removed the fear of irrationals by demonstrating that they were a pretty everyday

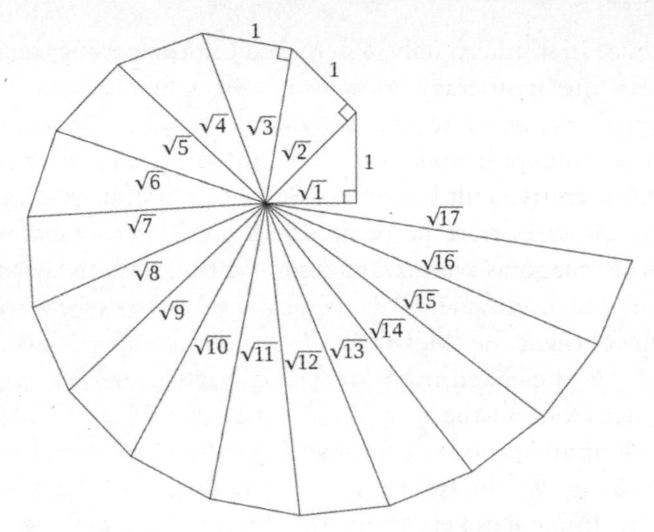

Figure 2.14

phenomenon. He discovered that $\sqrt{3}$, $\sqrt{5}$, $\sqrt{6}$, $\sqrt{7}$, $\sqrt{8}$, $\sqrt{10}$, $\sqrt{11}$, $\sqrt{12}$, $\sqrt{13}$, $\sqrt{14}$, $\sqrt{15}$ and $\sqrt{17}$ are all irrationals. In fact, the square root of any whole number that is not a natural square (such as 4, 9, 16 and so on) is irrational.

In exploring irrationals, Theodorus produced a spiral of right-angled triangles, starting with an isosceles triangle with two sides of 1 unit and one side that is the square root of 2. The next triangle has a hypotenuse that is the square root of 3, and so on. It is said to have been a mystery why his spiral stopped at 17, but as the diagram shows, it may be simply because the next triangle would overlap the previous ones (see Figure 2.14).

Theodorus also saw that the fraction $\frac{1}{7}$ is an exceptional irrational because its decimal sequence not only goes on for ever, it also repeats: $\frac{1}{7}$ in decimal form is 0.142857 or, to be more accurate, 0.142857 142857 142857 with the same set of six digits repeating forever... But this quirk means that it is not a true irrational number: true irrationals have non-repeating sequences.

Pythagoras' Legacy

According to what we know, besides his mathematical works, Pythagoras also engaged in politics, and even had some influence in southern Italy, where his political leanings tended towards the aristocracy. However, during his lifetime ideas and ideologies began to change, and the notion of encouraging a more democratic and equal society took hold. Pythagoras and his secretive cult began to lose favour, and he and many of his followers were persecuted. About 10 years before his death, Pythagoras was exiled from Croton, and his disciples scattered far and wide.

One of them was **Alcmaeon** (born around 500 BC), whose life was also cloaked in mysticism and false beliefs. He may have been one of the first people to dissect corpses to learn how the human body worked. Among his discoveries were the optic nerve, the tube that connects the ear to the mouth, and the differences between veins and arteries.

Parmenides (515 to about 450 BC) was the first Italian-born philosopher. He rejected Pythagoras' ideas of reincarnation, choosing instead reason as the cornerstone of his learning. He lived in Elea, close to the home of Pythagoras' clan, but he founded a school of philosophy (led by Zeno – more on him later) based on a more logical view of the world.

Philolaus, who was born around the time Pythagoras probably died, and who also had to flee Croton to escape persecution, was the first person to write down Pythagorean maths. This was no easy task, as the Greeks had no symbols for numbers, and used their alphabet instead. The first nine letters – alpha, beta, gamma and so on up to theta – represented the units 1 to 9; the next 10 stood for the 'tens' (10, 20 and so on); and the rest of the alphabet stood for the 'hundreds' (100, 200 and so on) until they ran out of letters. There was also no clear way of denoting whether a letter represented itself or a number; to confuse things even more, they didn't put spaces between letters to signify the ends of words or sentences. Nevertheless, despite the potential confusion, it was from Philolaus and another disciple, **Empedocles**, that a generation later (about 428–347 BC) Plato (see Chapter 3) learnt a great deal about Pythagoras, and spread the word.

Pythagoras was the first to truly realise the vast scope of mathematics, once remarking that 'All is Number', and clarifying this by splitting the subject neatly into two sections and four all-encompassing categories:

Mathematics			
Discrete		Continued	
Absolute	Relative	Stable	Moving
Arithmetic	Music	Geometry	Astronomy

For Pythagoras, arithmetic was absolute, because for any problem there is always a right answer. Music can be many things, from exquisite accomplishment to hideous noise, but all relative to our own tastes and appreciations. Geometry is solid and stable, ideally used by architects and builders to create mathematical permanence. Astronomy is always moving, not just because the Sun and planets continually sweep across the sky, but because our knowledge of the heavens is ever growing. This last assertion is by any standards an impressive prophecy: even today our knowledge of space, time and the Universe is still growing exponentially.

So, anti-establishment dreamer, geek, philosopher, mathematician, ground-breaking geometer, forefather of musical understanding, fanatical leader of a dubious cult – all these titles would fit Pythagoras. Which is perhaps why his legend lives on, even today.

CHAPTER THREE

The Great Age of Grecian Geeks

Nowadays, anyone who steeps themselves in a subject like maths or science is often referred to as a geek. Between the years 550 BC and AD 100 in Greece, a remarkable number of individuals emerged who sought a better understanding of the world and all that's in it by devoting their lives to study and learning. This chapter, following on from Thales and Pythagoras, features many, although by no means all, of those wonderfully original 'geeks' of the great Grecian age.

More Clues That the Earth Is Round

The Earth wasn't officially established as round until the third century BC, but several Grecian geeks were speculating on the planet's shape long before then. According to Thales, as a ship approaches land, the way the land rises into view suggests that the Earth is round, but few people bought that idea (the early Greek model of the Earth was a flat disc set in the sphere of the heavens). Ancient Egypt had no need for ships larger than those that sailed up and down the Nile, until King Necho (610–595 BC) began to fight battles far from Egypt, and he realised that being able to navigate the seas was an essential skill.

With that in mind, he attempted to link the Nile in southern Egypt to the Red Sea with an east/west canal, but the task proved too big and he had to abandon it. So he hired some Phoenicians – then the finest sailors around – to explore on his behalf, and they set sail from the Arabian Gulf, heading south.

According to the Greek historian **Herodotus** (484–424 BC), the Phoenician sailors always kept the shore on their starboard (right) side, and sailed around the continent he called Lybia, but which we now know as Africa. They returned via the Pillars of Hercules (two promontories on either side of the Strait of Gibraltar) at the entrance to the Mediterranean.

Herodotus reckoned the voyage took three to four years, during which time the sailors would put in to shore, plant crops, harvest them, stock their ships and move on.

The most remarkable story the Phoenicians brought back was about the behaviour of the Sun. As they sailed further south, they claimed, the Sun at first appeared directly overhead, and then, for some time, to the north. This, they deduced, could only mean that the world was in fact round, and that the continent they had sailed around straddled the Equator – something that even Herodotus struggled to accept.

This tale has often been confused with another legend derived from information in a *periplus*, an early Greek almanac of ports and coastal landmarks. This told of a Phoenician navigator called Hanno, who was born around 530 BC in Carthage in modern-day Tunisia. Hanno, the story goes, sailed with an armada of about 60 ships and 30,000 men and women, although these numbers may have been exaggerated. According to some claims the ships had each 50 oars, but assuming an average of about 500 people on each ship, the vessels would have been both heavy and cramped, and moving them at a reasonable speed with just 25 oars on each side would have been a mighty task, even in calm seas.

It's far more likely that Hanno sailed from Carthage, west through the Pillars of Hercules, then down the west coast of Africa, keeping land to the port (left) side. Apparently, on their travels they encountered men who could run as fast as horses, and who translated the local dialects for them. We have a pretty good idea that this was the west coast of Africa because of the other things they found: crocodiles and hippopotamuses, and gorillas – which they at first thought were hairy people, so savage and violent that they named them after guerilla warriors, known for their tough and violent tactics.

Hanno and his crew also claimed to have seen a mountain that spouted fire, with lava that ran into the sea, and which they named the Chariot of the Gods. This was probably Mount Cameroon (still an active volcano), which stands at about 4,000m. The travellers reported that making headway here was difficult, which gives us another clue as to their location: at that point the coast is affected by strong sea currents from the south.

Despite both these legends, however, most Greek scholars of the time still maintained that the Earth was flat, a point of view that they held for another couple of hundred years.

Philosophers a-Plenty?

After Pythagoras, dozens of original thinkers followed his example, and between them brought about the golden age of Greek philosophy. Although most of them were schooled in maths, many moved beyond its boundaries, contemplating such weighty subjects as how the Universe works, or what things are made of. Although many of their ideas were fanciful, and stemmed from little more than pure guesswork, slowly they began to devise sounder concepts, as well as robust maths to explain them.

We'll begin with three of Pythagoras' disciples. The first, **Philolaus** (around 470–385 BC) – whose name has the same root as the word 'philosopher' – was notable because he was the first to write down at least some of the Pythagoreans' ideas. What's more, Philolaus implied that some of these ideas were his own, including the notion that the Earth travels through space and around the Sun. Struggling to understand how the Earth didn't fall out of its orbit, he came up with the rather odd idea that there was an identical Earth on the opposite side of the Sun that kept things balanced.

A late disciple of Pythagoras, **Archytas** (428 to around 350 BC) lived in Tarentum, in southern Italy, across the bay from Croton, the last surviving centre of Pythagoreanism (see Figure 3.1).

Archytas is best known for improving on Pythagoras' basic musical ideas, devising a correct theory of harmonic sound progression. Providing further evidence that the Greeks were already dabbling with fractions, he explained that as you divide a musical string into fractional parts, the harmonic progression – the intervals between notes – gets shorter: 1, ½, ⅓, ¼ and so on.

Figure 3.1

Archytas maintained that higher pitched sounds are caused by faster vibrations – that suggestion was correct – and that they also travel faster than lower pitched sounds, which sounds logical but is actually wrong: it's the frequency that changes, not the speed. But measuring frequency was still some 2000 years away.

The third Pythagorean disciple, and perhaps the most influential, was **Empedocles**, a Sicilian who lived between 492 and 432 BC. Like Pythagoras, Empedocles was the subject of many stories that suggest his logical thinking was soaked in mysticism. He once prophesied that on a particular day he would ascend to be with the gods, and on the appointed day he climbed Mount Etna, stood on the lip of its crater and jumped in…Or did he? Another version – less dramatic, but probably more likely – is that he secretly fled to Greece and died there.

But Empedocles did add some powerful arguments to the body of scientific explanation. He realised that moonlight is actually reflected sunlight, and that the centre of our bodily system is the heart, not the brain. (Many sayings familiar today – not having the heart, heartbreak, the heart of the matter – originate from this time.) He also suggested that most creatures, including humans, had evolved over a very long time, something Charles Darwin would demonstrate two millennia later.

But Empedocles' most widely known idea was that everything is made of four specific elements. His reasoning was neat:

Earth	Water	Fire	Air
Dry and Cold	Wet and Cold	Hot and Dry	Hot and Wet

He even put forward explanations for how the elements combine and how one transforms into another. By applying Fire to a log (Earth), for example, it gives off sap (Water), smoke (Air) and ash (Earth again). In line with his mystical bent, he also maintained that there were two forces at work in the world: love, which helps to combine things, and hate, which drives them apart.

Atoms and Paradoxes

Despite their mystical nature, Empedocles' elemental ideas carried enough weight to be taken up by a favourite of mine in the pantheon of great Greek thinkers: **Democritus** (470–380 BC), whose name has the same root as the Greek word for democracy. He found great humour in man's faith in superstition and mysticism, and their lack of desire for true understanding, which earned him the nickname 'the laughing philosopher'. Democritus was also the first to suggest a theory of atoms – essentially that if you split anything into smaller and smaller bits, eventually you get down to a particle that won't divide any more (*atom* is the Greek word for 'indivisible').

Democritus related the elements Fire, Earth, Air and Water to four of the five regular geometric solids that

Pythagoras and his followers had identified. 'Sharp' Fire atoms, he suggested, might be like a tetrahedron; Earth fills space, so its atoms might be cubes that fit snugly together; Air atoms could be octahedral in shape because they too can be grouped together to fill space; Water atoms flow over each other, so might be like the smoothest solid, the icosahedron. The fifth solid, the dodecahedron, he suggested, filled space.

Democritus suggested two more celestial ideas: that the Universe is a swirling mass of atoms that clump together over huge swathes of time to form what we now call stars, and that the Milky Way is a vast group of such clumps. But not all the great minds of the time were happy with the apparent practicality of Democritus' reasoning.

One such thinker, **Zeno** of Elea (around 490–425 BC), really put the cat among the pigeons, by discussing an improbable race between a great Greek athlete and a ponderous tortoise. Zeno became famous for his paradoxes, or ideas that seem to contradict themselves. The most famous pits Achilles – heroic warrior and athlete – in a race against a tortoise. When asked who will win the race, most people would put Achilles down as a pretty safe bet. But Zeno put forward some propositions that make the outcome far from certain.

Let's suppose Achilles runs 10 times faster than the tortoise. The race would be unfair from the start, because Achilles would run 100m in the time the tortoise covers just 10. So let's give the tortoise a 90m start. Now who will win? The first thought you might have is that it will be a dead heat. But, said Zeno, even that isn't cut and dried.

Let's think about it in more detail. When Achilles has completed 90m, the tortoise has covered 99m. When Achilles reaches the 99m mark, the tortoise will have gone another $^9/_{10}$ of a metre. When Achilles makes the extra $^9/_{10}$ of a metre, the tortoise will have gone a further $^{90}/_{100}$ of a metre, and so on. By dividing time and distance infinitely, Achilles will never quite catch the tortoise, let alone pass him (see Figure 3.2).

Zeno devised nine similar paradoxes that all divide time, distance or space into ever smaller pieces. In the Fletchers Paradox, for example, he talks about a speeding arrow:

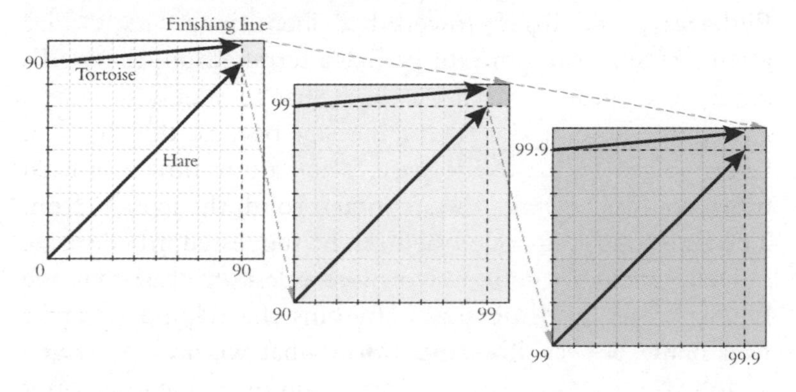

Figure 3.2

An arrow takes up a certain space when it is at rest. But it takes up the same amount of space at any one time when it's flying through the air. At any moment, the arrow is taking up that particular amount of space, so it can't be moving. Therefore, Zeno suggested, all movement is an illusion!

Are you happy with Zeno's paradoxes? Democritus certainly wasn't: for him, thinking about dividing time or anything else into infinitely smaller pieces gets us nowhere and only confuses things. Hence his concept of the indivisible atom, the limit of smallness that keeps things rational.

Down the years, however, many mathematicians and philosophers have realised the importance of Zeno's thoughts, culminating perhaps in the work of Werner Heisenberg and his uncertainty principle, devised in 1927. According to this now renowned theory, which concerns electrons flying around in atoms, if you know the speed an electron is travelling at, then it's impossible to work out where it is at any given moment; on the other hand, if you know exactly where it is at any given moment, you can't deduce its speed. In essence this is Zeno's flying arrow paradox restated about 2,400 years later; arguably, it also marks the threshold of the modern discipline of subatomic particle physics.

Almost all we know about Zeno comes to us courtesy of Aristotle, who proved to be the most influential of all the early Greek philosophers. But before I discuss him and the things he got up to, we must briefly trace the history of two other well-known Greeks, in the order in which they lived.

The Wisest Greek of All

Socrates was born in Athens sometime around 470 BC, and died there about 70 years later in 399 BC. We know exactly how and when he died, but pretty much everything else we know about him comes from the writings of others. Apparently he was short and stout with a pug nose – indeed quite ugly – but he became a legend, so clearly had something going for him: probably his wit, sense of humour and brilliant conversation. When I was a teenager my dad assured me that it was these qualities – rather than good looks and a fine physique – that won out with the girls! Socrates was also quite hardy – according to some reports he could sit all day in the snow, meditating.

Socrates was a fantastic speaker – Xenophon, a historian and chronicler of the time, invented the first shorthand system so he could record his speeches. Socrates was a fearless soldier and a shrewd and wise politician, and it was impossible to sway him from his judgement when he'd made up his mind about something. He also despised money, and generously spent it when he had it – apparently to the despair of his wife.

A master of debate, he even developed his own debating style, still known as *Socratic irony*: starting the discussion by pretending ignorance, but then, by questioning and probing his opponents, revealing their ignorance, and undermining their stance or opinion. Clever though his method was, however, making others look foolish in this way did not endear him to many people.

When the Oracle of Delphi declared him 'the wisest of all Greeks', Socrates laughed, admitting that 'If I am the wisest, it is because I know that I know nothing.' He rejected explorations in astronomy, preferring to seek out, as Thales did, the more down-to-earth virtues of real life. He did not dabble much in mathematics, but reportedly believed that it was possible to construct a model of the Earth using 12 pentagons sewn together in the shape of a dodecahedron – amazingly, just like a modern soccer ball. But it was his influence on moral rather than scientific study that perhaps explains why the Greeks of this period progressed so slowly on scientific fronts.

Sadly, Socrates' genius, attitudes and success made him many enemies. In 399 BC he was charged with atheism, treason and corrupting the young. But he also had many devoted friends, and had he attempted a defence of any kind he would probably have been acquitted, thanks to their influence. As it was, he goaded the jury (made up of 500 men), highlighting their collective foolishness, until by a slim majority of 280 to 220, they delivered the death sentence.

He then spent a month in captivity, during which time he could have probably escaped at any time, again helped by his many friends. But at 70 he considered himself to have lived a good life, so on the appointed day he calmly took his punishment: drinking hemlock. In a particularly cruel irony, the poison, while slowly paralysing first his extremities and finally his lungs, left his brain unaffected until the end.

The First Academic

Socrates' finest disciple was a man called **Aristocles** (427–347 BC), from which we get the modern word 'aristocracy'. He earned the nickname 'Plato' thanks to his athletic physique (*platon* meaning 'broad in the shoulder' in Greek), and it stuck. Socrates' death, and its manner, came as a terrible shock to Plato. He once remarked that 'society will never really improve until kings become philosophers and philosophers become kings' – but then being related to royalty he probably would say that.

As a young man Plato visited Italy, where he absorbed the work of Pythagoras and his followers, and Africa, where he studied under Theodorus before settling in Athens. There, around 387 BC, he set up the world's first university – the Academy – in grounds previously owned by Greek hero Academus, from whom the institution got its name. The Academy largely epitomised Plato's belief that students should gain knowledge to strengthen the soul, although he loved mathematics as a tool for explaining and proving things. The institution lasted about 900 years, until it was closed in AD 529 by Roman Emperor Justinian.

According to Roman author and engineer Vitruvius, who lived about 300 years after him, Plato is the possible source of a maths puzzle that is still popular today. Plato and Socrates asked an ignorant student how a square window could easily be doubled in size. Plato's aim was always to simplify and demystify mathematics and highlight its importance through its capacity to empower. So he explained: divide the square into four triangles by drawing two diagonals. Now add a mirror image of each triangle to produce a diamond shape – this is exactly twice the size of the original square (see Figure 3.3).

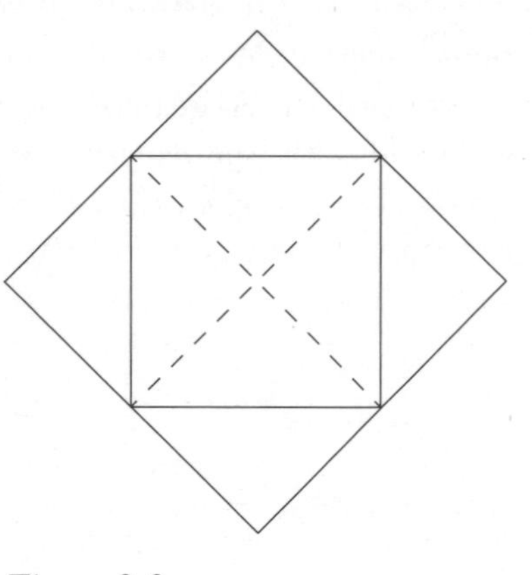

Figure 3.3

Over the door of the Academy was the phrase 'Let no one ignorant of mathematics enter here.' For Plato, maths was also a tool for exploring and explaining the heavens: he was a big fan of Pythagoras' Music of the Spheres, and the idea that each planet may have its own distinct note. He also absorbed Democritus' idea of linking the elements to the five regular geometric solids – shapes that we now call Platonic solids after him.

And as well as suggesting that Fire atoms are tetrahedral, Earth atoms are cubes, atoms of Air are octahedral and Water atoms are icosahedral, Plato also followed Philolaus in suggesting that space was filled with a fifth element, *aether*, made up of atoms that could be dodecahedral in shape. It was, of course, pure conjecture. Today we use the term 'platonic friendship', suggesting that, like the solids, two people might have many things in common but still maintain their differences.

Plato listed all his ideas in a dialogue entitled *Timaeus*, in which Socrates and Plato's favourite pupil **Theaetetus** (417–369 BC) discussed maths and philosophical concepts. Among Theaetetus' achievements, he was the first to 'prove' that there could be no more than five regular solids. He also advanced the understanding of irrational numbers, by showing that they are neither rare nor ugly, but highly plentiful and essential in the overall number system, and indeed our understanding of numbers.

In *Timaeus*, Plato explored an idea that, about 1,000 years previously, a Greek island had been destroyed in a giant earthquake; he then added a moralistic legend about an almost perfect society living there – a realm he called Atlantis. The lost island of Atlantis was almost certainly a work of fiction – a fact that has never dissuaded a host of romantics over the generations from spending vast sums of money hunting for it, or writing ceaseless books and film scripts about it. On the other hand, the Aegean Sea off mainland Greece is so uniquely scattered with islands that the idea that some of them are the remnants of a gigantic earthquake may well contain some truth (see Figure 3.4).

Figure 3.4

The Misconceived Father of Science

Aristotle (384–322 BC) studied under Plato at the Academy until his teacher's death in 347 BC, after which he left Athens, complaining that too much mathematics was being taught at the school, and not enough philosophy. While away he studied what turned out to be his greatest passion, biology. After observing that a dolphin – like humans and other animals – gives birth and suckles its young, he concluded that it must be a water-dwelling mammal rather than a fish.

King Philip II of Macedon hired Aristotle to teach his 14-year-old son Alexander (who later became Alexander the Great). In 336 BC, however, the king was assassinated, and his 20 year-old son took over, declaring that he had no time for or interest in further education. Aristotle returned to Athens, where he established his own school, the Lyceum. It was often referred to as the 'peripatetic' or 'walkabout' school, because Aristotle was keen on lecturing while walking in the grounds. While at the Lyceum he established a huge collection of manuscripts that eventually formed the core of the Library in Alexandria; Aristotle's lectures alone accounted for 150 volumes, making him a one-man encyclopaedia. The reason why his name and works are so widely known in Europe today is largely because many of his books were found in a pit in about 80 BC by the Roman general Sulla, then taken to Rome and copied.

Although Aristotle applied philosophical thought to everything we might call scientific, he never checked the validity of his ideas using experiments. Still, his 'scientific' concepts were handed down and taken as gospel for the best part of the next 2,000 years, until Galileo finally questioned them. Aristotle's ideas on logic and logical thought endured even longer, until the mathematician George Boole, in the nineteenth century, converted them into pure mathematics, which ultimately led to the dawn of the computer age.

When Alexander the Great died in 323 BC, many were strongly opposed to all he had stood for, and Aristotle, fearing that his status as Alexander's teacher would leave him open to victimisation, retired to his mother's home town, where he died the following year.

Aristotle had no love of maths for maths' sake, but it clearly influenced his teaching. Learning of the Golden Ratio he applied it to philosophical thought, describing it as the desirable middle between two extremes: excess and deficiency. So courage is a virtue, for example, but taken to excess it becomes recklessness, while if it's in short supply the result is cowardice.

The Golden Ratio could also be seen as an attribute of beauty and truth. According to Aristotle's teachings, beauty has three 'ingredients': symmetry, proportion and harmony. This triad of principles infused Aristotle's life: for him we love something for its beauty – whether in architecture, education or even politics. Aristotle judged life by these principles and was willing to accept only what pleased him about them.

This idea was by no means restricted to the West: a similar concept, the Doctrine of the Mean, was propounded in Chinese philosophy, by Confucius; Buddhism has a similar concept, 'the middle way'.

The trio of great Greek philosophers – Socrates, Plato and Aristotle – required little actual mathematics to help them formulate their philosophical ideas. In that case, which individuals, in those times, were responsible for taking maths several giant steps forwards in the ensuing years?

The First Golden Age of Greek Mathematics

The tales of Thales and Pythagoras are truly the stuff of legend, but the lack of evidence around the men themselves makes it hard for us to be sure what actually happened during their lifetimes. In fact, any kind of certainty in Greek mathematics had to wait until the second half of the fifth century BC, with the work of Hippocrates of Chios.

There were, in fact, two Hippocrates who lived around the same time, both named after the horse, which was regarded as a noble creature in ancient Greece. Both made names for themselves, leaving legacies that survive to this day.

The most famous of the two was Hippocrates the physician (460–370 BC), who lived on the island of Kos, where he set up what was then the finest centre for medicine in the Grecian world. Even today, medical students, after qualifying and before becoming fully fledged practitioners, swear a version of the Hippocratic Oath.

The name Hippocrates shares the same root as the word 'hypocrisy' (*hypocracy* in Greek means to simulate honesty or virtue). So it might perhaps be hypocritical of me to celebrate Hippocrates' words when we can't be sure that the Oath was actually written by him. But he was certainly a great man: the first eminent member of the Athenian school of Greek philosophers, followed by Socrates. It might also be more hypocrisy on my part if I were to claim that, of the two Hippocrates, I think his achievements were the most important. Because, being a lover of all things mathematical, I have to honestly say that I hold the slightly older **Hippocrates of Chios** (470–410 BC) in greater esteem.

A Master of Simplicity

It seems that Aristotle did not rate this Hippocrates as highly as I do today. Apparently he thought of him more as a hypocritic oaf. Hippocrates arrived in Athens, having been attacked and robbed by Athenian pirates, and his plan in the city was to claim all of his stolen possessions back through the courts. Knowing the wiles of the criminal fraternity, Aristotle and others laughed at him for even trying and, sure enough, after spending more money and getting absolutely nothing back, Hippocrates faced ridicule again. From then on, he concentrated his life's efforts on maths, at which he was far more successful.

Hippocrates, in fact, laid down basic and specific rules governing certain areas of mathematics. We do know that he wrote a book called *Elements*, which sadly has been lost. However, it may have been the original inspiration for the

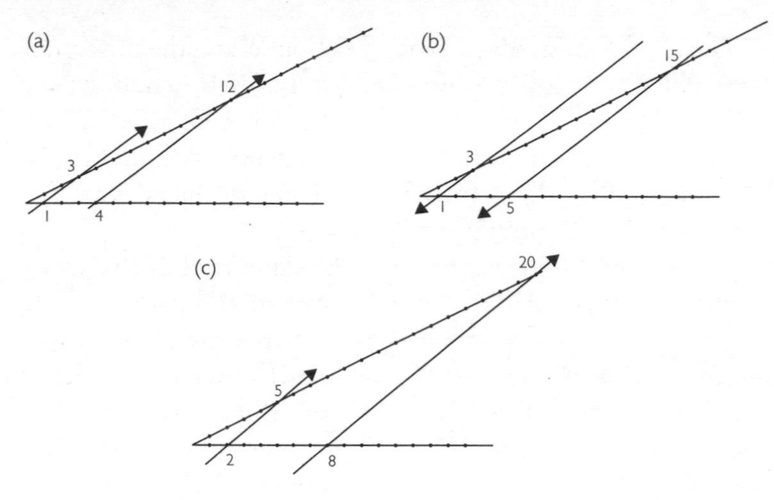

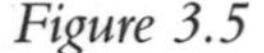
Figure 3.5

greatest book of mathematics ever, also called *Elements,* which was written just over a hundred years later, and which we encounter at the end of this chapter.

Hippocrates was perhaps the first person to realise the sheer power of a couple of straight lines, and how they could be used for many different types of basic mathematical calculation. With two lines set at an angle, it's possible to multiply any two numbers, divide any two numbers and transfer similar ratios from one line to another, a useful tool sometimes known as the *rule of three.*

The Mesolabe Compass

Try this: first draw two lines at an angle to one another – any angle will work, but about 40 degrees is fine. With the compasses set at a constant distance, mark off equal units along each line. You should now have produced what was called the *mesolabe compass* (see Figure 3.5).

Every journey begins with one step, and the secret to using this amazing calculating device lies in always starting with *unity* – or the number 1. Let's see how it works…

Multiplication: To multiply any two numbers – 3 times 4, say – first find 1 on the bottom line and 3 on the top line and draw a line through the two points. You now have a picture of 'one three'. (Maths in pictures? What a wonderful idea!) But we want four threes, so next draw a line parallel to the 1–3 line that passes through the 4 on the bottom line. This line will pass through the 12 on the top line, illustrating very neatly that 3 x 4 = 12 (see Figure 3.5a).

Successive parallel lines will give 2 x 3 = 6, 3 x 3 = 9, and so on. The simple beauty of it all is jaw-dropping.

Division: To divide a number by any other, simply reverse the process. To divide 15 by 5, for example, draw a line that goes through 15 on the higher line and 5 on the lower one. A parallel line through the 1 on the bottom line will pass through the 3 on the upper one, showing that $^{15}/_{5}$ = 3 (see Figure 3.5b).

Transferring ratios: Although the Greeks, like the Egyptians before them, were aware of fractions, they normally only used those with 1 as the numerator: ⅓, ¼, ⅕ and so on. (Decimals were still some way off in the future.) They almost always did calculations by comparing the ratio of two numbers to one another. This nifty tool – known as the rule of three – was rediscovered many times over the years (see Figure 3.5c).

Try this problem. As 2 is to 5, so 8 is to … what?

The mesolabe compass makes things easy. Draw a line through 2 on the lower line and 5 on the upper one. Now, if you draw a line parallel to it that passes through the 8 on the lower line, it will pass through the 20 on the upper line. So 2 is to 5 as 8 is to 20. Job done.

This simple method is so extraordinarily powerful. In the Wow Factory Mathematical Index we show how to divide any line of unknown length into exactly equal portions. But for Hippocrates this was only the start...

The Square Root of Any Number?

Many people find searching for the square root of just one number daunting enough. But is there a simple method of finding the square root of any number? 'Yes' is the answer – and what's more the method also seems to date back to around the time of Hippocrates of Chios, and is essentially an extension of Thales' Theorem, which we explored in Chapter 2.

Figure 3.6 shows half a circle with diameter AC. DE is a random line perpendicular to AC. Now complete the triangles ADE and EDC. We can see that the two triangles are similar – all their angles are the same, which means that their sides will be in the same proportions. So if EC is 3 units long and DE is 9 units, then AE will be 27 units long, because 3:9 = 9:27.

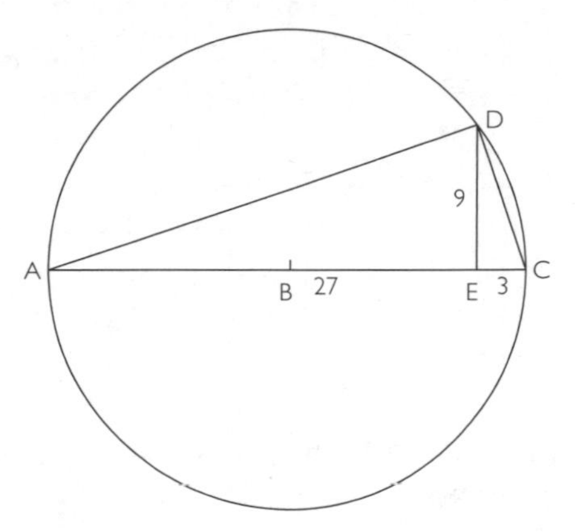

Figure 3.6

But if you multiply AE x EC you always get a square number, and the line DE is the square root of that number. In this case 27 x 3 = 81, which equals 9^2. Incredibly, it works every time.

But the magic really happens if you give EC a length of 1, or unity. Now, whatever length AE is, DE will always be its square root.

So, say you wanted to find the square root of 9? It's an easy one, I know, but it illustrates the principle nicely. Simply count 9 along a ruler or a line divided into equal units. Now add 1. Why? Because that's what makes it work!

Your line is now 10 units long. Find the centre point at 5 units and draw a semicircle with a radius of 5 units from the zero over to the 10 point. Now draw a perpendicular line from the 9 point. Where that line touches the curve, its length will measure exactly 3 units, or the square root of 9 (see Figure 3.7).

So, if we constructed the diagram accurately, with EC as 1 unit and AE as 'any other number', the length of DE will always be the square root of AE. Amazing or what?

These two examples were so powerful that René Descartes, some 2,000 years later, used them both to introduce his great work, *The Geometry* (see Chapter 12). His linking of geometry and algebra inspired both Newton and Leibnitz, and was an indirect cause of the Industrial Revolution: without the maths to underpin it, the technology of steam engines and all that came afterwards could not have progressed.

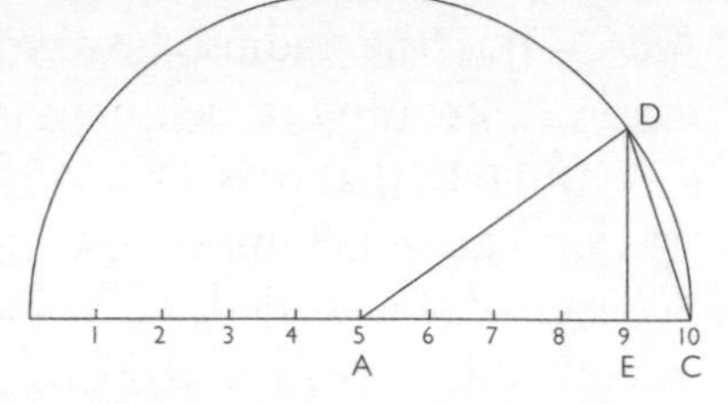

Figure 3.7

In the Wow Factor Mathematical Index, we show how an eminent 19th century scientist used this square root method to accurately measure the diameter of the earth.

Magical Lunes

Sadly, almost all of Hippocrates' work has been lost; nevertheless it is very well documented that he made an amazing discovery while playing with a pair of compasses and reflecting on Pythagoras' Theorem.

What people often miss is that Pythagoras' Theorem doesn't just apply to squares on the sides of triangles. It can work for triangles, rectangles – or even camels, set in proportion on each side. Hippocrates didn't use camels as far as we know, but in toying with a pair of compasses he discovered the area of a *lune*, or moon-shaped segment.

First, he drew an isosceles right triangle ABC inside a semicircle (see Figure 3.8). He then drew another triangle, AOB, which was a mirror image of the first. Then, taking O as the centre

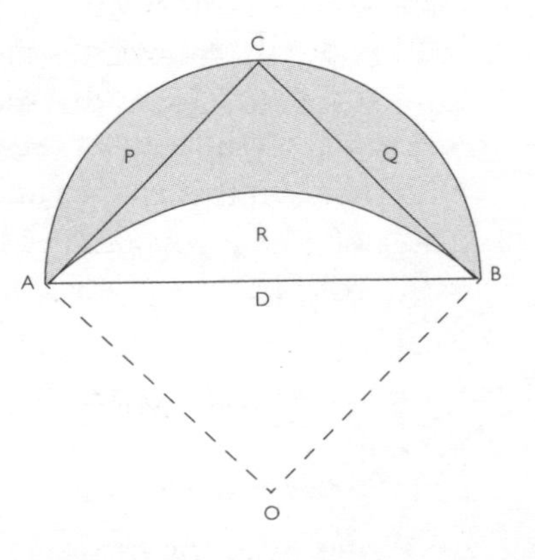

Figure 3.8

and AO as the radius, he drew an arc inside the original triangle, creating a segment of circle that we'll call R. Recalling Pythagoras' Theorem, Hippocrates saw that the area of segment R must equal that of segments P plus Q. So the area of the shaded lune must therefore equal the area of triangle ABC.

This was the first time somebody had managed to calculate an area bounded totally by curves. It wouldn't become possible to calculate any curved area until the advent of calculus in the seventeenth century.

Using ruler and compasses, Hippocrates discovered other lune areas using different methods, and we have no way of knowing which came first. It is clear that he simply loved playing with a pair of compasses and letting his mind do the rest. When I was at school, playing with compasses was a relief from the usual mathematical world of numerical calculation.

Hippocrates found that by constructing semicircles on the sides of right-angled triangles, he could calculate the areas of many differently sized lunes. Figure 3.9 shows a right triangle in a semicircle, with a vertical line dividing the triangle into two smaller, differently sized triangles. By constructing a semicircle on each of the sides, we form two new lunes. The area of each lune is equal to the area of the triangle directly below it.

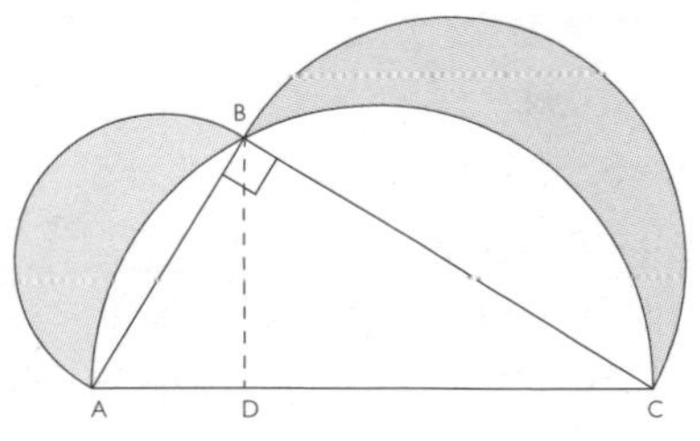

Figure 3.9

So where can this all this take us? Figure 3.10 is a beautiful example of where you can get to with a pair of compasses and a luney attitude.

The Three Classic Maths Problems

1. Duplicating the Cube

Hippocrates was the first person to try to put geometric theorems into a logical progression, and using his trusty

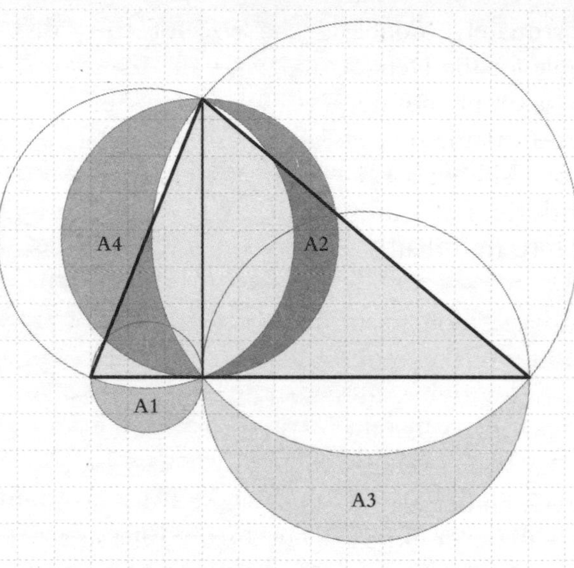

Figure 3.10

straight edge and compasses, he made several new discoveries. Then he attempted to tackle the three unanswered mathematical questions of his day. He began with 'duplicating the cube'.

This problem is usually known as the Delian problem, after the Greek town of Delos. To ward off a plague, the story goes, the townspeople of Delos erected a statue of Apollo. After consulting the Oracle, however, the town's elders decreed that the Gods were unhappy with the meagre size of the plinth, so they commissioned one twice as large. Maybe, on the face of it, this seems like an easy problem to solve. But oh, dear me, no!

It's not as simple as doubling the length of a side. If you start with a cube with sides of length 1, its volume is 1 x 1 x 1 = 1. If you then double the length of the sides, the volume of the new cube is 2 x 2 x 2 = 8. Eight times the volume of the original. If you triple the length of the side, you get 3 x 3 x 3 = 27 times the size of the original cube, as we know from exploring the Rubic Cube.

Unfortunately, doubling the size of the cube proved impossible for the Delians, and everyone else, because Greek maths was simply not up to the job at that time.

Using a calculator you can quickly find the approximate length of a side of a cube that's twice as large as a starting cube with sides of 1 unit. It took me just a few minutes, for example, to find that $1.26^3 = 2.000376$, which is very close to 2...

On a basic (non-scientific) calculator, to cube a number you can press the multiplication key twice, followed by the equals key twice. So if you punch in 1.26 x x = = you get 1.26^3, which is 2.000376. Better still, punch in 1.259921 x x = = and you get 1.9999997; or 1.259922 x x = = to get 2.0000045. Either way, you're getting close to a result of 2, but you will never quite reach it, because the square root of 2 is irrational.

The ancient Greeks had no calculators, and knew nothing about decimals, which hadn't been discovered yet. They were even shaky with fractions. So Hippocrates attacked problems like this using proportionals and simple geometry. But he never found a solution to doubling the cube. In fact, despite claims from many cranks down the ages to have solved the problem, it's impossible to state with absolute accuracy: the cube root of 2 is irrational.

2. Trisecting an Angle

The second problem the ancient Greeks wrestled with sounds simple enough. Using just straight edge and compasses, trisect – or divide exactly into three equal parts – any given angle. Hippocrates discovered that this was impossible too.

However, with a simple tool (and a bit of cheating) it gets quite easy. The tool – a carpenter's square – is a right-angled 'L'-shaped device used in ancient Greece, and which can still be found in modern tool boxes (but we don't know that Hippocrates discovered this solution).

First, looking along the short outer edge of the tool, call the corner X and mark two points, one that corresponds with the width of the longer arm (Y), and the other twice that distance along (Z). (Already this breaks the rules of using an unmarked straight edge.)

Now, to trisect the angle ABC, first place the long arm of the carpenter's square along one of the angle lines AB and use it to draw a line DE parallel to AB, as shown. Place the tool so that the upper side of the long arm passes through B and the corner X lies on the new line DE. Slide the tool along until point Z lies on the line BC. Mark the point on DE where the corner lies and call it X; also mark point Y. The lines BX and BY trisect the angle ABC perfectly (see Figure 3.11).

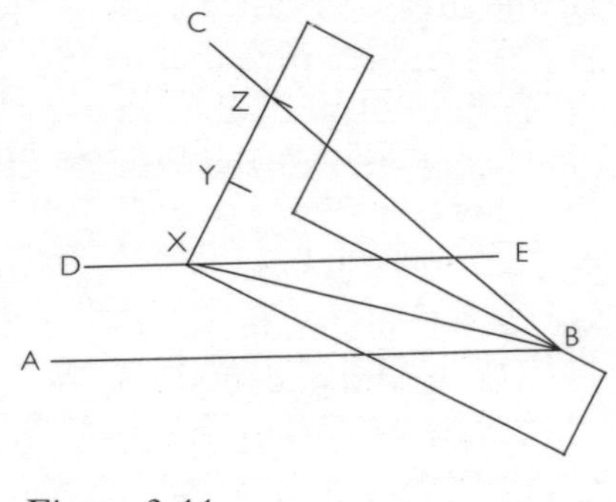

Figure 3.11

Of course, it is possible to draw the trisectors of a known angle using a protractor – if an angle is 60 degrees, then the trisectors would each be 20 degrees apart. In 1899 Frank Morley, an English mathematician who spent almost all his academic life in the USA, developed a theorem that involves trisectors of angles, and which still carries his name today. If you draw the trisectors from the three angles of any triangle, ABC, and then mark where the trisectors closest to each side meet, the three points will always form a perfect equilateral triangle (see Figure 3.12).

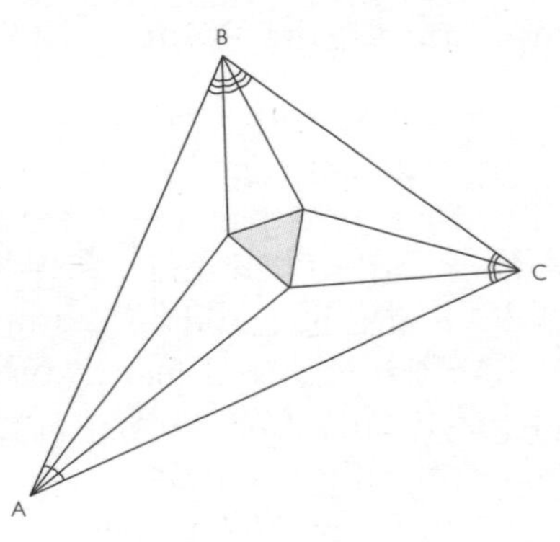

Figure 3.12

3. Squaring the Circle

This third classic problem is sometimes known as the *quadrature problem* ('quadrature' means 'area finding'). The aim is to create a square and a circle with equal areas.

The Egyptians had already found a reasonably accurate solution to this problem, as we saw in Chapter 1. They worked out the area of a circle as $\frac{8}{9}$ of the diameter squared. So the area of a circle with a diameter of 9 units would be 8 x 8 = 64 square units. To calculate the area of the circle using the value for π suggested in the Rhind Papyrus (3.16), we get πr^2 = 3.16 x 4.5 x 4.5 = 63.99. It is close to 64 but not quite. In fact, it was 1882 before German mathematician Carl Louis Ferdinand von Lindemann proved that π is not only irrational, but also transcendental: it cannot be found exactly, and transcends a solution using algebraic equations.

These days it is possible, using algebraic formulae, for a computer to appear to produce a square and a circle with identical areas. The flaw is that the image is made up of pixels, which are small squares, from which you cannot make a perfect circle. So while the image makes it look as if the two shapes have identical areas, in reality they don't.

Studying to the Point of Exhaustion

Hippocrates' work was extended by another influential Greek mathematician called **Eudoxus** (408–355 BC). He was born in Cnidus (now part of modern-day Turkey), but moved to Athens and studied under Plato. Being very poor, this wasn't easy, because he could only afford cheap lodgings in the docks area of the city, and had to walk several miles to and from Plato's Academy every day. But, as is often the case, early hardship only strengthened his character. Eventually, quite late in life, in 367 BC, he is believed to have headed the Academy while Plato was in Sicily.

Eudoxus also travelled to Italy and Sicily, where he met Archytas; in Egypt he studied astronomy and produced a map of the heavens, using lines of latitude and longitude for the very first time. Later, he used the same method to produce a map of the known Earth.

At first he accepted Plato's idea that the planets orbit the Sun in perfect circles, but according to his measurements this didn't seem quite right. So he suggested that the centre of each circle was itself moving in a different circle, thus preserving Plato's idea of circular orbits.

The Method of Exhaustion

Eudoxus also developed the idea of seeking mathematical solutions using the 'method of exhaustion': at each stage, you achieve a more accurate figure. To find an increasingly accurate solution to the square root of 2, for example, he produced a ladder of numbers.

Starting with 1 and 1 as the first row, he then added those numbers together to start the second row – so 1 + 1 = 2. He then added this number to the one above it to get the second number in the second row: 2 + 1 = 3. And so on. As the ladder builds, the ratio of each pair of numbers grows ever closer to the value for the square root of 2.

1	1	1/1 = 1
2	3	3/2 = 1.5
5	7	7/5= 1.4
12	17	17/12 = 1.416666
29	41	41/29 = 1.4137931
70	99	99/70 = 1.4142857

Square root of 2 = 1.4142135.

An unusual property of this sequence is that if you double the square of the first number it is always 1 more or 1 less than the square of the second number (so 5 x 5 x 2 = 50, while 7 x 7 = 49).

Eudoxus also explained something of supreme importance in mathematics: never include zero when working with multiplication or division. If you do, the whole method collapses, because if you multiply or divide anything by 0, the answer is always 0. So calculating 1 x 2 x 3 x 4 x 5 x 6 x 7 x

8 x 9 might take a little while, but calculating 1 x 2 x 3 x 4 x 5 x 6 x 7 x 8 x 9 x 0 is trivial – the answer must be 0.

Eudoxus also explained why Zeno's paradoxes tend to cause such confusion by dividing time into infinitely small pieces and why, in reality, Achilles – running 10 times faster than the tortoise – must always pass it eventually.

Eudoxus was a great teacher, and one of his students, Menaechmus, was the first to discuss the four shapes formed when you slice a cone, which differ according to the angle of the slice. This work was later taken up by Archimedes and Apollonius (who we meet in Chapters 4 and 5). Apollonius developed his first four books on *conics* from Menaechmus' work, after which he added four books of his own ideas to expand the subject.

Alexandria – a Great New Centre of Learning

Halfway through the fourth century BC, Athens was the centre of the Greek empire, and the most culturally established city in the known world. But very quickly great changes took place, brought about mostly by one great individual.

Alexander III of Macedon was born in 356 BC, the son of warrior king Philip of Macedon and his wife Olympia. As a teenager, Alexander complained that the process of learning mathematics was too slow, to which his teacher – Aristotle – replied, 'There is no royal road to mathematics!' This clearly didn't suit the young Alexander, however: within his short lifetime things moved very quickly indeed.

Aged only 18, under the direction of his father, he led a successful Macedonian cavalry charge, a victory that marked the beginning of the end of Athens as a centre of Greek power. At the age of 20, following his father's assassination, Alexander became King of Macedonia and Greece, and from then on his short life became almost one continuous fight. After conquering what is now Turkey, he fought his way around the eastern coast of the Mediterranean to Egypt where, in 332 BC, now aged 24, he founded the city of Alexandria.

Over the next eight years, Alexander rightly earned the title of 'Great': he established the greatest empire the ancient world had ever known, stretching an average of 1,000 miles north to south, and some 2,500 miles east to the Punjab in India.

But everything happened so quickly for Alexander that holding it all together was impossible. On his return to the Mediterranean, he faced treachery and rebellion, from the very leaders he had left in charge. He dealt with them quickly and brutally, slaughtering all but the most faithful. Still undaunted, he then began to draw up plans to take on Arabia. One day, however, quite tragically, this mighty warrior caught a fever and died – and he was still not even 33 years old.

Alexander's legendary exploits still feature in the history of every country he conquered, and even in their religions, and he is remembered as some kind of superhero. And although at heart he was always a warrior, Alexander's two greatest achievements (apart from his military ones) were spreading the influence of Greek culture far and wide, and founding the city of Alexandria – soon to become great too.

Alexandria sits majestically at the mouth of the mighty Nile, in northern Egypt. It is the African seaport nearest to Greece, and it also serves as a crossing point for the people of North Africa and the Eastern Mediterranean. However, in Alexander's time the city developed very little and its greatness was still to come.

Following Alexander's death, his empire was split into three, and Ptolemy I, one of his generals, took control of Egypt. Ptolemy set to work fulfilling his master's dream by developing Alexandria into the greatest centre of Greek learning there was, surpassing and quite rapidly eclipsing Athens itself.

Although Ptolemy started work on the Pharos Lighthouse, it was eventually finished by his son Ptolemy II, in about 280 BC (it was destroyed by an earthquake in the fourteenth century). It stood about 130m tall, second only in height to the Pyramids of Giza, and took its rightful place as one of the Seven Wonders of the Ancient World. During the day, a moving mirror at the top of the lighthouse reflected the Sun's rays around the horizon; at night a fire blazed within.

More important by far than the lighthouse, however, although it was never considered one of the Seven Wonders, was the Royal Library of Alexandria, which Ptolemy I started in about 300 BC. Scrolls from Greece – and all the known world, in fact – were gathered there, until it was reputed to hold some 700,000 volumes. The library was, in effect, a university, and one of its first teachers was to produce a set of 13 books that even today has had more editions printed than any other book in history except the Bible. Surprisingly, perhaps, the work was a maths textbook, listing everything that was known on the subject at that time. Its author was a man named Euclid.

A Master, Truly in his Element

Euclid (325–270 BC) was born close to Alexandria but was almost certainly educated in Athens. He clearly knew of Hippocrates and his work; following his lead, Euclid called his own 13-volume maths opus *Elements*. In these books he succeeded in collecting together the great wealth of mathematical knowledge that those before him had produced, distilled into 467 propositions.

In fact, in my own small way, *Elements* was more a research project than an original work. To this day, we are not sure that it includes even one idea that Euclid originated. His great achievement was to collect and edit the work of all the previous Greek mathematicians into one comprehensive masterpiece.

In fact, in my own small way, that's what I've been trying to do in this book so far, and I've done it as briefly as possible. So, given that I've covered so much material already, there is no need for me to go into great detail over what Euclid's *Elements* actually contained. Phew!

Across its 13 volumes, Euclid made *Elements* as comprehensive as possible. Starting with very basic ideas and progressing through to quite complex maths, he left out nothing he had gleaned from the mathematicians who had gone before.

The *Elements'* Thirteen Books

Books I and II deal mostly with lines, and plane and regular figures that are Pythagorean in origin (Pythagoras' Theorem features in Book I). Book II looks at how to find square roots geometrically, echoing Hippocrates, and Book III is mostly about circles, while Book IV features polygons inscribed in and around circles.

Book V features proportion, and can be traced back to Eudoxus; this leads to more advanced explanations in Book VI. Books VII, VIII and IX cover arithmetic. They discuss the theory of numbers, prime and composite numbers, the lowest common multiple (LCM) and the greatest common divisor (GCD), which once again seem to have come mostly from the Pythagoreans. In these volumes Euclid introduced the concept of perfect numbers (such as 6, 28 and 496), for which all possible divisors add up to give the number itself (28, for example, can be divided by 1, 2, 4, 7 and 14, which added together equal … 28). Larger perfect numbers are rather harder to find: the ninth in the series is 37 digits long.

Book X of *Elements* is mostly concerned with what it calls *incommensurables* or irrational numbers, and once again Euclid echoes Eudoxus, and his method of exhaustion. Book XI discusses three-dimensional geometry, while Book XII proves Hippocrates' theorem that the area of a circle is πr^2. It also discusses cones, pyramids and cylinders, and how the volume of a cone is one-third of the volume of the cylinder it fits inside.

Euclid's intention was that the final volume of *Elements*, Book XIII, should cap all that had gone before. Sure enough, in it he covered the five Platonic solids, introduced by Pythagoras and extolled by Plato and Theaetetus. He placed them all inside spheres, an innovation that would influence Johannes Kepler some 1,900 years later. The book finally arrives at the ultimate platonic solid, the dodecahedron, which by Euclid's time had become the geometric symbol for the Universe itself.

Examples of Euclid's Maths

To achieve his goal of collating all known mathematical knowledge up to that point, Euclid had to start with the simplest of mathematical ideas. For him, this was the statement,

'The shortest distance between two points is a straight line.' But he needed to be even more explicit than that, so he added 'A line has no width' and a 'Point has no magnitude.'

He also stated concepts that were obvious, but only to those who applied their minds to considering and understanding them – take, for example, the statement 'Two sides of any triangle must always together be greater than the third.' (see Figure 3.13). One modern critic has suggested that this would be obvious to a donkey, but in truth it only really becomes obvious when you 'think about it'.

For students down the ages who were strongly influenced by Euclid, however, it was invariably his revelations in geometry that inspired them. The very first proposition of Book I shows exactly how to construct an equilateral triangle using a pair of compasses and a straight edge. It's easy – try it?

First create a circle with centre A. Then create another circle, choosing the same radius and a centre (B) at any point on the first circle. Now draw lines to join the two centres and either point where the two circles cross, which we'll call C, and there it is – a perfect equilateral triangle, because AB = AC = BC (see Figure 3.14).

Now, using C as the centre, draw another identical circle, then another, until the original circle is surrounded by six others. The six points around the original circle are equidistant; join them and you get a perfect hexagon. But the whole pattern is a joy, showing that seven circles make a flower head arrangement (see Figure 3.15).

Pythagoras's Theorem – Euclid's Proof

By far the widest known of Euclid's propositions is actually good old Pythagoras' Theorem. As we've already seen (in

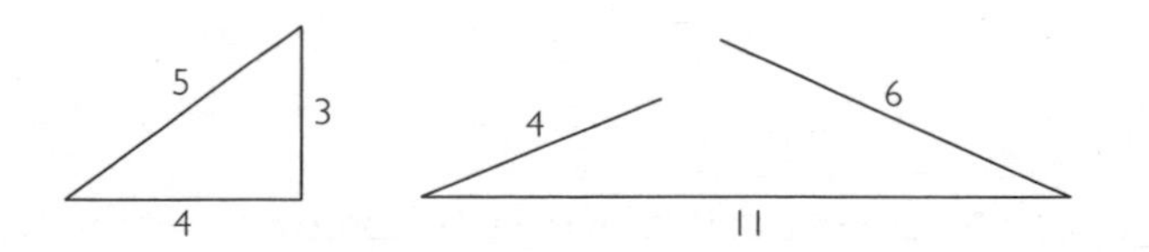

Figure 3.13

Chapter 2), the Pythagoreans themselves demonstrated the theorem by manipulating squares and triangles. Euclid's version, however, is the well-known but more formal one.

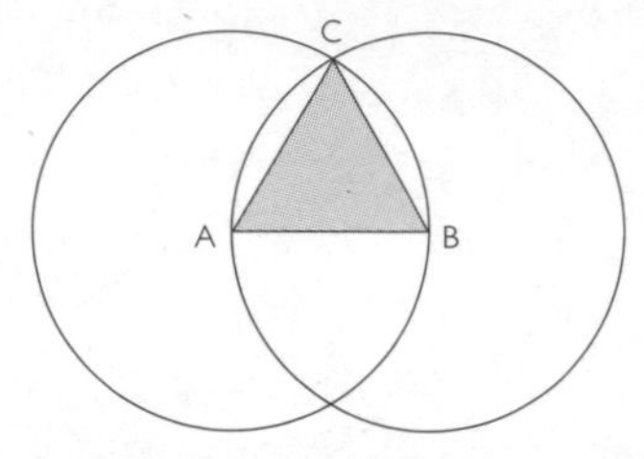

Figure 3.14

By this time, Euclid had already stated that any squares or rectangles with the same base, and set between two parallel lines, must have equal areas. This is easy to prove, because you can transform one shape into the other simply by lopping off one end and placing it at the other end.

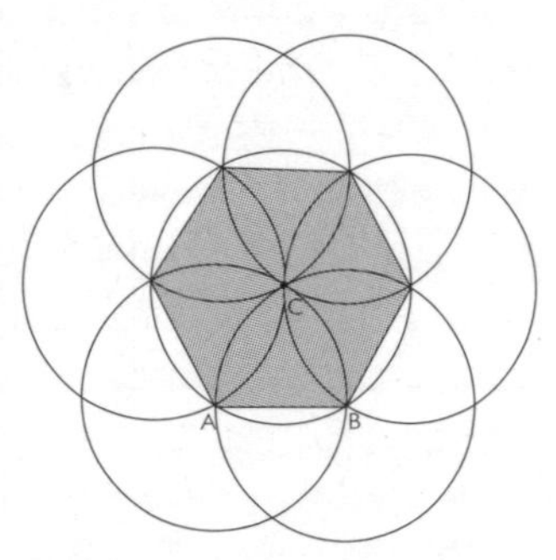

Figure 3.15

Euclid also stated that all triangles set on the same base, and with the same height (governed by a parallel line) have the same area, which is exactly half that of the square or rectangle on that base and with that height.

It is these facts that Euclid used for his rigorous proof of Pythagoras's Theorem. He drew a line from the right angle of the triangle, running parallel to the side of the larger square, and dividing the figure in two.

Now, taking one side of this line at a time, he found two identical triangles (shaded), each containing one side of the large square and one side of the smaller square. As these identical triangles must have equal area, any square or rectangle that contains them (governed by lines parallel with the same base), must be twice that size, and also equal (see Figure 3.16).

So the area of the small square must equal the area of the rectangle to the left, in the large square. The same must also apply to the third square and its corresponding rectangle. So the area of the square on the hypotenuse must equal the area of the squares on the other two sides. QED.

Today QED is an abbreviation for *quod erat demonstrandum,* meaning 'What was to be demonstrated'. But that phrase is actually a Latin translation of the original Greek, which was

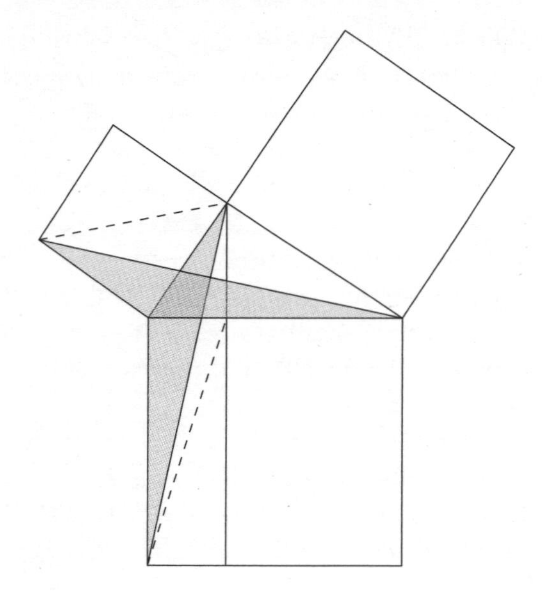

Figure 3.16

'The very thing it was required to have shown,' which I like rather better.

Amazing Facts About Lines That Cross Circles (Chords)

Elements is an impressive and hugely influential work, yet few people nowadays trawl through all the maths in its 13 volumes. But there are some absolute gems in the work that really should be more widely known, largely because they're so beautifully simple. How about this example: Book III, proposition 35...

Draw a circle and select any random point, P, within it. Now draw *any* line AB across the circle that passes through P. This line is called a *chord*. Measure AP and PB and multiply them together. Now draw any other chord, CD, that also crosses the circle and passes through P. It is an absolute certainty that CP x PD equals AP x PB, every time, even though P and the two lines were randomly chosen (see Figure 3.17a). Isn't it beautifully amazing?!

This proof dates right back to Thales, the first great Greek mathematician, and his Theorem. If you remember from Chapter 2, this states that any angle formed by two lines from the ends of a chord, and which meet on the circle arc on one side of the chord, will always be the same (see Figure 2.3).

So in the diagram we have now joined A to C, C to B and B to D. We can see that the angles CAB and CDB must be equal (as Thales explained). Similarly, the angles ACD and ABD are equal, and the two triangles APC and BPD are similar; their sides must therefore be in the same proportion. So AP/CP must equal PD/PB, and therefore AP x PB must always equal CP x PD.

But let's not stop there. Let's look at proposition 36 in Book III. Choose any random point outside a circle, which we'll call P (see Figure 3.17b). From P draw a line that passes through the circle in a random direction of your choosing, and call the points where the line crosses the circle A and B. Now measure and multiply PA x PB. Then select any other line starting at P that passes through the circle, and call the points where it crosses the circle C and D. Once again, in every case, PC x PD will always equal PA x PB. Is this a wow moment or what?!

There is another case where the line from P might reach the circle at its very edge, forming a *tangent* at point E. Now measure PE and square it (PE x PE); the result will be the same as PA x PB. As long as P is on the same plane as the circle, it could be absolutely anywhere and the theorem still works. It's staggeringly beautiful, don't you think?

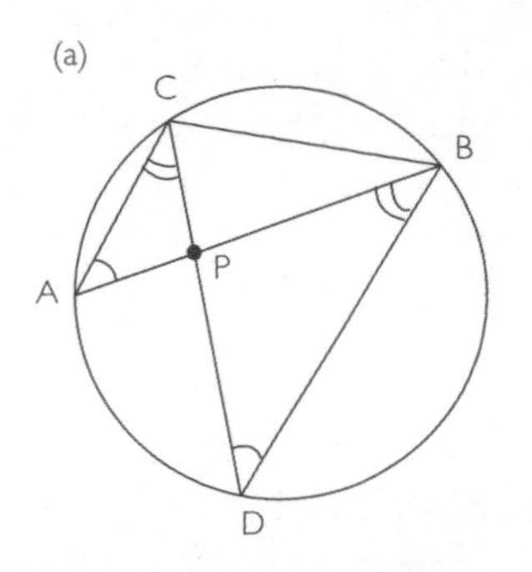

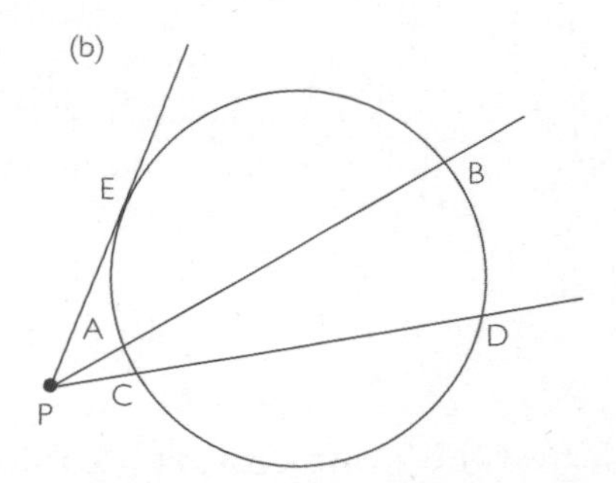

Figure 3.17

The Spread of Euclid's Influence

In the 2,000 or so years since it first appeared, more than 1,000 editions of *Elements* have been published. A scrap of papyrus showing part of an early copy was found in Oxyrhynchus in AD 100.

We know that Adelard of Bath, an English monk, travelled in the Middle East dressed as an Arab, so he could secretly learn Arabic Maths. He translated *Elements* from Greek to Latin in AD 1120. The first paper edition in Greek appeared in Venice in 1482, and was then translated into Latin in 1505.

The first version in English appeared in 1570, and a Chinese version, translated by Italian printers, followed in 1607. English mathematician Isaac Barrow produced a version that influenced Isaac Newton, but by that time Galileo and Kepler had already indicated how much it had helped them. The translation published by Thomas Little Heath in 1908 is still in print, covering three volumes.

Euclid also inspired many great personalities of history, including Abraham Lincoln, who always carried his Euclid in his saddlebag. Ada Augusta Byron, Countess of Lovelace, was saved from drugs and gambling by an instant love of Euclid, and eventually became the world's first computer programmer. Einstein called it his 'holy little geometry book'.

The great attraction of Euclid's work is in the way secrets are revealed in a progression that at first seems to have no form, but then – rather like Hercule Poirot or Miss Marple piecing together clues – the collective ideas build, allowing you to make further, more complex deductions.

Without knowing that it had originated with the Greeks via Euclid, I soaked up geometry at school like a sponge, and far more easily than any other subject. It was so inspiring that I suddenly found I loved mathematics, and would always have it as part of my armour, throughout my life.

The One Long-lasting Problem With Euclid

Throughout the centuries, Euclid's mathematics has been passed down to us as thoughts and ideas that can be relied upon 100 per cent. Except, that is, for one, which would cause mathematical arguments for the next 2,000 years. He stated the *parallel postulate*:

> *If a straight line falling on two straight lines, makes the interior angles on the same side less than two right angles, the two straight lines if produced indefinitely, will meet on that side on which the interior angles are less than two right angles (180 degrees).*

Modern mathematicians have agreed that this proposition – which is correct for two-dimensional planes – defines 'Euclidean geometry'; there can be other geometries where this postulate doesn't hold. To me it has always seemed a silly argument, because Euclid was talking about plane flat surfaces, and it's only on curved surfaces that the postulate is not certain – but on these curved surface geometries, lines can never be thought of as straight in the first place.

Henry Ernest Dudeney, Britain's greatest ever puzzle creator, explained it perfectly when he asked, 'If 200 guardsmen on parade stand perfectly to attention, how many are standing parallel to each other?' The answer is none, because if they are perfectly upright, the lines passing down through each of them will eventually all meet at the centre of the Earth.

What's more, if they are in perfectly straight rows, then those rows must be on great circles that, if continued, would divide the Earth exactly in half. This means that, in truth, even their rows are slightly bowed with regard to one another and not quite parallel, because their parade ground is on the curved surface of the Earth.

Euclid's Other Works

Euclid wrote many more books besides *Elements*, but sadly all have been lost. Included among them were *Catoptrics* – a study of mirrors that included concave and even spherical mirrors – and *Phaenomena*, an early guide to 'spherical astronomy'. It provided the maths required to study the entire heavens if we imagine that all the stars are set on a huge sphere always the same distance from the Earth.

Euclid's book on optics was the first to consider perspective, and explained why the Moon seems bigger near the horizon than when it is high in the sky. He also tried to explain sight by claiming that light rays come from the eye and spread out in all directions (we now know, of course, that it happens the other way round). It's believed that he also produced a book on mechanics, which included a study of levers and the circular movement created when they're attached to wheels.

The loss of all these books is a tragedy but, as we will see later, Euclid – more than anyone else of that time – began a tradition in Alexandria that heralded the city's golden age. Before we go there, however, we will meet arguably the greatest of all Greek mathematicians and scientists. By choosing to spend most of his life on the island of Sicily, far from Alexandria, he set himself apart physically and intellectually.

CHAPTER FOUR

Archimedes – the Greatest Greek of Them All

Archimedes was, without doubt, the greatest mathematician, engineer and scientist of the ancient Grecian age. Today he is rated on a par with Newton and Einstein, but for me, considering the age in which he lived, his achievements were even greater than theirs. We have more examples of Archimedes' exploits than we do for any other ancient mathematician, including Thales, and most of them are directly linked to his phenomenal achievements in mathematics and engineering.

In the most famous Archimedes legend, he jumped out of the bath and ran down the street in the nude. Mind you, when you list all that he achieved in his lifetime, you might wonder how he ever found time for a bath?

The Eureka Streaker

King Hiero II (who was Archimedes' cousin) hired two jewellers to make him a crown from a bar of pure gold. When he saw the result he was thrilled, until someone whispered in his ear that the jewellers were not completely honest and may have substituted less valuable silver for some of the gold, which they'd kept for themselves.

The problem was, how could he prove it without bashing the crown back into its original gold bar shape? Hiero turned to Archimedes for help, but even he had no idea how to solve the conundrum – until one day, as he climbed into the bath, water slopped over the sides…

Get in a bath brim-full with water, and you sink just like you ought-er,
But what happens to the water?
As you sink in more and more, the water slops upon the floor.

(Song: *'Eureka I've found it, and it wasn't even lost'* by Johnny Ball)

In a flash, Archimedes was out of the bath and running down the street, crying 'Eureka', while everyone else cried 'You streaker!' Eureka means 'I have found it!' in Greek. But what had he found? The act of his body pushing water out of the way and over the side of the bath was the clue to solving the crown problem. Inspired, he took the crown, a bar of gold identical to the one used by the jewellers and an urn full to the brim with water.

The urn is full right to the brim, take the bar and lower it in,
The water slopping o'er the brim,
Must equal the volume of the bar, that's the answer, there you are.

Archimedes immersed the gold bar in the water, forcing some of it to flow over the brim. When he took out the bar again, the water level in the urn was much lower.

So, quickly, your highness, your crown, the very finest,
Let's ascertain how honest, these 'honest' jewellers are?
We'll now dip the crown in, and if the crown is genuine,
The water will rise to the brim, as it did with the bar.

So the crown is lowered in, water rises to the brim – Stop!
But there's still more to go in? Ahha!
Our experiment has showed, the crown cannot be solid gold!

The crown displaced even more water than the bar of gold, so it had to have a greater volume and must therefore contain a metal lighter than gold, but more of it, to bring the crown up to the same weight as the bar of gold. The king confronted the jewellers who confessed and as well as losing the gold, they lost their heads – ooh, nasty!

The king was thrilled – but not as pleased as Archimedes, who had done much more than solve just one problem. He had discovered the secret to solving thousands of problems using the principles of *hydrostatics*, which he summed up in three simple laws:

Law 1. 'An object immersed in water will displace its own volume of water.'

Immerse an object in water and the water displaced equals the object's volume. Jewellers have been using this method ever since. Gold is measured in *carats* (denoted by the letter K): pure gold is 24 carats; 18 carat gold is 18 parts gold and 6 parts of another metal or metals, and is three-quarters pure. Adding other metals can affect the weight of a piece of gold, but it also helps to achieve the desired colour or hardness.

Law 2. 'An object weighs less in water – the loss being equal to the weight of the displaced water.'

Feel the weight of a stone in your hand. Lower your hand, holding the stone, into water, and notice how the stone feels much lighter, even though it's still supported by your hand.

Law 3. 'An object less dense than water will sink until it has displaced its own weight, and will then float "weightless" upon the water.'

This explains why a ship floats. Its weight pushes the water out of the way until the weight of the ship equals the weight of the displaced liquid, and the ship floats – effectively weightless – on the surface. Today's heaviest oil tankers have a dead or unladen weight of around 275,000 tonnes. However, they can hold about 400,000 tonnes of oil, and in order to float must displace almost 700,000 tonnes of seawater. They are so big that they cannot pass through either the Suez or Panama canals – in fact, the biggest can't even pass through the English Channel.

Archimedes and the Mighty Five

Archimedes also explained the physical laws that govern the five simple machines or devices used by ancient engineers, which were known as the Mighty Five.

1. **The ramp** was used in Egypt in the building of pyramids and temples. A road winds up a mountain, seeking a manageable path. It's a ramp. If a ramp is five times as long as the height it reaches, then it's five times easier to move a load up the ramp than to lift the load straight up (although you have to

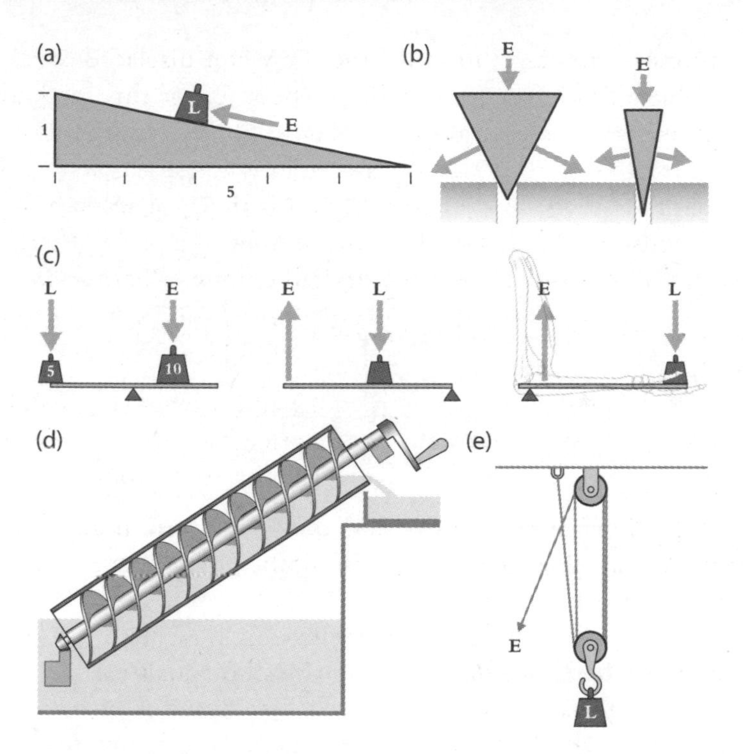

Figure 4.1 a–e

overcome the friction between load and ramp). (See Figure 4.1a.)

2. **The wedge** is really a double ramp. An axe splits wood by forcing it sideways up two ramps. The sharper and thinner the axe, the more efficient it is (see Figure 4.1b).
3. **The lever**, the most versatile of the ancient tools, has three components: the *pivot* or *fulcrum*, the *load*, and the *point of leverage*. There are three classes of lever, depending on which component is in the middle (see Figure 4.1c).

   The first-class lever has the fulcrum in the middle, like a seesaw. For it to balance, the weight times the distance from the fulcrum must be the same on each side. So it will balance with equal loads equal distances from the fulcrum, while a load that's 2m from the fulcrum on one side will balance a load that's twice as heavy but 1m from the fulcrum on the other side.

First-class levers mostly use an off-centre fulcrum. If one side is five times longer than the other, then a force applied at the long end can move a load five times greater on the short end. This is what Archimedes was alluding to when he said, 'Give me a lever long enough, and a firm place to stand, and I could move the Earth.' (How long would that lever need to be? See the Wow Factor Mathematical Index for my quite startling estimate!)

The second-class lever has the fulcrum at one end and the load in the middle, like a wheelbarrow. Through experience one quickly learns that the nearer the load is to the fulcrum, or the longer the lever, the easier it is to lift and move the load. The maths is the same as for the first-class lever.

The third-class lever has the fulcrum at one end and the load at the other, with the point of leverage somewhere in the middle. This is how the human forearm works – and it doesn't seem to be well designed at all. The distance between the elbow and the tendon linking the forearm to the biceps muscle is very small, while the distance to a load held in the hand is much further.

But our muscles have developed to cope with this imbalance well. Modern hydraulic cranes and diggers use the same principle, forcing oil at great pressure through narrow tubes over short distances to gain a huge mechanical advantage at the other end of a long arm.

4. **The wheel** had been around for perhaps 6,000 years by the time of Archimedes, although the Egyptians missed it. Wood rots away over time, so early examples of wheels are very rare, but some have been discovered in Eastern Europe that date back to between 6500 and 4500 BC, coinciding with the domestication of the horse.

A tree is naturally strong in an upwards direction, but not across the grain, so a round slice of tree does not make a good wheel, because it fractures easily. The first wheels were made from two layers of cross-planking – the biggest problem, however, was getting them to revolve, which required a fixed axle and a suitable lubricant.

The spoked wheel didn't arrive until somewhere between 2200 and 1550 BC, coinciding with the invention of the chariot. As is often the case, technological advances frequently occur to solve problems in warfare.

Archimedes saw the wheel as another form of lever, revolving around a central point. A modern ship's capstan, turned by a motor, is used to draw in long and heavy ropes. In Archimedes' time the capstan was operated by several strong men, and was used to raise heavy loads, or to create tension in catapults and other missile-throwing devices.

5. **The screw** Archimedes' screw was used to lift seawater from ships. A leather tube wrapped around a post was inserted into the water at an angle. A crank at the top turned the post, lifting the water (rather like a conveyor belt would), then pouring it away (see Figure 4.1d).

   The screw was an important device in Archimedes' day: ships were getting bigger and required better and more powerful technology – which leads us neatly to the sixth mighty machine.

6. **The pulley** was Archimedes' addition to the traditional Mighty Five machines. The pulley may have been used first by Archytas in around 300 BC, although Archimedes is generally credited with it. However, there is now evidence that the Sumerians were using a simple pulley to raise water in around 1400 BC (see Figure 4.1e).

In this particular pulley, a rope is fixed to an overhead beam, then travels down and around a wheel attached to a hook, which in turn is connected to the load to be lifted. The rope then travels back up and over another pulley wheel and down to the operator.

Because the rope travels the distance from the fixed beam to the load three times, the operator has a mechanical advantage of three to one, and the load feels only one-third of its actual weight. By attaching the first system to a second similar system (making a compound pulley), the operator would have a 9 to 1 advantage. Add another pulley system to the chain and the advantage would be 27 to 1, making the job of moving the load appear effortless. But to move the load 1 metre, 27 metres of rope would have to be pulled through the mechanism, and friction would lessen the advantage to some degree.

According to the Greek/Roman scholar Plutarch, Archimedes used a compound pulley system to draw a heavily laden three-masted merchant ship onto a beach. Now using a pulley system I can believe – but to pull such a ship onto a beach, as Plutarch suggested, seems to make little sense, especially if the ship in question happened to have been the mighty *Syracosia*.

The Mighty *Syracosia*

I don't think exaggeration is warranted when talking about the *Syracosia*, because the facts as we have them are truly remarkable. It was built in Archimedes' lifetime, for King Heiro II, as the flagship *quadrireme* of a fleet that was supposed to number 60 ships in total. A chap called Archias built and designed her using timber from the Italian mainland and hemp from Spain, with tar from the Rhône Valley to make her watertight. Three-hundred tradesmen and helpers came from far and wide to build the ship. The cedar hull took one year to lay down, before being covered in canvas, then tar, and finally soldered lead sheet to protect it against the voracious attacks of shipworm, which could eat through an unprotected hull very quickly.

To protect the ship, an iron fence ran around the main deck, and the ship carried 60 fully armed marines. It also had three masts, and eight towers containing lifting devices and weapons to lob rocks at anyone coming too close. The hull housed a freshwater tank and a seawater tank containing live fish, as well as a mill for grinding flour and fuel for the galley stoves, so the crew could cook and bake bread. There was also ample storage space for cargo, plus stabling for 20 horses, as well as quarters for the riders, their slaves and the marines.

The middle deck contained 30 four-cline cabins on each side for passengers and officers, each big enough to sleep four people on couches or the floor (a 'cline' was a sleeping space). The ship commander's quarters had an area of 15 clines. Homer's *Iliad* was displayed in its entirety throughout the ship, as a mosaic on the floor and in paintings on the walls and ceilings. The upper deck contained a gymnasium, as well as gardens with fresh plants (including kitchen vegetables and vines) watered by an automatic system of pumps and pipes, while ivy formed a shaded promenade. On this deck there was also a temple to Aphrodite, with polished gemstones on the floor and walls of cypress and cedar. Alongside the temple was a library, with ivory-inlaid doors, and a bathroom with three bronze baths and marble hand basins. The centrepiece of the library was a huge sundial, which presumably revolved so that it could be set to north, whatever direction the ship was facing. Statues and urns were everywhere.

These details come down to us from the writer and raconteur Athenaeus, and we now know that such extravagance and luxury would not be seen in ships again until the great liners of the early twentieth century. So what became of the *Syracosia*? It sailed to Alexandria and was offered as a gift to King Ptolemy III, and as far as we know never sailed again.

Archimedes Played With Mathematics

Above all else, Archimedes was a mathematician, and throughout his long life he played with maths in many ways, devising games, puzzles and mathematical conundrums. Here are some of the most famous that we know of.

The Ostomachion

Archimedes is known to have dabbled with the *Ostomachion*, and may even have invented it. Today it is often called the *Stomachion*, which, according to some scholars, means 'stomach turner'. I personally question that, though. Maths haters – then and now – might well refer to a maths problem as something that could turn your stomach, but I doubt Archimedes would have used such a term!

The word 'Ostomachion' is derived from the Greek for 'bone fight' – which is far more likely to be its true name – and the puzzle was probably invented before the time of Archimedes (see Figure 4.2).

British mathematician W. W. Rouse Ball (no relation of mine) suggested that it might be a game in which two or more players take turns to fit 14 accurately fashioned pieces of

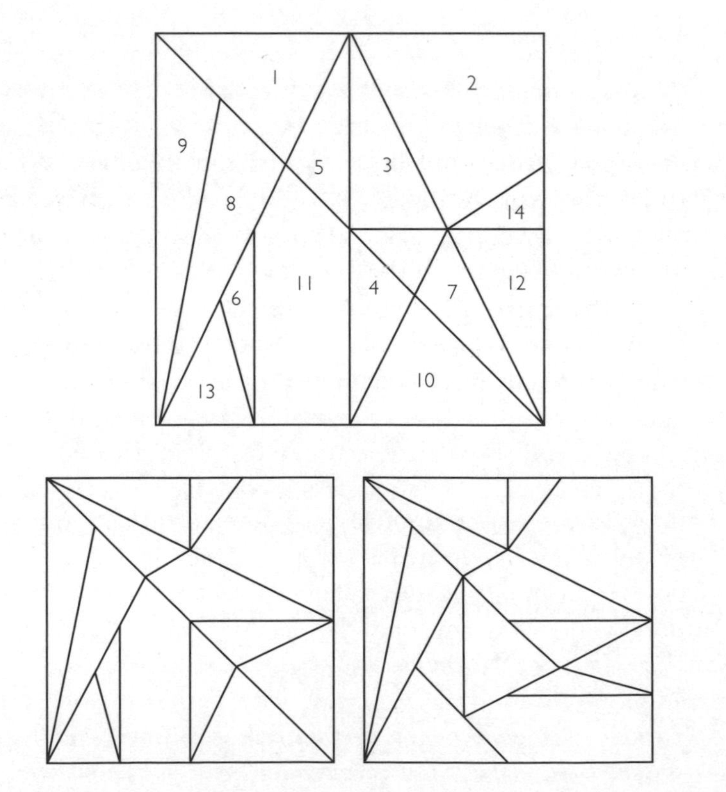

Figure 4.2

bone into a box in the shape of a square. In fact, it would make for a very good game: there are 17,152 different ways to form a perfect square with all the pieces, assuming you're allowed to flip some of them over.

It was designed on a 12 x 12 template; of the 14 pieces two are repeated: those on the upper left and lower right of the centre line. From the illustrated solution, other solutions can be achieved simply by flipping the pieces left and right or up and down across either the centre line or the long diagonal. Each half-section to left and right also has a long diagonal, so those pieces can be flipped too. Then again, if you measure the sides of each piece it soon becomes apparent that some shapes will only ever fit with their present neighbour. As in chess, by learning how to assemble smaller sections through trial and error, it would be possible to gain an advantage over your opponents.

The Sand Reckoner

Archimedes made some excursions into big numbers just to prove that anything and everything should be countable, and that one day we would need a system of numbers to cope with the concept. So he suggested to Gelon, then King of Syracuse, that we might want to know how many grains of sand it would take to fill the whole Universe.

The Wow Factor Mathematical Index shows in more detail how Archimedes proved that we need a more powerful number system to solve such a problem.

The Cattle of the Sun

Perhaps in another attempt to get others to think about very large numbers, Archimedes wrote a letter to Eratosthenes (who we meet in the next chapter) containing a complex puzzle he may have invented. It was called 'The Cattle of the Sun', and it began something like this:

> *Compute the number of cattle of the Sun, oh stranger, and if you are wise, apply your wisdom and tell me how many once grazed upon the plains of the island of Sicilian Thrinacia.*

He then described four herds of differently coloured bulls and cows, never giving actual numbers, just fractions of one set of bulls or cows compared to the others (the full puzzle is listed in the Wow Factor Mathematical Index).

Solving the bull and cow equations produces a total of 50,389,082 cattle. But in that solution the white bulls and the dark-skinned bulls, when added together, do not produce a perfect square, as required in the correct answer.

It wasn't until 1880 that an actual solution was found – by A. Amthor, who gave the answer of $7.76 \times 10^{206,544}$ cattle. This is far more than could possibly fit in the observable Universe, let alone Sicily. If Archimedes had a solution it has not survived, but some historians suggest that the whole thing might well have been a huge Archimedean leg-pull.

The Semi-regular Solids

Archimedes studied the five Platonic solids and took the whole concept a giant step further by finding all the *semi-regular solids*: solids with two different-shaped faces, but edges of the same length, identical vertices and regular plane figures for faces. Once he had found them, he also proved that there are – and can be no more than – 13 possible variations (the 13 shapes and their nets are detailed in the Wow Factor Mathematical Index).

Did Archimedes Treat War as a Game?

Down the centuries, when the threat of war looms, states have always looked to mathematicians and scientists to develop weapons or defence systems to beat the enemy. But when Syracuse was threatened by Rome, Archimedes – now in his sixties – seems to have taken to the task of defending it almost as though it were a game.

Following the outbreak of war between Carthage and Rome, Carthaginian general Hannibal marched through Spain and southern France, then famously took 38 elephants, 38,000 men and 8,000 cavalry over the Alps to enter Italy from the north. The severe weather conditions on the journey reduced his forces by almost half; many of the elephants also perished. But he surprised and out-flanked the Romans, and at Cannae in 216 BC he wiped out 48,000 Roman troops. For Rome, this was extremely serious.

Geographically, Syracuse, in Sicily, was right between the opposing sides – Carthage across the Mediterranean to the south and Rome to the north. Its new king, Gelon II, decided to side with the Carthaginians, much to the annoyance of Rome: if the armies of Carthage moved into Sicily, with Hannibal to the north, Rome would be trapped in the middle. So the Roman general, Marcellus, laid siege to Syracuse, aiming to conquer and hold it at all costs.

It was at this point that Archimedes' exploits raised him to truly legendary status. The city already had huge defensive walls, plus caves large enough to hold the whole population if required. To improve the situation, however, Archimedes let his inventive mind run riot. Most of what he actually achieved comes to us from Roman sources and cannot be wholly reliable, but it is clear he had an enormous and most impressive impact on the Romans.

He built catapults and devices for hurling rocks (some weighing a quarter of a tonne – about the same as a small family car). He built so many different devices that the Romans claimed he could hit their ships from close range to anything up to 200m away – in itself an amazing feat. The range of a catapult is in direct proportion to its mass or weight, so to double its range you have to make it twice as wide, long and tall, making its mass eight times as large.

If Archimedes really could hit a ship irrespective of its range, some feel he must have completely understood trajectories. He definitely knew about parabolic curves, as we shall see later. But did he know that this is the path a projectile always takes as it flies through the air? Galileo is normally credited with working out the parabolic flight of cannonballs and other missiles, but that didn't happen until 1,800 years after Archimedes lived.

Most scholars think that Archimedes – by then getting on in years – may have honed his accuracy with catapults by trial and error, rather than by discovering the true maths of parabolic flight. Personally, I doubt it: although we have more evidence of the workings of Archimedes than of any other ancient scientist, the evidence is still very skimpy, and there are large amounts of data that we have no record of.

What we do have, however, demonstrates his genius. So I believe, without evidence to the contrary, that by that time he would have worked out how missiles fly, and would have been able to achieve what the Romans commentators wrote about him: hitting a target at any possible range. Whatever the truth of the matter, he was certainly very good at it. According to the Romans, missiles fell like rain around them, battering and destroying both men and ships.

Legend has it that when they moved out of range, Archimedes positioned mirrors to focus the Sun's rays on a particular ship until it burst into flames. To achieve this he would have needed to employ his work on parabolic curves (which we look at later), but the concept had been written about even before his time.

Several scientists have tried to reconstruct this device, but have found that it's impossible with a single hexagonal mirror – which is what Archimedes employed, according to the original legend (recounted by Lucian of Samosata). In 1973, in an experiment conducted by a Greek scientist, a large number of sailors used flat mirrors about 3 x 5ft in size to set fire to a ship a bow-shot away.

But there are other Roman legends about Archimedes that perhaps try to explain and excuse why it took the Roman forces so long to take Syracuse, which was a fairly small city. According to some reports, when a ship sailed under the walls of Syracuse, Archimedes' Claw – a huge grappling device – either descended from the harbour ramparts or rose up out of the water. Either way, it took the ship in its grip, lifted it out of the water, tipped out the crew and dashed it against the rocks.

For the Romans, attacking by sea was proving futile. Surely their land forces could do the job? They had already conquered Leontini, a city just a few miles north. All they had to do was march south and set up an attack from there. Amazingly, Roman land forces feared Archimedes just as much as those at sea. According to one story, as they approached the city walls, a long pole with a rope attached to it appeared over the battlements. Instantly, someone shouted, 'Oh, no – here comes another of Archimedes' weapons!,' and the entire Roman force turned and fled.

Nevertheless, as Syracuse was besieged from land and sea, it was only a matter of time before it was forced to give in, as hunger became an ever greater problem. So it seems strange that, according to Roman writers, the city's inhabitants held a carnival to celebrate Artemis, goddess of hunting, during which the guards at one gate apparently got very drunk. Someone tipped off the Romans, and they broke in, ending the siege.

The Romans found Archimedes on the beach, unaware of the situation, and drawing geometric figures in the sand. A soldier approached and commanded that he attend Marcellus. Archimedes either refused or, more likely, just carried on working at his geometric problem. The soldier lost his cool and simply ran Archimedes through with his sword. Archimedes was 75 years old.

On hearing the news, Marcellus had the soldier flayed and skinned alive. The general himself dressed in sackcloth and ashes as penitence for being party to the destruction of so great a man. But that is not the end of our story. We haven't yet covered the most important work that Archimedes achieved in his lifetime: his revelations in mathematics.

Archimedes' Mathematics – Pure and Simple?

Archimedes' engineering and scientific exploits are legendary, but his discoveries and innovations in mathematics are of phenomenal importance, because he invariably tackled problems in new and original ways.

The Search for PI

Let's start with his work on circles. The Egyptians had suggested that π was $16^2/9^2$ or $256/81 = 3.16$, which was a very good estimate. Archimedes decided to try to improve on that. Earlier mathematicians had used the method of exhaustion – in which every new calculation moves closer to an accurate solution. Archimedes followed a similar procedure in his search for π.

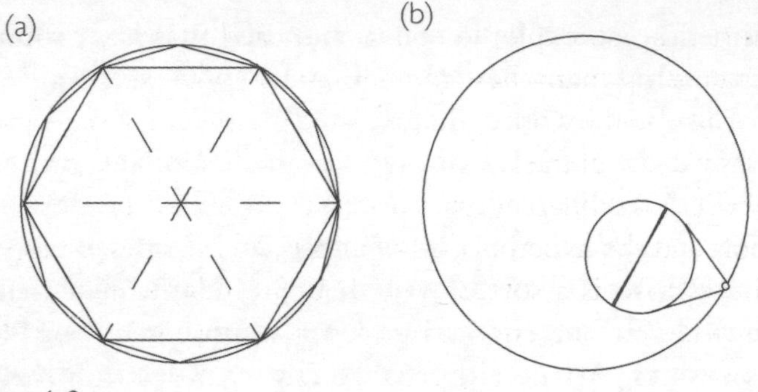

Figure 4.3

First, using his compasses at a fixed setting, he drew a circle and divided its perimeter into six segments. Then he produced a hexagon divided into six equilateral triangles. This figure is 2 unit sides across and 6 around. So the perimeter of the hexagon is three times the circle's diameter. But as the circle curves outside the hexagon, it's clear that its circumference must be slightly greater than 3. But by how much? (see Figure 4.3a)

Archimedes then drew a small triangle on each of the hexagon's six sides, fitting into each outer segment of the circle. The result was a *dodecagon*, or 12-sided figure, and the comparative distances around the dodecagon and the circle were now closer. He repeated the process to produce a 24-sided figure, then 48-agon and finally a 96-agon, which is so very close to an actual circle with each side only 3.75 degrees wide. Even enlarging our image, the difference between side and circle is hardly noticeable (see Figure 4.3b).

Next Archimedes circumscribed a hexagon around the outside of the circle, and kept doubling its sides until he had circumscribed a 96-agon. Now he could measure the distance around both figures; the true distance around the circle had to be somewhere between the two. After many calculations and measurements, he concluded that π was greater than 3 10/71 (3.140845), but less than 3 10/70 (3.142857) – the latter one is actually 22/7, which is the very adequate approximation we mostly use today.

Chronic Conic Problems

The most celebrated ancient scholar of the mathematics of cones was Appollonius, who we meet in the next chapter. But

Archimedes also explored cones, and used them as a route to understanding many new mathematical concepts.

He had read the work of Eudoxus and his student Menaechmus, who (as we saw in the last chapter) was the first to slice up a cone to discover four different types of curve: the circle, the ellipse, the parabola and the hyperbola, all of which Archimedes explored.

The **ellipse** is a sort of egg-shaped circle, except that an egg is wider at one end and tapers at the other, rather like a drip of water. A true ellipse is doubly symmetrical: the left side is a mirror image of the right and the top half a mirror image of the bottom.

Although we often talk about circles, we seldom speak of ellipses – and yet we see far more ellipses than circles every day of our lives. A cup has a circular hole, but unless we are directly above it, when we look at it we don't see a circle – we almost always see an ellipse. And whereas a circle has one centre point or focus, an ellipse has two *foci* (although this term wasn't known until the days of Kepler). The definition of an ellipse is 'a collection of all points whereby the combined distance from the two foci is always the same'.

An easy way to draw an ellipse is to stick two pins in a piece of card, then loop some string loosely around them. Use a

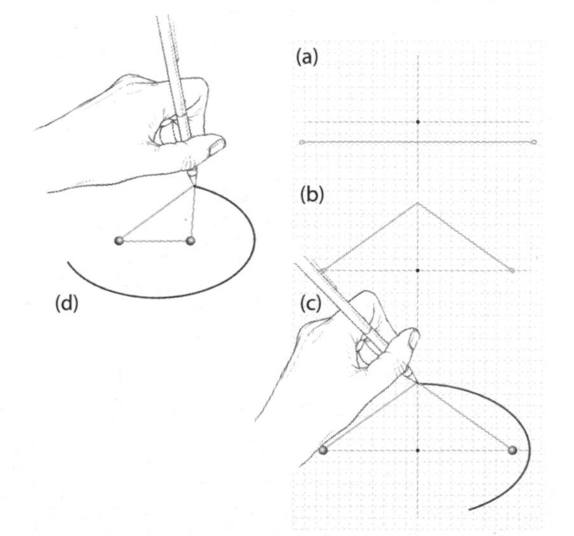

Figure 4.4

pencil to pull the string taught, then draw round the two pins to create a perfect ellipse. But this method is not helpful if you're trying to draw an ellipse of desired dimensions. Say, for example, you want to draw an ellipse that's 20cm wide by 12cm high…

First draw a cross on your piece of card, and from its centre measure a horizontal line 10cm left and one 10cm right, followed by a vertical line 6cm above the cross, and one 6cm below. Now take a piece of string and make a tiny loop at each end, but make sure that when it's stretched out it's exactly the length of the horizontal line: 20cm. Fold the string exactly in half and place its centre point at the higher 6cm mark on the central axis. Stretch one half of the string down and left, and the other half down and right, and mark precisely where they just touch the horizontal line.

Insert your two pins at these points, making sure they pass through the end loops of the string. Then repeat the pencil and stretched string routine described earlier, and you have your desired ellipse (see Figure 4.4).

Finding the area of an ellipse is slightly more complex than finding the area of a circle (which is πr^2). An ellipse has a horizontal and a vertical radius, and we need both for the calculation. The neat formula for finding the area of an ellipse is π x ½ h x ½ w – or, in the case of our 20cm by 12cm ellipse, π x 6 x 10cm.

Elliptical mirrors have amazing repeating properties similar to parallel ones. When I made *Johnny Ball Reveals All* for Central TV in the early nineties, we created a small elliptical wall using flexible mirror card.

We then placed a small coloured flag at each of the two foci. The TV camera peeped over the wall, focusing on one flag, then started to track sideways, always keeping the flag centre-screen. Through the entire camera movement, a reflection of the second flag appeared directly behind the first flag, and another image of the first flag could be seen behind that, and so on.

The **parabola**. Surprisingly, parabolic curves are everywhere. The path of water from a fountain, the flight of

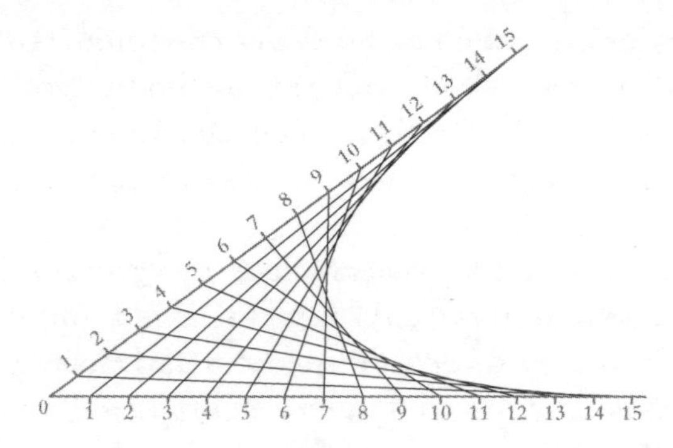

Figure 4.5

a soccer ball, a leaping dolphin, or the descent of a high-board diver: all form natural parabolic curves.

You can make parabolic curves using the mesolabe compass (see page 72). Start by joining, say, point 15 on the top line to point 1 on the lower line. Now repeat with straight lines from 14 to 2, 13 to 3, 12 to 4 and so on until you join 1 on the upper line to 15 on the lower one. Your lines will have formed a beautiful parabolic curve, in which every line is a tangent to the parabola, touching but never crossing the parabolic line (see Figure 4.5).

I find the parabolic curve the most fascinating and empowering of all mathematical figures, because although it appears complex, it explains beautifully a wealth of maths that confronts us in everyday life.

So I have always found it very sad that parabolic curves are introduced to students so late in the mathematics curriculum. I have yet to meet a nine year old (in my experience the age at which kids start to take this on board) who hasn't been fascinated by parabolas, if they've been explained simply.

In modern car headlights, a dipped or focused beam is achieved by varying the geometry of the reflective surface behind the light source. In the Second World War, searchlights swept the skies over England, looking for enemy bombers. Their reflective surface was in the shape of a parabola, so every ray from the light source hit the surface and reflected directly forwards. The beam neither narrowed nor widened as every ray was parallel to every other.

At Odeillo in the French Pyrenees there is a solar furnace that – possibly inspired by Archimedes – comprises a field of mirrors that direct the Sun's rays onto a huge parabolic mirror. This then directs the rays onto a receiver, which can reach temperatures of about 3,500 degrees Centigrade.

For visitors they demonstrate the power of a parabolic reflector using a rusty old Second World War German searchlight, but in reverse. First the searchlight is pointed directly at the Sun. Then someone holds a long stick so that the far end is at the focal point of the reflective lens, where the light source originally came from. The stick immediately starts to smoke, and in 10 to 20 seconds it has burst into flames.

The Greek who first wrote about using parabolic curves as mirrors was Diocles, who was born around 350 BC (a hundred years before Archimedes). A student of Aristotle, he was said to be the greatest early mathematician after Hippocrates (who was born a hundred years before him). Realising that calculating how sunlight reflects on a curved surface like a parabola is complex, Diocles imagined lines lying at a tangent to a parabolic curve. At the point where curve and tangent touch, he drew a line perpendicular to the tangent. He saw that a ray of light hitting this point from any angle inside the parabola would reflect back at the same angle on the other side of the line.

You can create your own parabolic curve in a couple of minutes. On a piece of A4 paper, halfway along one of the long sides, make a dot a few centimetres in. Fold the bottom edge of the paper so it meets the dot, and make a crease. Now repeat the process a good few times, always bringing a new point on the long edge of the paper up to the dot and making a crease. When you have a dozen or so creases, you'll start to see a curve forming. Each fold is at a tangent to the parabolic curve, and the original dot is its focus, showing neatly how the position of the focus affects the shape of the curve (see Figure 4.6).

This trick reveals more about this special curve. For every parabolic curve, there is a line (in this case the edge of your sheet of paper), which is called the *directrix*. Draw any line at right angles from the directrix to the point at which it reaches

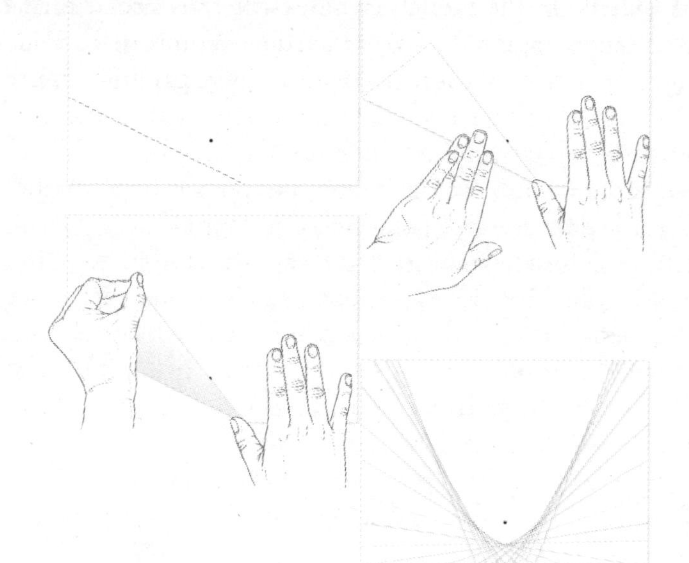

Figure 4.6

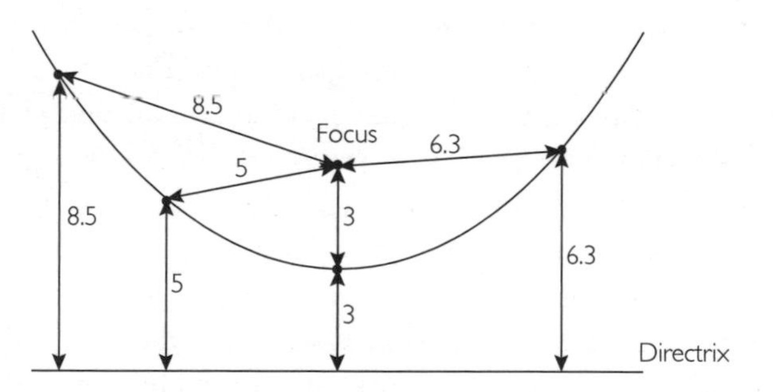

Figure 4.7

the parabolic curve. Now draw a line from that point of contact to the focus of the curve (your original dot). The two lengths are always exactly the same, which is why the folding trick works (see Figure 4.7).

The Area Under a Parabola

Archimedes was very concerned about parabolic curves and how their areas compare with the areas of rectangles either inscribed within them or circumscribed around them. He already knew from Eudoxus and Menaechmus that a cone is exactly one-third of the volume of a cylinder it fits into, just as a square-based pyramid is one-third of the volume of the cuboid it fits snugly inside.

So it seems pretty likely that Archimedes might have drawn a parabola inside a rectangle and asked himself, 'How do the two areas compare?' (see Figure 4.8).

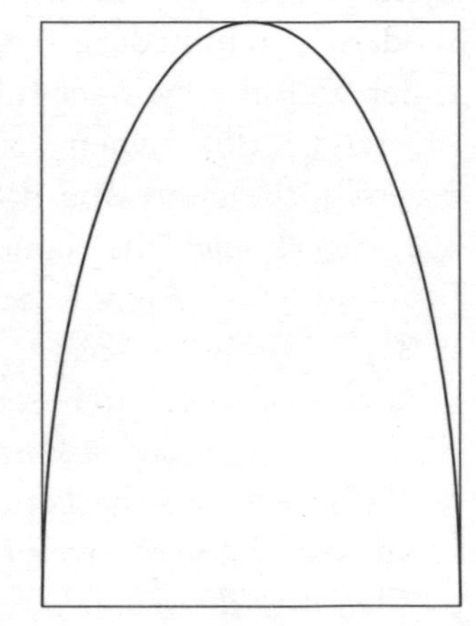

Figure 4.8

Clearly the parabola's area is not one-third of the rectangle's, as is the case with cones and cylinders. But mathematics keeps producing beautiful answers. If the area is not one-third, then what might it be? Could it be two-thirds?

Now I have no proof, and certainly cannot read the thoughts of a man who lived 2,200 years ago – but I do know that this is the way engineers behave when they develop new car engines, aircraft, bridges, computers and so on. Although they have the maths as part of their background training, they often find the quickest way to a solution is to make an educated guess and say, 'I bet it's so and so!'

So it seems quite possible, if not probable, that Archimedes might have drawn a parabola in a rectangle on a piece of vellum, cut out the parabola and weighed the pieces, which would prove the two-thirds relationship. This wouldn't be a mathematical proof, but Archimedes was not averse to finding solutions using any approach he could, as we'll see when we explore his Method, which is featured in the Wow Factor Mathematical Index. Following it is not child's play – but it's well worth the effort.

The Cylinder, Sphere and Double Cone

An egg timer is a remarkable thing: it's the timepiece with the greatest number of moving parts. The timepiece with the fewest moving parts, of course, is the sundial: the only moving part is the Sun itself. Figure 4.9 shows a modern Archimedean egg timer with the cylinder full of sand. But when it's reversed, the same sand fills the sphere *and* the cone, proving – as Archimedes was first to do – that the volume of the cylinder equals the combined volume of the sphere and the cone. Here's how he may have proved it.

Figure 4.9

The mathematician David Singmaster explained this to me as what might have been Archimedes' thought process when he explored these three shapes, and I have happily demonstrated it ever since, because it explains so much. We have no proof that Archimedes did it the following way, but knowing so much about these shapes as he did, he must surely have thought about them along these lines.

Figure 4.10 shows a cylinder, a sphere and a 'double-napped cone' (two cones joined at their tips). All three are the same height and width. As the volume of a cone is one-third of its base area times its height, the two double-napped cones have the same volume as one cone with the same overall base and height.

If we take a horizontal plane (A) across the top of the three figures, what do we see? The cross-section of the cylinder equals the end cross-section of the cones plus zero at the

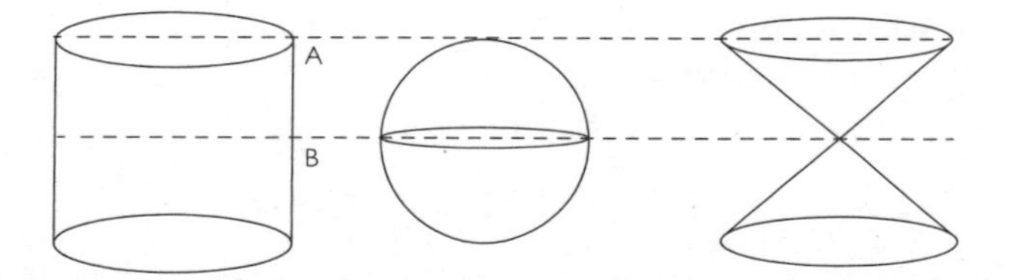

Figure 4.10

north pole of the sphere. Now if we take a similar horizontal plane (B) through the centre of the three figures, the cross-section of the cylinder equals the cross-section of the sphere at its equator plus zero at the single point where the two cones meet. So, can we assume that if we took any random horizontal plane through the three figures, the cross-section of the cylinder would always equal the cross-section of the sphere and cones combined?

Archimedes already had a method to prove this might be so, using hydrostatics. He could have immersed the cylinder in water and measured the water level. After taking it out, he could then have immersed the sphere and two cones. If the water rose to the same level, then the volume of the cylinder must have been equal to the combined volume of the sphere and the cones. If you try it, that's exactly what happens. QED.

We cannot know Archimedes' actual method for revealing the relationship between a cylinder and a sphere and two cones. But he could have done it using simple logic, demonstrating how powerful mathematical thought can sometimes be.

We now know that the volume of the cones plus the sphere equals the volume of the cylinder. As the cones are one-third of the cylinder's volume, the sphere must be two-thirds of the cylinder's volume. From this we can now deduce any and all of the following:

The volume of one cone is ⅓ of the half-cylinder.
The volume of a hemisphere (or half-sphere) is therefore ⅔ of the half-cylinder.
The volume of the half-cylinder is the end section: $\pi r^2 \times r = \pi r^3$.
The volume of the whole cylinder is $2\pi r^3$.
The volume of the two cones is ⅓ of the volume of the whole cylinder, or $\frac{2}{3}\pi r^3$.
So the volume of a sphere is ⅔ of the volume of the cylinder, which is $\frac{4}{3}\pi r^3$.

This complex exploration of volumes produces amazing mathematical results, all so neatly wrapped up. But that is only the start! What about the surface areas of these solids?

Surface Areas of Cones, Spheres and Cylinders

Let's take the surface area of the cone first. For a right circular cone (one with the upper point directly above the centre of the base), the slope, height and radius form a right-angled triangle. Using Pythagoras' Theorem, we can work out the length of the slope: $s^2 = r^2 \text{ x } h^2$ or $s = \sqrt{r^2 \text{ x } h^2}$.

We need to know the length of the slope to find the surface area of the cone, because the formula for the area is: curved sloping area (CSA) = πrs (where s is the slope length). It's as simple as that (see Figure 4.11).

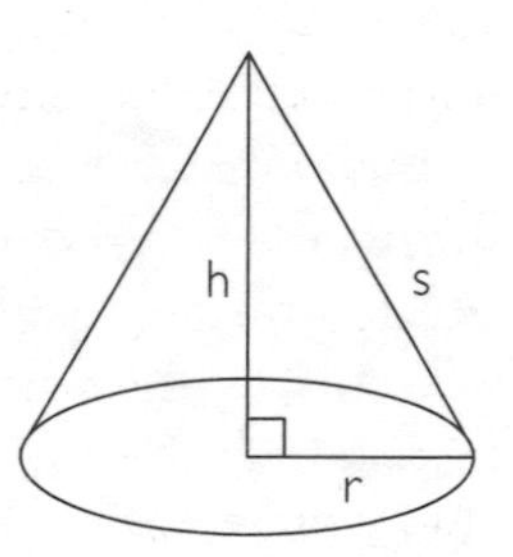

Figure 4.11

To calculate the cone's whole surface area, we now add the area of the base.

So $\pi r^2 + \pi rs = \pi r\ (r + s)$. It's very neat.

But the surface area of the sphere and the cylinder, and their close connection, was Archimedes' major discovery – and it's quite amazing in its simplicity.

Let's start with the sphere. The Moscow Papyrus (see page 25) tells us that the circular cross-section of a sphere is πr^2. Egyptian basket weavers had estimated the surface area of a half-sphere to be twice this amount, or $2\pi r^2$. We're not sure that Archimedes knew this, but it seems he did 'prove' for himself that the total surface area of a complete sphere is four times that of its cross-section, or $4\pi r^2$, which confirms the above.

The surface area of the cylinder wall is the distance around the cylinder times its height. So the area of the cylinder wall is $2\pi r \text{ x } 2r = 4\pi r^2$. But that is exactly the same area as the surface of a sphere.

Many centuries later this was to prove a great boon to mapmakers in calculating the area of segments of the surface of the Earth (see Figure 4.12).

Calculating the surface area of the Earth between, say, latitudes 30 degrees north and 60 degrees north is a complicated task. The distance around the sector at 60 degrees north is much smaller than that at 30 degrees north, and we are now attempting to measure complex curved

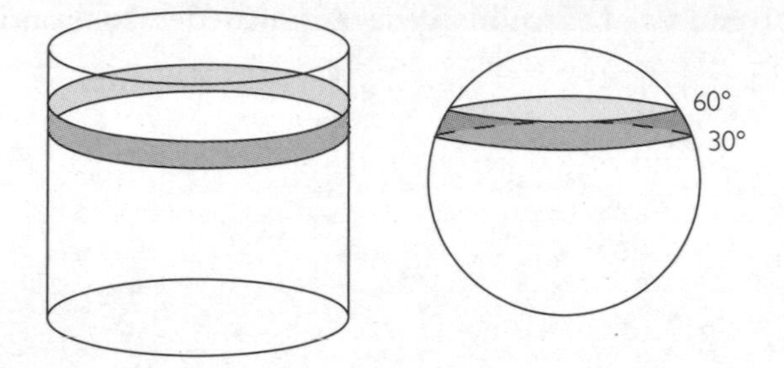

Figure 4.12

areas. But if we compare this area with a cylinder of the same height and width, the calculation becomes a lot simpler.

We know the distance around the cylinder is $2\pi r$. So if we multiply that by the parallel height of the sector, we have the area of the band on the cylinder. But that is also the area of the sphere's surface between latitudes 30 and 60 degrees north. How brilliantly simple!

It's also a simple matter to measure the area of the Earth around the North Pole out to, say, 80 degrees north. Draw a horizontal line through the point on the sphere that is 80 degrees north, and continue that line across the cylinder. The area of this narrow band on the cylinder is also the area of the sector of the sphere around the North Pole that's above 80 degrees north.

Lastly, the surface area of the entire cylinder is the area of the cylinder wall ($4\pi r^2$) plus the area of the two ends, each of which is πr^2. So the surface area of the entire cylinder is $6\pi r^2$. But just as the volume of a sphere is two-thirds of the volume of its matching cylinder, so the surface area of the sphere, at $4\pi r^2$, is two-thirds of the surface area of the matching cylinder.

It is all so beautiful and neat – and for me it probably explains why Archimedes requested that a simple drawing of a sphere contained in a cylinder be carved on his tomb when he died. This was duly done, but because the Greeks at that time (as far as we know) had no conception of perspective drawing, the image was simply a circle in a square. The

Roman general Marcellus gave Archimedes an honourable burial in 212 BC, but the actual site of the tomb was lost until 75 BC, when the Roman orator Cicero, then Governor of Syracuse, found one with the tell-tale design. Since then the site has once more been forgotten, but the name of the great Archimedes will surely live forever.

CHAPTER 5

The Glory That Was Alexandria

By establishing Alexandria close to the mouth of the Nile, Alexander always intended that it would become the greatest city of the age, dominating the Eastern Mediterranean. Its glory lasted about 300 years, until the Roman Empire took over and changed everything. Until then, however, Alexandria – echoing its founder – became truly great.

Seeing the potential for a significant port, Alexander sited the centre of the city opposite the Island of Pharos. To shelter the enormous harbour he built a *mole*, or long connecting causeway, which he called the Heptastadion, because it was seven *stadia* (about 1.25km long).

From 303 BC, Egypt, including Alexandria, was ruled by the Ptolemaic dynasty. Ptolemy I was actually from Macedonia, and had been one of Alexander's bodyguards. Nevertheless, the Egyptians accepted him and his children as their new Pharaoh dynasty.

Between 280 and 247 BC, Ptolemy II Philadelphus commissioned the huge Pharos Lighthouse on the end of the Heptastadion. The lighthouse was between 103 and 118m tall (historical accounts vary), but it was one of the tallest man-made structures in the world for many centuries, and was one of the Seven Wonders of the Ancient World (see plate section). Its crafty architect Sostratus had 'Ptolemy II' etched in plaster over the entrance; generations later, however, when the plaster fell away, the name 'Sostratus' was revealed permanently carved in the stone.

The base of the lighthouse was a tower $30m^2$; the lower stones featured interlocking grooves filled with lead to lock them together so they could withstand the action of the waves.

About two-thirds of the way up the lighthouse became octagonal shaped, then circular at the top. At the very top, a rotating mirror reflected sunlight, and when the Sun went down a huge fire was lit inside. Using Thales' mathematics (from Chapter 2), we can work out that the light could be seen almost 50km (30 miles) away.

The commanders of the mighty *Syracosia* would have first seen the Pharos Lighthouse when the ship arrived from Syracuse as a gift to King Ptolemy III. We don't know how much he appreciated the gift, but there is no record that it ever sailed again. Nor were any more ships like it ever built, although 60 had been planned.

In any case, Alexandria around that time was a major centre for naval warfare, and a ship perhaps even more amazing than the *Syracosia* was built there.

In those days, fighting ships were swift-oared vessels crewed mostly by local young men who trained hard and took pride in their performance and skill – rather like local football teams today. These ships were designed for fast action and could turn quickly, and their crews could out-manoeuvre those with lesser skills, before ramming their ships and boarding them. However, like today, military generals were always looking to build something bigger and better than their enemies could come up with – and in Alexandria, they did just that.

They built a catamaran comprising of two large hulls joined by a huge deck, on which were pavilions and tents for the generals and their staff, and weapons to lob missiles or rocks at anyone who approached. The huge two-hulled ship had three banks of oars down its four sides. It was a *quinquereme*, in which the two lower oars had two rowers per oar, but the upper oar had a single rower. Although quadriremes and quinquiremes did exist, three banks of oars was usually the maximum, but with more than one oarsman on some oars.

There are many historical references to triremes, quadriremes, quinqueremes and even heptaremes. There is even John Masefield's celebrated verse – *Quinquereme of Nineveh from distant Ophir* – but there is clearly poetic licence here, as Nineveh was a city on the eastern bank of the Tigris, more than 300 miles from the sea. It would have known heavy river traffic but few seagoing ships.

How many oarsmen could a single ship possibly have? The mighty catamaran I've just described was designed for 4,000 oarsmen. There were three layers of 200 oars powered by 1,000 oarsmen along each of its four sides. And although the ship was apparently taken on sea trials, it was never used in battle – I should imagine she was a bit sluggish when cornering.

But this ship and the *Syracosia* do give some idea of the technological expertise the Greeks had in around 250 BC, and which they had been developing for more than 200 years. Being on the Mediterranean they were in the perfect place for this to happen – although it can get very stormy, it's not usually as rough as the open ocean.

Calculated Rowing

When I was writing my early TV shows in the 1980s, I discovered – surprisingly – that no one knew exactly how the ancient Greeks had made their oared ships. The only evidence of ships existing 2,000 years ago comes from designs on pottery and wall paintings. So half a dozen major universities around the world set about unravelling this lost technology, and built ships as they imagined the ancient Greeks had. They even raced them off the Grecian coast (see plate section). But getting the dimensions right was a major headache.

In a rowing boat, the rowlock or fulcrum on which the oar pivots is a safe distance above the surface of the water, just high enough to prevent waves slopping over the sides and into the boat (see Figure 5.1). Our picture shows oars where the length to the rowlock is about a quarter of the length of the whole oar, giving a mechanical advantage of about 3:1.

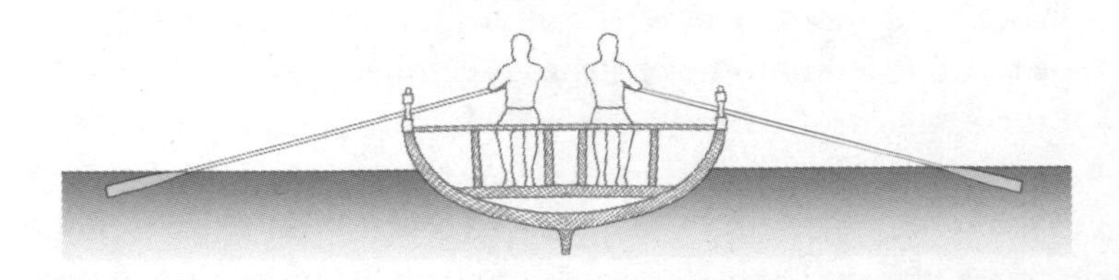

Figure 5.1

Modern oarsmen find the ideal ratio for maximum power and speed is slightly more than 2:1, with just under a third of the oar in the boat. But in planning for several banks of oars, ancient Greek designers had many more problems.

With three banks of oars, each set one above the other, rowing at the highest level would be difficult. The oars were staggered, so that each higher rower was positioned diagonally between the two below. But now the poor lower rowers often found that, as they bent forwards, their noses were very close to the backsides of the rowers above. So for each bank of oars the rowers were offset with the oarsmen on the highest level, seated furthest out, on an outrigger deck overhanging the side of the ship. The second row of oarsmen was positioned slightly more inwards, and the lowest-level oarsmen were even further in. In this way, all the oars could be of a similar length, and their leverage was about equal too.

The rowing on cargo ships was done by slaves, to a rhythm set by a drumbeat. On the lighter fighting triremes, however, a young girl faced the rowers and set their rhythm by playing a tune on a flute or pan's pipes, repeating a high-pitched note for each actual stroke. So the rhythm might go '*peep* and a two, and a *peep* and a two, and a *peep*' and so on. The regular piercing note would ping through the noise and turmoil of battle, keeping the rowers in perfect time even when they were under attack. And if the captain demanded more speed to ram the enemy, it was the girl who would increase the tempo.

The Great Men of Alexandria

What do we know of the scholars and engineers who made Alexandria great, some 2,000 years ago? They designed and introduced a host of new technologies, and developed a far greater understanding of all things mathematical. The names of the greatest of them (as far as we know) have come down to us through the centuries.

Though schooled in Athens' Lyceum, the great scholar **Strato** (340–270 BC) – who was close to Ptolemy I and taught his son – helped to establish Alexandria as a scientific centre.

Strato later returned to Athens to become the third director of the Lyceum, and produced scientific ideas more advanced than Aristotle before him.

He is best known for observing exactly how water drips fall from a roof. He saw that successive drops got further apart, which meant that the lower drops were falling faster; from this he deduced that falling objects accelerate. This was confirmed when he saw that the splash created by a drop falling a short way was much smaller than that created by a drop falling a long way (see Figure 5.2). There is no evidence that he worked out the mathematics that governs falling bodies – for that we would have to wait some 1,400 years, when Galileo began a new scientific revolution.

Figure 5.2

Aristarchus (310–230 BC) may well have been Strato's most famous pupil – in Alexandria he became an important (although not very accurate) astronomer, well thought of by Archimedes. Aristarchus was the first known astronomer to suggest that the Sun was at the centre of things, with the Earth and other planets revolving around it, and that the Moon was the only heavenly body revolving around the Earth. Perhaps because the Moon clearly travelled around the Earth, other astronomers concluded that everything else – the Sun, other planets and indeed the heavens – did too, and that the Gods had placed the Earth at the centre of the Universe.

Aristarchus also calculated the size of the Moon, by observing the curve of the shadow the Earth cast on it during a lunar eclipse: his conclusion was that the Moon was a third of the width of the Earth. We know now that it is actually just over a quarter of the Earth's width. The Sun, on the other hand, is more than 100 times as wide as the Earth.

Aristarchus' measurements of the Moon weren't too far off, but his comparisons of the relative distances to the Earth from the Sun and the Moon were more wildly out. He waited until the half-moon, when the Moon is exactly half in light and half in shadow, and forms a right angle between the Earth and the Sun. Using Pythagoras' Theorem he then tried to calculate their comparative distances, but his maths was not really up to the job.

He calculated that when the Earth/Moon/Sun angle was 90 degrees, the Moon/Earth/Sun angle was 87 degrees. In fact, it's 89 degrees and 52 minutes, so close to 90 degrees because comparatively the Sun is so far away. He suggested that the Sun is 19 times as far from the Earth as the Moon (it's actually 390 times). But even though he assumed the size of the Sun to be bigger than the Earth, and that it was much further away than the Moon, Aristarchus could not conceive of it travelling around the Earth every single day. So he suggested that the Earth must be spinning at one revolution per day – but that in turn posed an unanswerable question. If the Earth spins every day, then the land under our feet must be travelling very fast – in which case, why do we not feel that movement, and why are we not all flung off into space? The conundrum puzzled mathematicians for another 2,000 years, until it was eventually solved by Isaac Newton.

Despite his wayward calculations, however, Aristarchus had started the ball rolling. Astronomers who followed, also questioning how the Universe works, gradually improved on his estimates. Indeed, some 1,800 years later, Copernicus – finally establishing the truth – credited Aristarchus as his inspiration, although he then had second thoughts and took sole credit himself.

A couple of generations after Aristarchus, around 190 BC, **Seleucus** was born in Seleucia on the Tigris. Like Aristarchus, he felt that the Sun must be at the centre of things. No one else listened, however – in fact, all other astronomers seem to have ignored both men, including the greatest of all the Greek astronomers, Hipparchus (who we meet later).

The First Great Alexandrian Engineer

As a young man, Archimedes had been schooled in Alexandria, at the Museum – so called because it was dedicated to the nine Muses. Someone he would surely have known during his stay there was **Ctesibius** (285–222 BC), a remarkable engineer of about the same age. Ctesibius began honing his engineering skills as a youth by helping his dad – a barber – solve a particular problem.

In those days mirrors were made of polished bronze and gave a pretty good reflection. In their shops, barbers had one or maybe two mirrors so that customers could approve their haircuts all the way round. But holding up heavy bronze mirrors was a chore, something that Ctesibius' father complained about. So his son decided to help.

Ctesibius got a rope, and attached two pulley wheels to the ceiling, one above where his father required the mirror, and one in the corner of the room. He then had a square section stone cut that was about the same weight as the mirror.

He hung the stone in the corner, and the mirror at the other end of the rope just above head height. Now when Ctesibius' dad wanted the mirror, he simply reached up to pull it down and it stayed in place, balanced by the stone. When he'd finished, he pushed up the mirror, the stone went down and the two – still balanced – stayed put. Dad was overjoyed – for a while…

When Ctesibius returned a few days later, his father complained that the stone was knocking the plaster off the wall. 'No problem,' said Ctesibius and, taking two planks, he neatly boxed the stone into the corner. His father now pulled the mirror down, but when he tried to raise it again, the rope went slack. The stone fitted too snugly in the corner, trapping the air, and refused to descend. Taking an auger (a type of drill), Ctesibius bored a hole in one of the planks, quite close to the floor, to let the air out.

Now the mirror worked again, because the stone could rise and fall in its square corner container. But every time Ctesibius' dad raised or lowered it, a whistling sound emerged from the hole. The note lowered as he pulled the mirror down (raising the stone) and got higher as he pushed it up again.

Quite by accident, Ctesibius had invented what we know today as the Swanee Whistle. The stone was the first ever piston, sucking or blowing air through the hole and creating the whistling sound, which rose in pitch as the column of air became smaller, and lowered in pitch as the column grew larger. Ctesibius was so inspired by this accident that he went out and designed the world's first ever organ.

He started by making a bigger version of pan pipes – which had been around for a long time – with individual pipes of varying length. Then he added keys that, when pressed, opened the lower end of a particular tube, allowing air to enter it to produce a specific note.

The most ingenious part of the invention was the way it delivered air into an open tube. Standing at the back of the organ was a slave, who repeatedly depressed a foot pedal with a spring that returned it to its original position (see Figure 5.3 – our design shows two pump levers). This action pumped air through the system and an equalizer tank at the bottom ensures a constant pressure. So no matter how vigorously air was pumped in, it always entered the pipes at the same pressure, and stayed at that pressure until the organist pressed a key to produce a note.

Ctesibius then invented a simple yet most ingenious device – the one-way flap valve. A simple leather flap was fixed over a hole. Air could pass in one direction through the hole, blowing the flap out of the way. But air coming in the other direction sucked the leather flap over the hole, closing it off.

Until electronic organs were invented in the twentieth century, this system changed very little. Around 1918, my dad, aged about 10, earned a few pennies on Sundays by sitting behind the local church organ and pumping the bellows to ensure there was always enough air in the pipes to produce the hymn tunes to accompany the congregation. In old organs it was common to hear a drone emerging, as excess air escaped so that the right pressure was maintained.

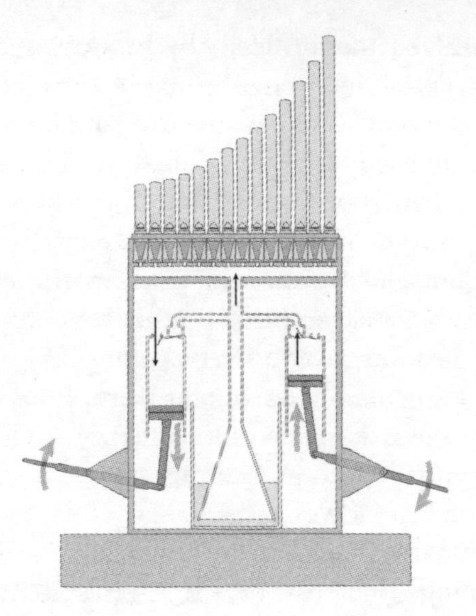

Figure 5.3

Ctesibius was now on the crest of an inventive wave. With a reed pipe and his flap valve, along with a sheep's bladder, and a tube and resin glue to seal it all up, he had everything he needed to invent the bagpipes (or so legend has it) – a device that today, it seems, the Scots are prouder of than the Greeks.

Ctesibius went on to create all kinds of novelty devices. Using fire to expand air, he made tiny figures spin round and emit whistling noises; he also designed the first known fire engines: huge barrels of water on wheeled carts. Long bars down each side of the cart were raised and lowered to pump water through hoses towards the fire. Similar engines were in use in America up to around 100 years ago, and 2,000 years after Ctesibius.

But his most famous and ingenious device was a water-clock: the *clepsydra* or 'water thief' (see plate section).

In those days, keeping accurate time was much more difficult than it is today, because the Greeks still followed the Ancient Egyptian system of dividing the 24-hour day into 12 night-time hours and 12 daytime hours. Alexandria is 30 degrees from the Equator, but as the Sun swings from 23.5 degrees north in midsummer to 23.5 degrees south in midwinter, the length of daylight hours varies considerably.

Ctesibius solved this problem by producing a rectangular chart with a central line drawn from left to right to represent midday. He then drew 12 lines above and 12 below the central line, to represent each half hour of daylight time. Each of these lines formed a hyperbolic curve, with the curves increasing as they moved from the midday line. He then drew a centre line from top to bottom to represent midwinter's day, when the daylight hours were at their shortest. Moving left or right, the hours would become progressively longer until midsummer's day, which appeared halfway around the cylinder. He assessed and calibrated these lines over the course of a year or two until he produced an accurate set of lines.

The clock had a clever system of gears to allow the water level, and a floating pointer, to rise slowly throughout the day, indicating the time. At the end of the day the water was siphoned off, and the marker dropped to the bottom of the tank, ready to start the next day. Extra gears ensured that the cylinder revolved just once in a whole year. The clock was quite brilliant – and it makes you wonder whether Ctesibius in any way inspired the most amazing Grecian device of them all, parts of which were found in a ship wrecked off the island of Antikythera in 1900 (see plate section).

The Amazing Antikythera Device

Divers searching near a shipwreck hauled a collection of bronze objects from a depth of 40m. These proved to be parts of a box with original dimensions around 30 x 17 x 10cm. On one face of the box was a double circular scale showing months of the year and signs of the zodiac. At one time there was probably also a pointer that revolved to indicate the date. The back of the box was inscribed with two more scales, and had central holes, possibly for more pointers.

Only four cog wheels were recovered, but it's now believed that originally there were more than 30, connected in a variety of ways, probably to indicate astronomical timings like the phases of the Moon. It would have been of immense value to travellers at sea or in unfamiliar places. Incredibly, though, it dates to the first century BC, so was probably made more than a hundred years after Ctesibius was alive.

Ctesibius was hugely famous in his day, and he eventually became principal of the Library at Alexandria, a post he held until 225 BC, when he was about 60 years old. The principal's post then passed to one of his contemporaries, who held it for almost 30 years. His name was Eratosthenes.

The Man Who Measured the World

Eratosthenes (276 to around 196 BC) was born in Cyrene, west along the coast from Alexandria, although at some point he was taught in Athens. Rather than follow one path of learning he chose to be a polymath, a master of many skills. However, because he never actually excelled in any one of them, his critics called him Beta after the second letter of the Greek alphabet – rather than Alpha, the first. His friends, though, rectified the situation, nicknaming him Pentathlos, after athletes who tried their hand at many events, but never specialised – in modern athletics we still have the pentathlon, heptathlon and decathlon, with five, seven and ten events respectively.

Eratosthenes became a poet, music theorist, and theatre and literary critic. Demonstrating that Grecians did not lack a sense of humour, he even wrote a treatise on comedy. As a historian, he attempted to date all major events from the Trojan Wars and the wooden horse episode in 1184 BC, when Troy fell to the Greeks. First and foremost, though, he was an astronomer and mathematician, and a great friend of Archimedes. As principal of the Great Library, he improved the running of the place, and read almost all of the many thousands of scrolls contained there.

As for mathematics, his aim was always to make the subject more inviting. He promoted the mesolabe compass (first used by Hippocrates) and other simple maths tools. He also studied number theory, and devised his 'sieve' method for finding prime numbers.

Eratosthenes was a great astronomer, and produced a star map with 675 stars. But he was also a geographer, creating

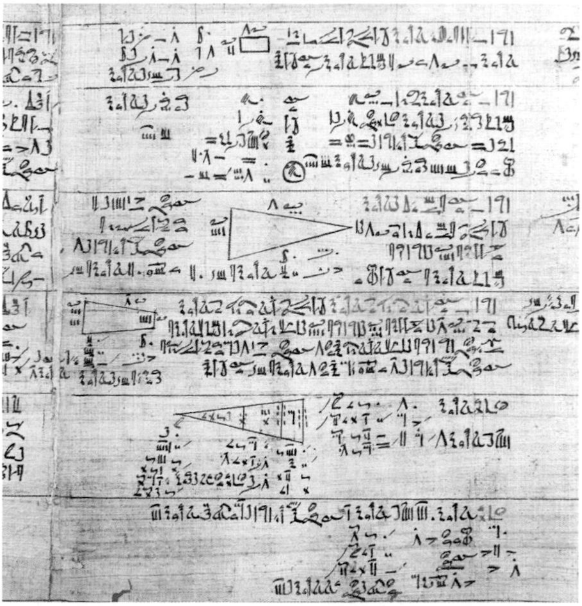

Left: A fragment of the Rhind Papyrus, from the British Museum in London (p. 14).

Left: Djoser's Step Pyramid. Egypt's first pyramid (p. 24).

Below: Pharos Lighthouse – an artist's impression (p. 118).

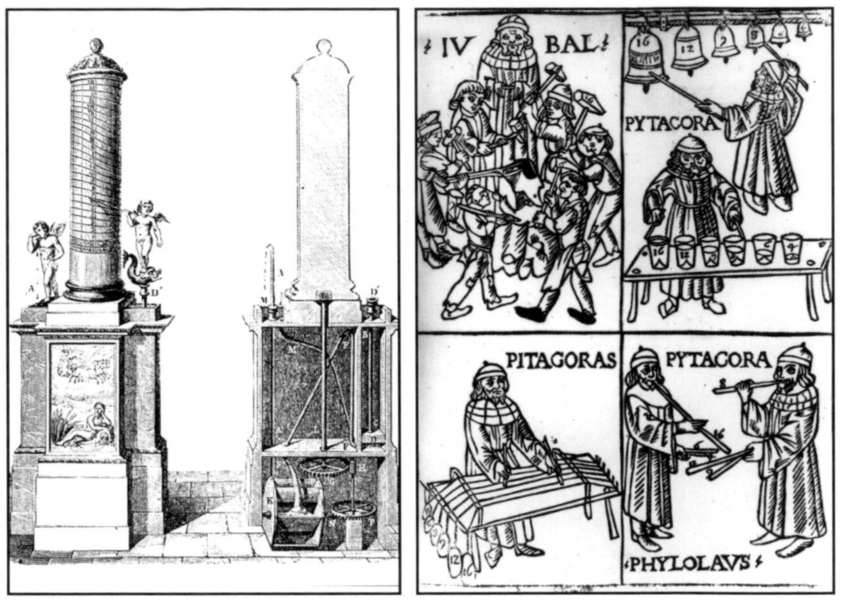

Above: Ctesibius' Clepsydra showing gearing mechanism (p. 126).

Above: Medieval woodcuts of Pythagoras' musical ideas (p. 42).

Below: Eratosthenes' world map – the first map ever to have lines of latitude and longitude (p. 131).

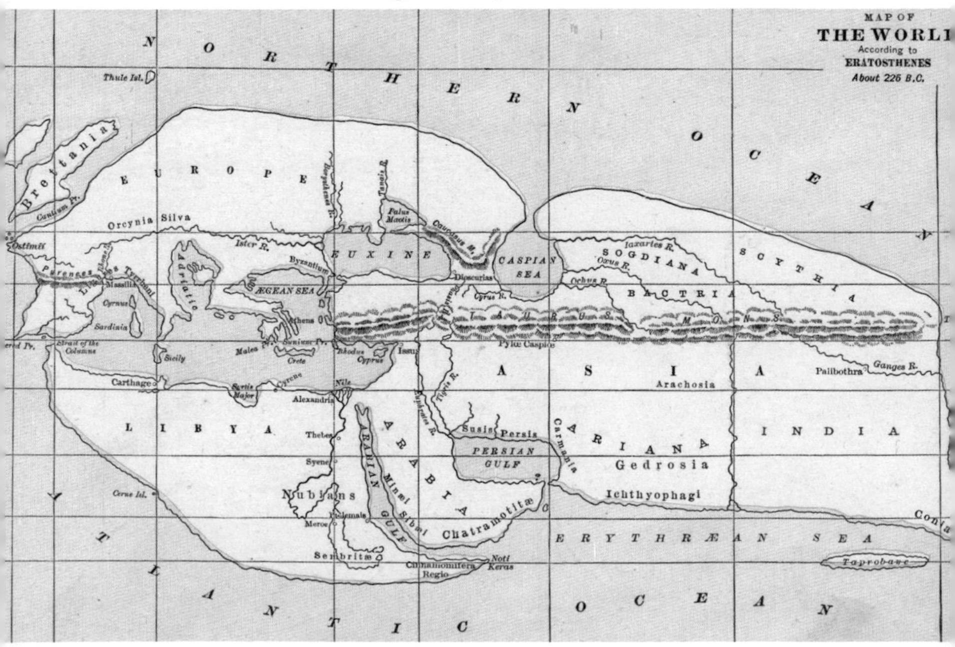

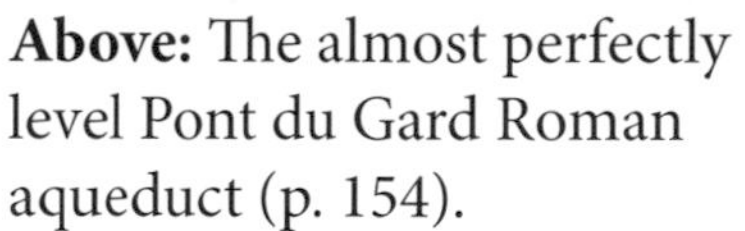

Above: The almost perfectly level Pont du Gard Roman aqueduct (p. 154).

Above: The Antikythera mechanism. One of just four bronze fragments found (p. 127).

Above: A quite detailed astrolabe (p. 136).

Above: The city of Gwalior's first example of a written zero (p. 202).

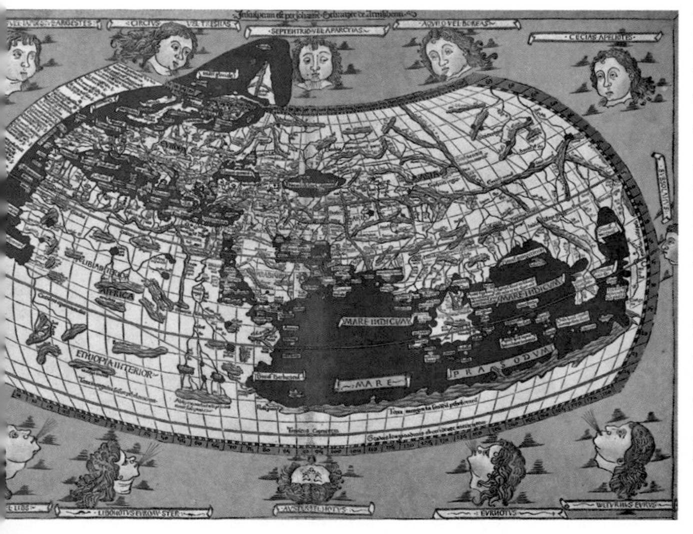

Left: Ptolemy's world map. The longitude lines are numbered from 0 to 180, implying that this was half the world (p. 171 and p. 248).

Left: The Monte Alban Great Square, with the angled observatory in the foreground (p. 206).

Left: A 19th-century illustration of a Greek trireme. The higher oarsmen overhung the lower ones (p. 120).

Below: A Peruvian quipu (p. 205 and p. 454).

Below: The Hsi and Ho brothers receive instruction from Emperor Yao (p. 183).

Left: A stone in a Peruvian wall with 12 perfectly-fitting angles, Cusco, Peru (p. 203).

Left: A 1000-year-old Olmec stone head (p. 205).

Left: Flying buttresses, Bourges Cathedral – even these essential supporting buttresses were designed with light elegance (p. 235).

Below: St Denis Cathedral, Paris, showing the birdcage-like tracery, allowing in light but maintaining great strength (p. 234).

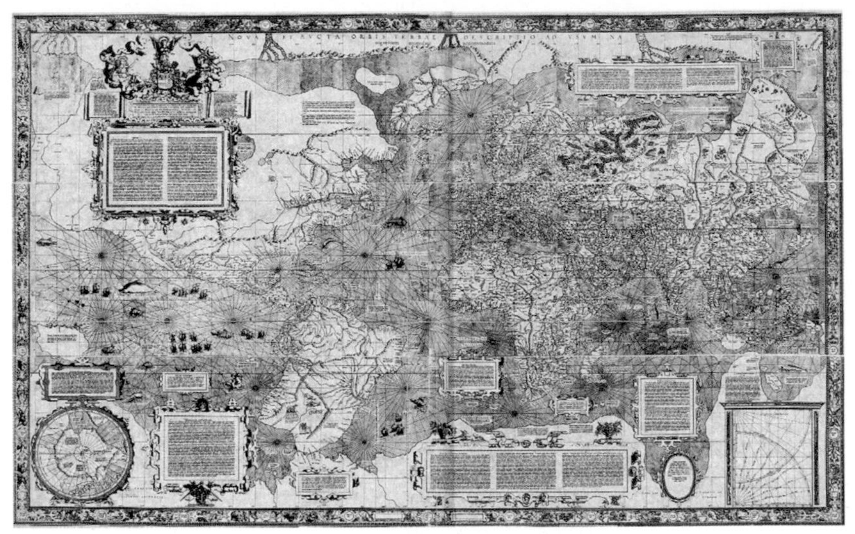

Above: Mercator's world map of 1659. He drew on information from his Spanish and Portuguese adventures (p. 256).

Left: The Beautifully decorated church of Santa Maria Novella, Florence, designed by Alberti (p. 272).

Below: John Harrison's incredibly accurate Chronometer Number 4 (p. 263).

Below: The Pacific Ocean covering half of the earth's surface (p. 253).

Above: Eilmer and a model of his wings at Malmesbury (p. 239).

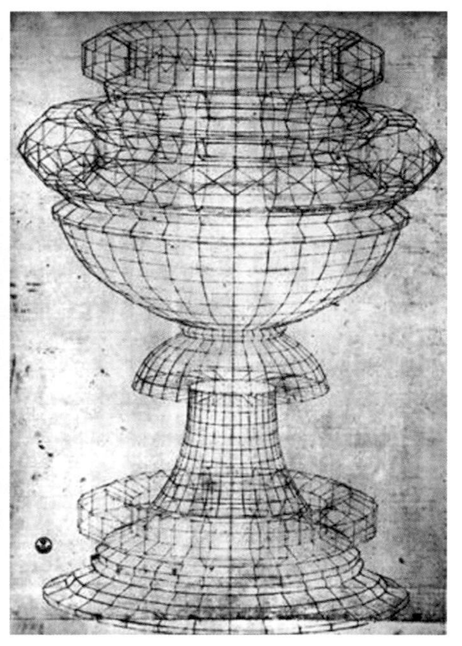

Above: Uccello's groundbreaking study in three dimensions, *c.* 1430 (p. 271).

Left: *The Flagellation of Christ* by Piero della Francesca, 1450, with great perspective depth (p. 267).

Left: Otto von Guericke's experiment – 16 horses failed to overcome the force of the atmosphere on the iron ball (p. 344).

Left: The plan of Tycho Brahe's Uraniborg Observatory (p. 304).

Above: The 1338 Astrarium of Giovanni Dondi, with seven faces to calculate the orbit of the Sun, Moon, planets, time and date. It took 26 years to complete (p. 290).

Above: Tycho Brahe's great armillary sphere and viewing ampitheatre. The sphere was invented by Hipparchus (p. 305).

The Sieve of Eratosthenes

Eratosthenes devised his sieve as a way of finding prime numbers. From the 100 numbers in a 10 x 10 square he began to eliminate all the numbers that were *not* primes (see Figure 5.4).

1	2	3	4	5	6	7	8	9	10
11	12	13	14	15	16	17	18	19	20
21	22	23	24	25	26	27	28	29	30
31	32	33	34	35	36	37	38	39	40
41	42	43	44	45	46	47	48	49	50
51	52	53	54	55	56	57	58	59	60
61	62	63	64	65	66	67	68	69	70
71	72	73	74	75	76	77	78	79	80
81	82	83	84	85	86	87	88	89	90
91	92	93	94	95	96	97	98	99	100

Figure 5.4

The definition of a prime is a number larger than 1 that's divisible only by 1 and itself. The number 2 is prime, and the only one that's even. First Eratosthenes cancelled out all the numbers divisible by 2, except 2 itself, or the entire even rows. Even numbers divisible by 3 had already gone, so next he took out the odd 3s: 9, 15, 21 and so on. Then he took out the odd 5s, as those ending in 0 had already gone.

The most complex number to remove was 7, but as the even 7s and those divisible by 3 and 5 had already gone he was left with only three: 49, 77 and 91.

Now, magically, all that remained were the 25 primes: 2, 3, 5, 7, 11, 13, 17, 19, 23, 29, 31, 37, 41, 43, 47, 53, 59, 61, 67, 71, 73, 79, 83, 89 and 97.

The sieve works even better if the numbers are listed in rows of 6 instead of rows of 10 (see Figure 5.5).

1	2	3	4	5	6
7	8	9	10	11	12
13	14	15	16	17	18
19	20	21	22	23	24
25	26	27	28	29	30
31	32	33	34	35	56

Figure 5.5

Now, with the exception of 2 and 3 in the top row, all the primes fall into the first or the fifth column, and are either one more or one less than numbers divisible by 6 (+1 or -1 *modulo* 6). This includes all the primes ever discovered, and those that have not yet been revealed.

much of the geographic terminology we still use today. And he was the first person to accurately calculate the distance around the Earth.

In a town called Syene, directly south of Alexandria and near present-day Aswan in Lower Egypt, a strange thing happened every midsummer's day. For only a few minutes either side of noon, the Sun shone straight down the wells in the town, casting no shadow on the sides of the well at all.

Eratosthenes, on hearing of this, waited for the next midsummer's day in Alexandria and sought out an obelisk which the Egyptians had set perfectly upright 2,000 years earlier. At noon precisely he measured the angle of the shadow cast by the obelisk, which was 7.2 degrees. Now he had all the information he needed to measure the entire distance around the Earth.

He reasoned that if both the obelisk and the well were extended downwards, they would eventually meet at the centre of the Earth. As the Sun's rays arrive at the Earth parallel to each other, he saw that the two lines meeting at the centre of the Earth would form the same angle as the shadow of the obelisk (7.2 degrees). This was a very convenient figure, because 7.2 divides into 360 degrees exactly 50 times.

In those days long distances were measured using the distance covered by a camel in a day: not a particularly accurate measure. Nevertheless, Eratosthenes knew that Syene was about 50 camel days south. More accurate distances were paced out by professional pacers, and grouped together in *stades* (the length of a stadium), a measure that Eratosthenes also used. But as stadia varied from place to place, we cannot be absolutely sure which stade he used.

He did give the distance from Alexandria to Syene as 5,000 stades and the distance around the Earth as 50 times that, or 250,000 stadia. Today we know that the distance from Alexandria to Syene is almost exactly 500 miles (or 800km), so the stade Eratosthenes used was probably about $\frac{1}{10}$ of a mile long.

So:
360/7.2 = 50 x 500 miles = 25,000 miles
or 360/7.2 = 50 x 800km = 40,000km

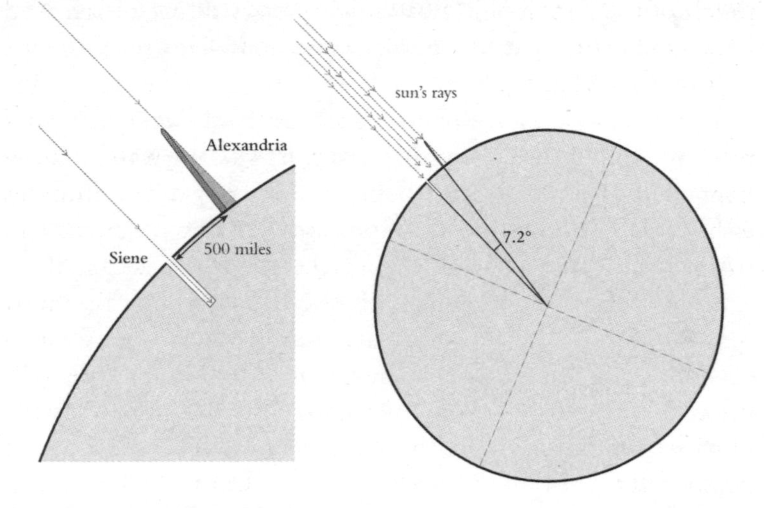

Figure 5.6a and b

When the French introduced the metric system in 1799, they decided that a metre should be 1/10,000,000 ($\frac{1}{10}$ millionth) of the Earth's quadrant at sea level, based on measurements made between Barcelona and Dunkirk. This gave the distance around the Earth at 40,000,000m, or 40,000km.

Eratosthenes' measurement seems to have been amazingly accurate for his time. Since the 1960s we have been able to measure any distance on Earth with mind-boggling accuracy, thanks to satellites that can look back at our planet from space (see Figure 5.6).

Having announced that he had discovered the size of the Earth, Eratosthenes then produced a map of the known world, stretching from the British Isles in the north-west to modern-day Sri Lanka in the south-east. His was the first land map with lines of latitude and longitude, based on the available geographical knowledge of the time (see plate section).

On seeing the map, however, most Greek scholars doubted that so much of the world could be undiscovered, believing strongly instead that the Earth must be much smaller than Eratosthenes had suggested. One hundred years later, the philosopher and astronomer Poseidonius, by studying star angles rather than the Earth itself, declared that the Earth was about two-thirds Eratosthenes' suggested size – a figure that was to cause much trouble, as we shall see.

But Eratosthenes didn't stop with the size of the Earth. Over the course of a year he measured the shadows cast by the Sun, finally declaring that the Earth's axis was tilted in relation to the Sun by 23.5 degrees. This explained why the Sun shone straight down the Syene wells at noon on midsummer's day. It was the furthest north that the Sun reached each year, and marks the line of latitude we now call the Tropic of Cancer. Similarly, Eratosthenes realised – without ever going there – that there was a similar southerly limit to where the Sun is overhead, which is 23.5 degrees south of the Equator, and which we now call the Tropic of Capricorn (see Figure 5.7).

By measuring time and angles, Eratosthenes saw that the length of a year was actually closer to 365 and a quarter days. So he suggested that every fourth year should have an extra day, and should be called a leap year. Sadly, Egyptian and Alexandrian conservatism prevented the change and it wasn't accepted for another 150 years.

Eratosthenes also calculated the distance from the Earth to the Sun to be 804 million stadia, or 80.4 million miles – pretty close to the actual figure of 93 million miles, especially considering the time in which he lived, and the equipment he would have had to hand.

Eratosthenes was revered by everyone, yet when he was about 80 years old, somewhere between 196 and 192 BC, he

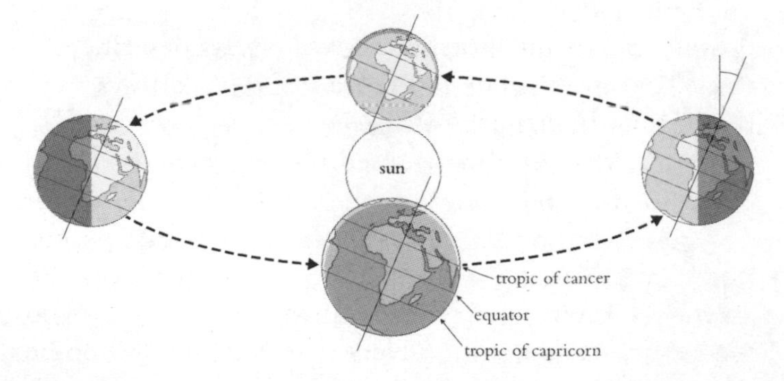

Figure 5.7

decided he could be of no further use. So he shut himself away in a dark room, seeking neither help, comfort, food nor water, until he passed away.

The Great Geometer

Apollonius (around 262–190 BC) was a contemporary of Eratosthenes, but was purely a mathematician. He was born in Perga (in present-day Turkey), and spent some time in Pergamum in North Africa, although he lived most of his life in Alexandria, where he was known as the 'Great Geometer.'

He is remembered for three mathematical discoveries that bear his name. He began by following the path of previous Greek mathematicians in using proportions to make amazing discoveries. For instance, he found what we now call 'the circle of Apollonius'.

The Circle of Appolonius

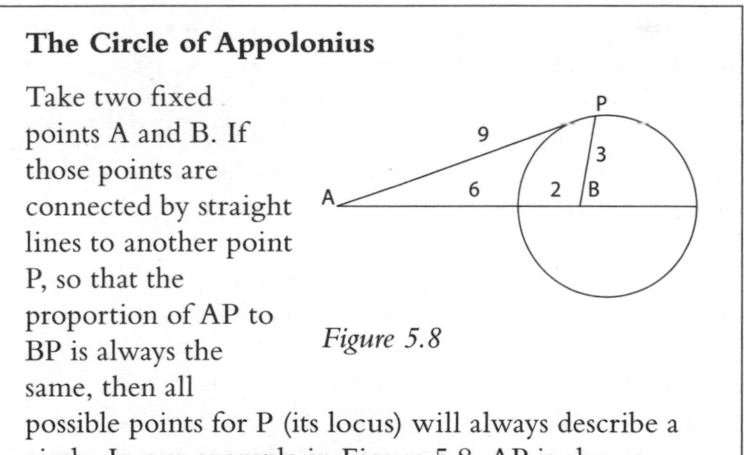

Figure 5.8

Take two fixed points A and B. If those points are connected by straight lines to another point P, so that the proportion of AP to BP is always the same, then all possible points for P (its locus) will always describe a circle. In our example in Figure 5.8, AP is always three times the length of BP.

It's odd that Apollonius used the term *locus* or *loci* to describe the collection of all points forming under certain conditions, but never *focus* or *foci* for focal points. For that terminology we had to wait until Pappus (in about AD 260) and Kepler (in about 1600), who formalised the maths. Given that the foci of conic sections and parabolic curves are very helpful in understanding and manipulating them, it's quite surprising that Apollonius and Archimedes seem to have missed them.

The Theorem of Appolonius

Apollonius also devised his own theorem, which is similar to Pythagoras' Theorem, but which works for all triangles. Finding it was sheer brilliance, but it is seldom featured in mathematics books today.

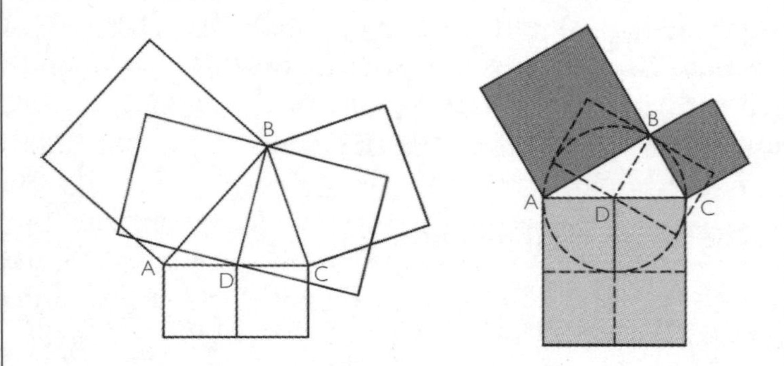

Figure 5.9

Take a triangle, ABC, that is not right angled. From B draw a line that meets AC at its centre point, which we'll call D. Apollonius' theorem states that the sum of the squares on sides AB and BC is equal to twice the squares on BD and AD (or DC).

In the right-angled triangle (right) you can see that the theorem is true. But looking at the left hand diagram, you begin to realise that if you make any adjustments or slight changes to the angle at B, the differences are reciprocal and cancel each other out.

Apollonius was also troubled by a circular problem – until, that is, he solved it. After that it became known as 'Apollonius' Problem', although for him – as he'd solved it – it had ceased to be a problem. Here it is…

Given three circles, can you produce a fourth that will touch all the others? There are eight solutions, and you can find them in the Wow Factor Index". Amazingly, Apollonius proved this using only the original basic mathematical tools – a straight edge and compasses.

Apollonius also investigated conic sections, or slices through a cone (see Figure 5.10). From these explorations he discovered how to accurately draw a tangent to a curve such as an ellipse, using equal proportions.

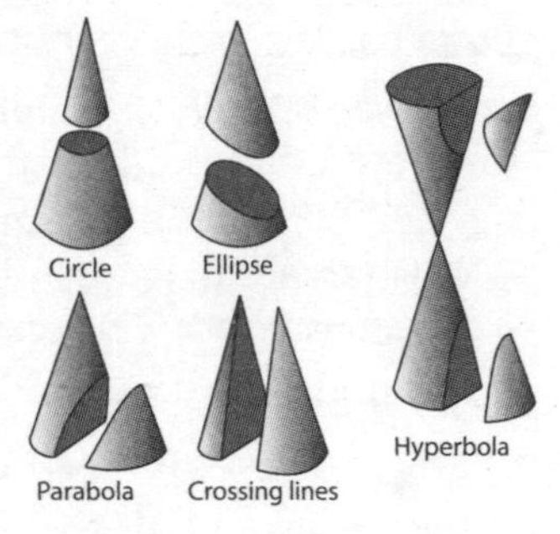

Figure 5.10

In Figure 5.11 we have an ellipse with an axis line AB running through it, and a point N that we have chosen so that, for example, N is 1 unit from A and 3 units from B (though this will work for any two ratios). Raise the perpendicular from N to meet the ellipse at P. Now extend the axis line and find a point T so that TA and TB are in the same ratio as AN and NB (in this case, 1:3). The line TP will be a tangent to the ellipse.

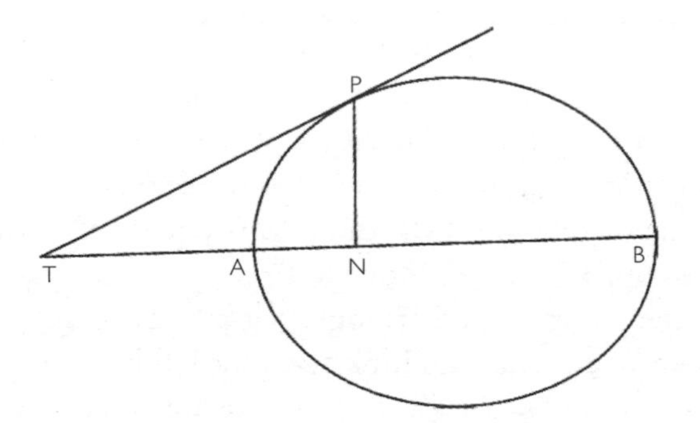

Figure 5.11

Apollonius also studied *hyperbolas*. The best place to find hyperbolas today is in our living rooms. Look at the way a circular lampshade casts light on a flat wall: the curved line between light and dark is a hyperbola.

Apollonius cut through a pair of double-napped cones (which we met in the previous chapter) to produce a pair of hyperbolas. There is more on hyperbolas in the Wow Factor Mathematical Index.

I have only been able to touch on a few of Apollonius' geometric discoveries, but you can see why he was called the Great Geometer. Although as far as we know he never considered that the Sun might be at the centre of things. In fact he spent a huge amount of time using his immense knowledge of the geometry of circles and conic sections to try to make sense of the movements of the Sun and planets around a central Earth.

However, all his work was superseded and improved upon by the greatest of all the ancient astronomers, who was born around the time of Apollonius' death. His name was Hipparchus.

The Ancient Star of the Heavens

Hipparchus (190–120 BC) was born in what today is north-western Turkey, and he seems to have acquired a great deal of knowledge from the ancient Sumerians and Babylonians. He moved to the island of Rhodes, where he observed the heavens for at least 35 years. There he also invented most of the gadgets used to observe the skies with the naked eye, which would be in use for the next 1,800 years, until the advent of telescopes. The most famous of these was the *astrolabe* (see plate section).

The basic astrolabe was a circular metal disc with 360 degrees marked around the outer edge and a ring attached to the top so it could dangle from your fingers. An index arm revolving around the centre could be lined up with the Sun or the Pole Star so that you could read the degree of elevation. We're not sure how much Hipparchus refined his basic idea, as over many years each new astronomer sought new ways to improve it.

An ecliptic circle was added later, which indicated the path of the Sun and the planets; it was different from the equatorial line because of the Earth's natural tilt. This helped users to measure the angles of the Moon or planets on their paths across the sky, and even calculate possible eclipses. North, or the *zenith*, was clearly marked, as were the major stars, which were marked on one side as part of a map of the heavens, while signs of the zodiac were displayed around the edge on the other side. The astrolabe became such a powerful tool to calculate how far north you might be, or to estimate the hours of day or night remaining, that soon no ship's captain or traveller left home without one, and it became the chronometer of its day.

Hipparchus could not believe that the Earth revolved once a day; rather, he believed that the heavens – in their entirety – revolved around the fixed Earth every day. But he also observed that the Sun, the Moon and the planets all revolved in a very complex way, and was determined to unravel and explain it, making it his life's work. As an example of what he achieved, he measured the lunar month to within about one second of the figure we use today.

He also accurately measured the difference between a solar day and a *sidereal* day. We measure our 24-hour solar day by the time it takes the Earth to turn one revolution relative to the Sun. But in that time the Earth has progressed about 1 degree in its orbit around the Sun. To be precisely facing the Sun again it has turned slightly more than one exact revolution. Hipparchus measured the time it takes for the Earth to come around so that it exactly faces a fixed star once again, and found it did this in very close to four minutes quicker. That is the sidereal day.

Hipparchus, remember, always assumed that it was the Sun that revolved around the Earth and not the other way about – but he still got the same result.

Hipparchus used 44 fixed stars to measure and confirm the sidereal day, and to calculate each night-time hour accurately. In doing this over a very long time, he eventually estimated that the heavens themselves progress west by about 1 degree per century – today we know that the figure is more like 1 degree every 72 years.

Hipparchus also suggested that Rhodes, from where he made all his heavenly measurements, should be on the prime meridian, and that all other places on Earth should be given positions east and west relative to it. It was a great idea, but it wasn't taken up until 1884. In that year, an international agreement accepted that a north/south line through the Greenwich Observatory just east of London (founded in 1675) should become the Greenwich Meridian, and that all times around the world should be fixed relative to distances east or west of it.

To help with this work in calculating the true angles of stars relative to one another, Hipparchus invented the *armillary sphere* (see plate section) – a model of the heavens made from metal bands, with the imaginary Earth at its centre. More importantly, however, he introduced a new branch of mathematics that was to prove a powerful aid to geometric calculation throughout the history of mathematics: trigonometry, or triangle measuring.

The Trick to Trigonometry

In observing the distances between stars, probably to place them on a map of the heavens or the surface of his armillary sphere, Hipparchus saw something unusual about right triangles. Let's assume we are looking at a right triangle ABC from an angle C (see Figure 5.12). Knowing C helps us calculate the ratio of the lengths of the sides of the triangle. As an example, if the angle is 45 degrees, then the sides AB and BC are equal, and their ratio to each other is 1.

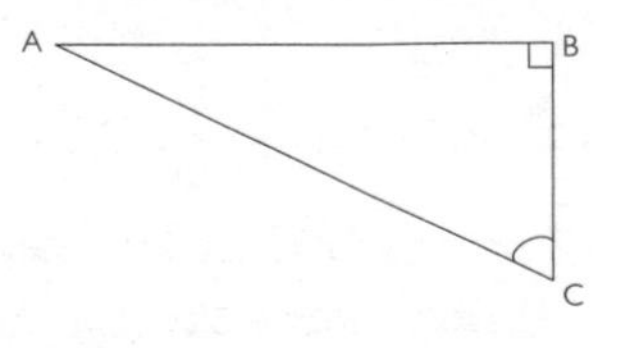

Figure 5.12

Hipparchus saw that there were several different possible ratios to consider, and he gave them names. The ratio AB/BC is called the *tangent*, AB/AC is the *sine* and BC/AC is the *cosine*. He even compiled a table of ratios (for more details, and some examples, see the Wow Factor Mathematical Index). Trigonometric values have been extremely powerful tools in mathematics ever since, but it is not certain that anyone used or extended the system for several centuries after Hipparchus.

In 134 BC Hipparchus noticed a star in the constellation of Scorpio that he had not seen before, and which was not marked on any existing star chart. This defied the religious belief of the time that the heavens, apart from the planets, were unchanging. This star had to be either one that had previously been too faint to see and was now brighter, or a completely new one.

Which of these was the answer, he could not tell, but he was determined that he wouldn't be caught out again. But how could one man possibly do that? Easy – he would create a star map containing every major star, with an accuracy, in terms of position, better than anyone had ever achieved before.

Hipparchus mapped out more than 1,000 of the brightest stars in the heavens. He called the brightest 20 'stars of first magnitude', and progressed through five more orders of magnitude to the sixth-magnitude stars, which can just be seen with the naked eye.

To accurately locate the stars on his map, Hipparchus added a grid of latitude lines across it, relative to his home in Rhodes, and longitude lines going up and down. He probably copied the idea from Eratosthenes, who had already used similar lines when mapping the Earth.

How High is the Moon?

One of Hipparchus' big achievements was to measure the distance from the Earth to the Moon using *parallax*. If you hold out a finger out in front of you, and look at it first with one eye and then the other, it seems to hop from side to side (see Figure 5.13). The effect is stronger in a moving car or train, if you compare nearby objects with those in the far distance. Parallax can be an effective way to measure distance, but the bigger the distances involved, the less reliable it is.

Using parallax, Hipparchus calculated – correctly – that the distance from the Earth to the Moon is 30 times the diameter of the Earth: the Earth is 13,000km (8,000 miles) wide and the Moon is 390,000km (240,000 miles) away. As

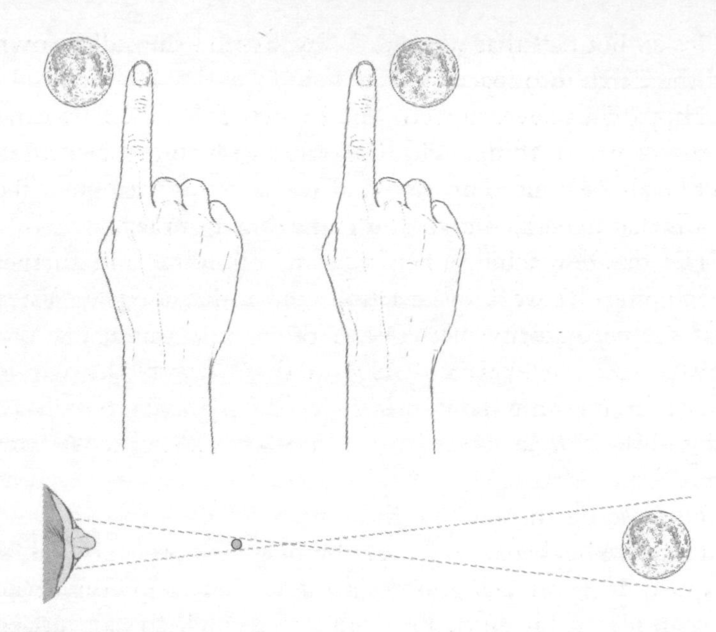

Figure 5.13

the planets in our Solar System are much further away than the Moon, we had to wait until telescopes were invented before we could use parallax to measure those distances.

Surprisingly, Hipparchus – and, for that matter, all the other early astronomers – seems to have missed a key clue to the workings of the heavens: he didn't consider how the brightness of the planets changes over time. This would have indicated that their distance from the Earth varied, suggesting that their earthly orbits were eccentric, and that their movements around the Sun would be far smoother and better related to natural orbits.

We might wonder today why the astronomers of antiquity didn't seriously consider the possibility that the Earth goes round the Sun. For them, however, the alternative was even more incredible and improbable. They had only recently taken on board that the Earth was a sphere and not flat. They could not now believe that the Earth was making a complete revolution every day. If you accepted Eratosthenes' calculations for the size of the Earth, then everyone living around the Mediterranean was constantly travelling east at about 800 miles an hour. If that was true, why weren't they all thrown off the Earth into space?

Hipparchus never moved from his belief that the Earth was at the centre of things with everything revolving around it. But he also set out to prove that it was so, by working out the maths that explained it all, with amazing accuracy.

His greatest achievement was in following and further developing the work of Apollonius, who had first suggested that the eccentricity of the orbit of each planet around the Earth could be explained using a system of (at least) two circles. Apollonius had suggested that a planet like Mars orbited in a circle about a centre point, and that it was this point – rather than Mars itself – that followed a circular path around the Earth.

Hipparchus began by describing the inner universe as comprising seven heavenly spheres, one each for the seven known planets, including the Sun. This, he believed, moved in a small circle about a point that in turn moved in a huge circle around the Earth. We now know that the change in the Earth's distance from the Sun is not to do with the Sun revolving in a small circle, but because of the eccentricity of the Earth's orbit around it.

The orbits of all the other planets were greater around some centre point that itself was orbiting the Earth. Eventually Hipparchus devised a system that seemed to work, and which enabled mathematicians to predict the position of any planet relative to the Earth at any point in the future. Hipparchus' astronomical maths was used to produce incredibly elaborate working models of the solar system for wealthy Romans – who had now taken control of most of the Grecian world, including Alexandria. The models worked extremely well.

The Times – They Were A-Changing

But things were changing fast. From perhaps 200 BC onwards, the power and importance of Ptolemaic Egypt declined and the Romans increased their domination of the region militarily, economically and politically. Being far less mathematically inclined, they were never going to produce mathematicians or scientists to match the Greeks. But they were happy to adopt all the science and technology they considered useful, and for their sons to be taught by scholars of the Greek tradition. One scholar who was a particular Roman favourite was **Poseidonius** (135–50 BC).

Poseidonius began his academic life in Athens, where he carried huge influence, teaching future Roman leaders such as Cicero and Pompey along the way. He dabbled heavily in astrology, and believed that heavenly bodies influence our lives, something that also went down quite well with the superstitious Romans. His astrological ideas had little foundation in fact, but he was correct when he asserted that the Moon influences the tides. He even travelled the length of the Mediterranean to study the relatively higher tides on the Atlantic coast.

Poseidonius also continued the work of Hipparchus in Rhodes, but like Aristarchus, his measuring methods left a lot to be desired. In fact he's well worth a mention precisely because of the effect he had long after his death, due to one enormous error...

Eratosthenes had accurately measured the Earth's circumference, and its size, but that implied that a huge part of the Earth – my rough estimate is between 75 and 85 per cent – was completely unknown to the Greeks. Many scholars just could not or would not believe this. Human vanity is often the cause of great errors of judgement!

Hipparchus had accurately calculated the hours of the night according to the revolving movement of the heavens. Surely, thought Poseidonius, by knowing where you were on the Earth and relating that to the turning of the heavens, you could calculate the actual size of the Earth?

Ignoring Eratosthenes completely, Poseidonius made many measurements of his own, and eventually calculated the size of the Earth at about two-thirds (0.71) of the size we now know it to be. The trouble was, possibly because of his influence with powerful Romans, everyone who mattered believed him, and rejected Eratosthenes' theory. This colossal error was to shape history dramatically.

The Nose That Changed History

Perhaps the most legendary tale of those times contains virtually no maths at all. But it is significant, as it marked a huge turning point in the history of the period, and the development of both mathematics and science.

As Rome's power increased, Egypt and Alexandria faded, although Alexandria retained some importance: the fertile banks of the Nile were the most productive farmland in the known world, providing the grain that fed much of the Roman Empire.

Twelve successive Ptolemys had ruled Alexandria, along with several queens, nearly all of them called Cleopatra. One of these – Cleopatra VII – attempted to make Egypt great once more. She was 18 when her father Ptolemy XII died; her brother Ptolemy XIII was just 10. As was the custom at the time, they were married and declared joint rulers. Cleopatra, however, refused to share the throne with the lad – but as the politicians backed the male heir, she was exiled in 48 BC.

Meanwhile, civil war had broken out in Rome. In September 48 BC, the Roman general Pompey arrived in Alexandria seeking sanctuary from Julius Caesar's far more powerful fleet, which was just three days behind him. From his mobile throne, the young Ptolemy watched as Pompey stepped ashore and offered a greeting, before being unceremoniously beheaded on the spot.

When Caesar arrived, Ptolemy, to display his loyalty, proudly presented Pompey's head. Caesar, however, showed no sign of gratitude – but then why would he? Although he was on the opposite side, Pompey was married to Caesar's daughter Julia, who had witnessed her husband's beheading. Caesar promptly banished Ptolemy and took total control of Egypt.

Conquerors often married into the families of those they conquered, strengthening and cementing ties for the future. Alexander and his generals had certainly followed that tradition. The Romans, however, had not generally adopted this habit – until now.

It's difficult to separate fact from myth, but it's pretty certain that Caesar ignored Cleopatra at first, despite her attempts to impress him, which included having her disgraced brother killed. Finally, she had herself smuggled into his presence, where she magically and seductively appeared from a carpet unravelled before him. In many ways, this would eventually lead to the unravelling of Caesar himself, and after him, Marc Antony.

Caesar was instantly captivated, and nine months later Cleopatra gave birth to his son, whom they named Caesarion or Little Caesar. Today we refer to a Caesarian birth, but that actually relates to the way Caesar himself was born, made necessary because of difficulties in his delivery.

Cleopatra had real scientific influence over Caesar, introducing him to her astronomer, Sosigenes (pronounced Sos-i-genees – not Sausagiknees…). The Romans followed the same calendar as the Egyptians and the Greeks, but interference from various religious leaders had made all calendars wildly inaccurate. Caesar adopted Sosigenes' ideas, including Eratosthenes' suggestion of a leap year every fourth year. The Julian calendar was introduced in Rome in 46 BC. But Caesar couldn't help messing things up again, claiming – and naming – the month of July for himself.

Meanwhile, Cleopatra's other young brother, Ptolemy XIV, also in exile, raised an army with the help of Romans opposed to Caesar. They besieged Alexandria, setting fire to Caesar's fleet in the harbour, and trapping Caesar and Cleo in their palace. According to later reports, on that night the great Library of Alexandria burnt down, but it's far more likely that some books stored on the quay went up in flames. The Greek geographer Strabo certainly studied at the library a few years later.

During the fire, Caesar dropped quietly into the harbour, and amid the confusion swam past the burning ships to join a vessel that had escaped. He immediately took charge of his forces, regrouped and destroyed the Egyptian force at the Battle of the Nile, after which Ptolemy drowned as his fleeing ship sank.

In 46 BC Cleopatra and Caesarion travelled to Rome to join Caesar, who proudly displayed them; he even erected a gilt statue of Cleo and set her up in a huge house. This, along with his many political quarrels and intrigues, pushed relations to breaking point and, while Cleo was still in Rome, on the Ides of March (15th March 44 BC), Caesar's enemies assassinated him.

Shortly afterwards, Cleopatra and her son returned to Egypt. She had earned her place in history for protecting Egypt and Alexandria by seducing a major Roman statesman. But she then proved that lightning can strike twice, by doing it all over again. Marc Antony had first laid eyes on Cleopatra when she was just 14, and according to one historian he was smitten even then.

Caesar had made his nephew, Octavian, his heir. Determind to avenge his uncle's death, Octavian formed a ruling triumvirate with Marcus Lepidus and Marc Antony (who married Octavian's sister to cement the partnership). They then hunted down and killed all of Caesar's assassins, after which Marc Antony was given control of the Eastern Roman provinces, including Egypt. On his arrival, he clearly couldn't resist the charms of Caesar's previous mistress, Cleopatra, for very long.

With Antony, she bore twins – Cleopatra Selene II and Alexander Helios – and another son, Ptolemy Philadelphus. She then revealed her opposition to Octavian, and turned Marc Antony against him, although this plan eventually backfired. Octavian sought total control of the Empire, and in Rome he began to discredit his brother-in-law Marc Antony, for 'going native' – the term used for having affairs on foreign soil – completely ignoring the fact that his god father Caesar had done exactly the same, and with the same woman.

It was all destined to end in tragedy, however, as another Roman civil war ensued. In 31 BC Octavian defeated Antony and Cleopatra at the Battle of Actium. This was one of the largest ever sea battles: even after they'd lost, Antony and Cleopatra fled back to Alexandria with 60 of their ships that had survived the fighting.

It was only a question of time before Octavian's forces arrived. Marc Antony committed suicide, followed some 10 days later by Cleopatra herself who, legend has it, held an asp to her breast. In fact, while she did indeed take poison, it was from something pre-prepared rather than the fangs of a snake. Her son by Caesar, Caesarian, was declared Pharoah Ptolemy XV, but his reign ended quickly after Octavian ordered his execution, decreeing that only a Roman should rule the Roman province of Aegyptus. Grecian control of Alexandria was at an end.

The mythical Helen of Troy's face had launched a thousand ships. Cleopatra had only turned two heads – but both appeared on the coins of the mightiest nation the world had ever known.

Around 1,700 years after this event, the French mathematician Blaise Pascal suggested that Cleopatra's beauty must surely have been something to behold, and had her nose not been quite so beautifully shaped, history might have turned out very differently. Rather oddly, perhaps, he remarked, 'Had her nose been shorter...?' Today we would probably associate beauty with a smaller snout, but as ancient Romans were already known for their pronounced snifters, our perception of female beauty may have changed since then!

Clearly though, Rome's military dominance brought about the demise of the great Grecian age of mathematics and scholarship. Other scholars, including Romans, would follow to carry on the baton, but mostly they would repeat the discoveries of earlier Greeks rather than advancing them. The Romans' legacy was less about scholarship than philanthropy, and carving a permanent niche in history with their magnificent architecture.

CHAPTER 6

Total Eclipse of the Greeks

In the first century BC, the Romans' military might gained control of the Mediterranean, and they now began to dominate both politically and economically. Intellectually, however, they mostly left mathematics, science and philosophy to Greek scholars, although some Romans were happy to learn Greek ideas.

One such chap was **Lucretius** (95–55 BC), who lived in Rome. Just before he died he published a book, *On the Nature of Things*, in which, in a poetic style, he stated his belief in the atomic theories of Democritus. He went further, though, suggesting that even our very mind and soul were made of atoms, though much finer ones than those that made up materials like iron or earth. He was the first to suggest that man had evolved and developed through a stone age, a bronze age and into an iron age, and that evolution would carry us forwards physically, biologically and socially.

The Builder of Marble Rome

Marcus Vipsanius Agrippa (64–2 BC) became a great Roman general while still very young. He had been at school with Julius Caesar's godson Octavian, with whom he had engineered the sea victory against Mark Antony and Cleopatra at the Battle of Actium, after which Octavian became the first Roman emperor, adopting the name Augustus.

Following Caesar's example, Augustus named a month for himself, and so that it would be no less significant than July, he gave August 31 days as well. So poor old February, which was then the last month of the year, was reduced to 28 days and 29 in a leap year, as is still the case today.

Together, the two men planned to make Rome a city of marble. Agrippa took charge, renovating major aqueducts to give all Romans access to the highest quality public services, and creating many baths, porticoes and gardens. Amazingly, the Romans used about three times as much fresh water per head per day as we do today.

Agrippa was a master of self-promotion who never felt the need to mention the people who had helped him with their expertise. This is possibly why another slightly older engineer, designer and architect had a good old grumble in his books about not being recognised for the great influence he had in many major Roman building and engineering projects of the time. He was probably justified in doing so. His name was Vitruvius.

Ancient Rome's Primary Engineer

Vitruvius (around 80/70 to after 15 BC) was probably about 10 years older than Agrippa, and had served under Julius Caesar in Egypt as a military engineer, providing weapons, siege engines, missiles and all the hardware required to conquer opposing armies and lay siege to cities. He would surely have experienced Alexandria and the home of his heroes, Ctesibius and Eratosthenes, and it is clear that he read all he could about the movers and shakers of the great Grecian Age.

Vitruvius wrote *De Architectura*, a 10-volume encyclopaedia that was to have a huge influence on the builders and architects of Renaissance Italy 1,400 years later. Leonardo da Vinci's celebrated image of a man in a square and a circle was inspired by Vitruvius' work (as we see in Chapter 10).

From his books, it is clear that Vitruvius was a truly accomplished engineer. So given that his name does not appear on a single building or engineering project created in his lifetime, it's hardly surprising he felt disgruntled.

It is from Vitruvius that we get the legend of Archimedes leaping out of the bath and running down the street in the nude shouting 'Eureka!', and the story of how Ctesibius discovered the piston while sorting out his father's mirror. He had clearly read all the Greek maths and technology he could lay his hands on. In fact, he goes back almost 400 years, crediting Plato with solving the 'doubling the square window' problem (see Chapter 3) by turning it into a diamond shape. He also loved – perhaps was even besotted with – geometric concepts.

Vitruvius used the 3, 4, 5 triangle in his architecture. He suggested that a whole flight of stairs could be based on the triangle, by dividing the height from ground floor to first floor into three, then taking four of those units along the floor. It is then simple to construct the staircase of similar steps with each riser 3 units high and with treads 4 units deep. Most domestic staircases today follow more or less the same dimensions. Mine does – check yours!

Of course, Roman civic buildings often had very majestic steps leading up to them. A better choice for those would have been the 5, 12 and 13-sided right-angled triangle, which would make the steps far more elegant, with a tread 12 units deep and 2.4 times the height of the 5-unit riser.

Vitruvius also showed how an architect can easily draw 1, 2 or 3-unit squares using the angles of the 3, 4, 5 triangle as a template (see Figure 6.1). For a single square you can use the triangle's right angle to make the corners. Better still, use

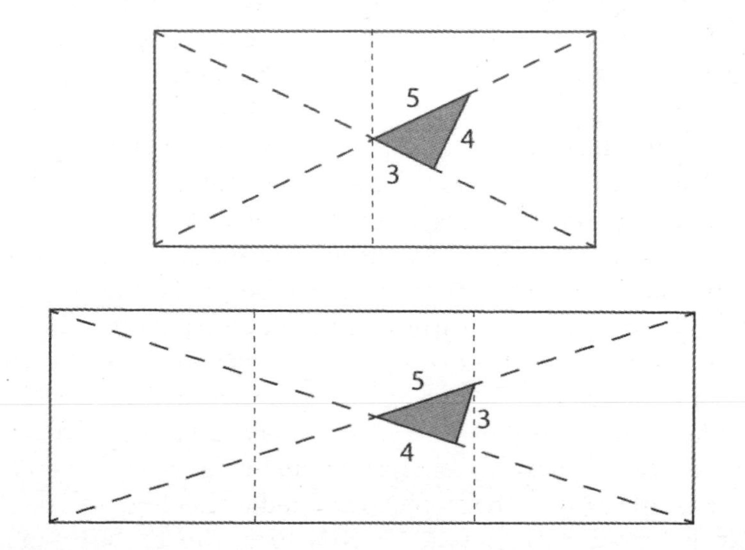

Figure 6.1

it to draw two diagonals first, then measure the same distance along each diagonal to produce the corners of a square of the desired size.

> To produce two squares side by side, use the 3/5 corner at about 54 degrees and draw lines along the 3 side and the 5 side. Measure the desired length along each diagonal from the crossing point to produce a double square of any desired dimensions.
>
> For a row of three squares, follow the same procedure, using the 4/5 corner at about 36 degrees to form two crossed diagonal lines. It's now easy to produce three perfect unit squares side by side.

A common design for medieval Christian churches was based on two squares with three diagonal squares – signifying the Holy Trinity – placed on top (see Figure 6.2).

Vitruvius saw that the isosceles triangle formed by diagonals coming from the very top down to the base had an upper angle of 36 degrees and lower angles of 72 degrees, and these were in a Golden Ratio to each other. He also saw that this triangle could be found by slicing a regular decagon in 10 equal pizza slices.

Vitruvius also wrote about how to achieve good acoustics in theatres. The Greeks had already established the semicircular shape of the auditorium; Vitruvius suggested that the front edge of all the rows of seats should follow one straight line

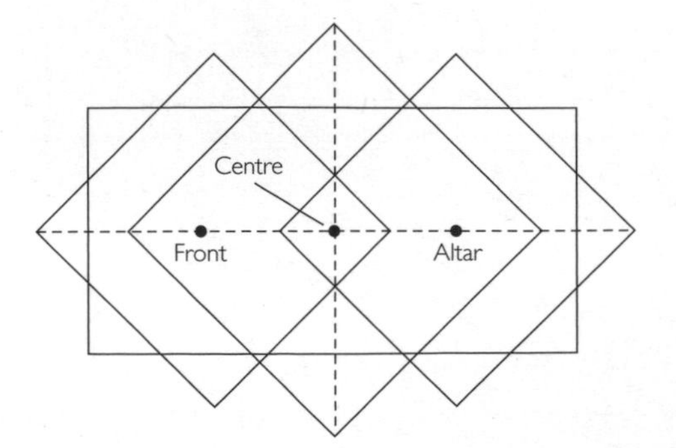

Figure 6.2

from the front to the back of the theatre. If wide gangways were needed, the back wall of the gangway continued this straight-line concept – so it wouldn't reduce the sound quality for those sitting further back.

Today, visitors to the ruins of Greek and Roman theatres are urged to stand at the focal point at the centre of the stage and whisper. Their voices can be heard by anyone in any of the thousands of seats around them (the largest stadia could hold between 40,000 and 60,000 people). And as the stage is usually hollow underneath, an impromptu soft-shoe shuffle will resonate clearly right to the very back row.

Vitruvius also wrote about all kinds of gadgets and tools that helped builders and architects with their tasks. Among them was the *odometer*, which he identified as something passed down from more ancient engineers (I wonder if Ctesibius wasn't originally involved here…?). The odometer was a wheeled instrument used to measure distance. Several discs were connected to the main wheel using cogs. Two discs had holes in them that aligned periodically, allowing pebbles to drop through to a lower container to measure a specific distance – a

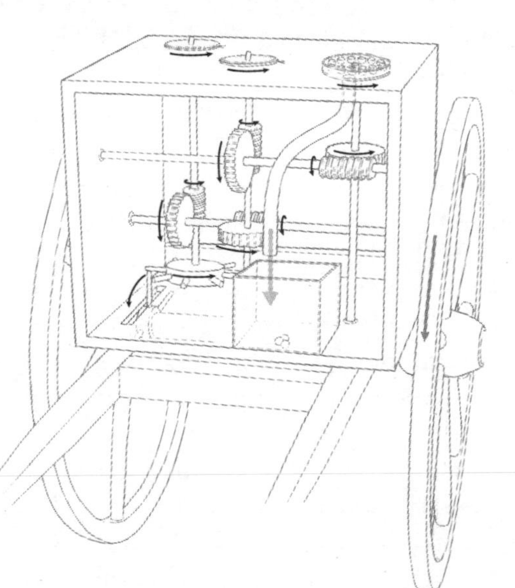

Figure 6.3

mile, maybe, or a stade. It could be a single-wheeled device, or could be mounted on a carriage or chariot, or even a wheeled taxi cart (see Figure 6.3).

Greek engineer Heron is sometimes credited with inventing it, but I am more keen to accept that it originated earlier, and in Rome rather than Alexandria. Rome was a truly bustling international city at the peak of its power, and many first-time visitors would have been unsure about how to get around. These were ideal conditions for a taxi service, especially one where both customer and driver had some proof that the right distance was being charged for.

A device like this for measuring accurate distances would have been a great advantage to road builders, but such measurements had always been made satisfactorily by professional pacers. Roman legions always set 'duty pacers', who counted each double pace. One counted tens and another hundreds, and 10 of them measured out 1,000 double paces, equivalent to one *mille*, or mile. It is said that wherever in the world the Romans pitched their camps, they always knew the exact distance back to Rome.

For a TV show, I once laid a tape measure on the floor and calculated my own double pace to be 63in, or 1.75yd. A thousand of those make 1,750yd, just 10yd short of the 1,760yd mile. With a tiny adjustment I could have become a fairly accurate pacer…

Vitruvius even suggested an odometer for ships, consisting of two light paddle wheels on either side of a vessel, connected by a long axle. As the ship progressed the wheels turned, and by counting the revolutions sailors could tell how far or fast it was travelling. Today surveyors use a simple device with a wheel that measures out exactly a metre or a yard; a counter ticks off each revolution and records the grand total with great accuracy.

We can see from his books just how powerful an engineer Vitruvius must have been. The eighth book in the series covers water and the construction of aqueducts. When a suitable mountain spring is found, Vitruvius explains, excavations need to be undertaken so that the water forms a reservoir or header tank in which impurities can settle out. From this, clear, pure water is allowed to flow over an adjustable sluice and down a very gradual slope into the aqueduct, at a regulated speed and volume. The Romans prevented moss and reeds from growing in aqueducts, which were covered in wooded areas to keep out leaves, debris and pollution. Their pipes were made of stone, or occasionally wood or terracotta; some earlier pipes were made of lead, although Vitruvius condemned lead as poisonous. We know, however, that some hundred years after Vitruvius was alive, the Emperors Caligula and Nero both took a liking to sweet food. Honey wasn't always available, and sugar as we know it had yet to be discovered, so it is said that they put shavings of lead on their food to sweeten it. Both became violent men, and eventually 'mad' – one theory is that lead poisoning was the cause.

Archimedes had said that you can't really measure horizontal levels with water, because the Earth's surface is curved. Vitruvius agreed, but not for short distances. So, to measure levels accurately at any particular point, he created a device called a *chorobates* – a plank of wood perhaps 2m long with a water trough in it.

Building aqueducts was a highly complex task. If water flowed too fast it quickly got out of control; if it flowed too slowly it tended to stagnate and collect debris. So aqueducts were built in separate manageable stretches between frequent reservoirs or header tanks. The Romans cut stone-lined streams across flat land, or supported them on arches to carry the water over uneven land. They only used semicircular arches, but the system saved them vast amounts of stone. However, if a valley was deeper than, say, 150ft, arches became impractical and they needed another system.

In those cases they built an enclosed stone tube running downhill, filled completely with flowing water that rushed down the tube very quickly. The force at the bottom was very strong, so the Romans bored an angled tube hole through a lump of solid rock that could withstand the constant rush of water. Another, rising tube naturally siphoned the water to almost the same height it had fallen from. It was then allowed to flow into a cistern before continuing on its way. This siphon system worked very efficiently, provided there were no leaks in the pipe (see Figure 6.4).

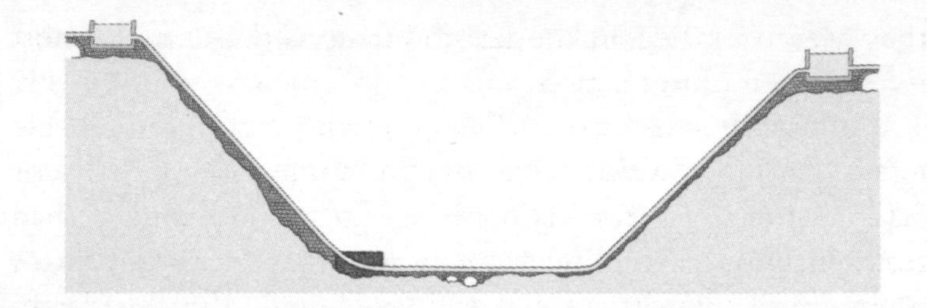

Figure 6.4

But water sometimes had to be delivered over long distances – more than 30 miles in some cases – so its descent had to be very gradual. Vitruvius advised that the minimum slope to ensure a smooth flow should descend at a rate of one unit in every 2,400 – just 10cm (the width of the human hand) over 8km (or 4in over 5 miles).

In 1814, Englishmen Humphry Davy and Michael Faraday investigated the ancient Roman aqueduct at the Pont-du-Gard near Nimes. This impressive Roman aqueduct has three tiers of arches: the lower two are each about 22m (75ft) high, while the top tier of arches, which carries the water, is quite narrow and controlled the slope 48m (160ft) up. The aqueduct still stands today, and the low end descends by a mere 2.5cm (1in) – a gradient over 75m of only 1 in 3,000 (see plate section).

In its heyday, the aqueduct carried about 200,000m^3 (44,000,000 imperial gallons) of vital water a day to the fountains, baths and homes of the citizens of Nimes. The steepest slopes used to deliver water to Rome, as well as other cities and military stations, fell at the rate of 1 in 150, although one – which was probably completely enclosed – apparently declined at a rate of 1 in 60, which seems pretty extreme.

Vitruvius wrote a great deal about town planning, and especially fortifications. In building city walls, he was keen to stress that main entrance approaches should be positioned in such a way as to force attackers to approach with their right sides close to the walls. This clever ploy was designed to maximise the damage done by archers attempting to repel the attackers, who would be firing at the soldiers' right-hand sides. As the soldiers were predominantly right-handed, if they weren't killed in the hail of arrows, they would most likely be put out of action.

The precision of Vitruvius' engineering skills is remarkable given that in his writings he gave the value for π at 3⅛th, or 3.125, when many of his predecessors whose work he had read, including Archimedes, gave more accurate figures of around 3.14. Nevertheless, 1,400 years or so after his death, his 10-volume work proved a major influence on the work of the builders and architects of the European Renaissance.

A Couple of Geographers

During this period, when Rome dominated the ancient world, two geographers are worthy of note, although neither of them was actually Roman. The first, **Strabo** (63 BC–AD 25), was born on the Black Sea coast of modern-day Turkey. Of the 17 volumes he wrote, all but one survive to this day, and from these we know that he was highly influenced by Eratosthenes, although his maths wasn't as good, and he was poor at making accurate land measurements on a spherical Earth.

Strabo boasted that he had travelled further and wider than any previous geographer. He followed the Nile to the borders of Nubia in 25 BC and encountered a climate hotter than anything he had previously experienced.

He suggested dividing the Earth into frigid, temperate and tropical zones, an idea picked up and improved upon by Pomponius Mela, a Spaniard who wrote just one book on geography, around AD 43 (he wrote it in Latin, though, making it popular with the Romans). Mela divided the Earth into north frigid, north temperate, torrid, south temperate and south frigid zones, claiming that only the temperate zones would sustain human life, which is in fact wrong. And as the southern half of the Earth was unknown territory to the Romans, Mela jumped to the conclusion that it would be similar to the northern half, and proposed the existence of vast undiscovered lands south of the Equator.

This idea influenced many great explorers for centuries, right up to Captain Cook (who was killed in Hawaii in 1779), but they discovered that Mela was wrong. In fact, south of the

Equator, the Earth's surface is 81 per cent water and contains only 32.7 per cent of the planet's land, making the earth remarkably similar to a spinning top, which spins better and longer with a heavier top half.

A Workaholic Curious to the Very End

Amazingly, in their 800 years of Empirical dominance, the Romans did not produce any great scientists or mathematicians, although a few Roman scribes repeated much that the Greeks had introduced. The greatest of these was the historian Gaius Plinius Secundus, or **Pliny** the Elder (AD 23–79).

Pliny had immense energy, and was a workaholic his entire life, considering both sleeping and walking a waste of time. To ensure he could be constantly reading or writing, even while travelling, he rode – or was carried – everywhere he went. He served as a military commander, mostly in central and northern Europe, and at one time was Governor of Gaul – but throughout his army life he studied and wrote continuously.

His 37-volume *Naturalis Historia* (*Natural History*) was gleaned, he claimed, from reading over 2,000 books by 500 writers. Sadly he could not discriminate between truth and fiction, and was gullible enough to believe in unicorns, mermaids and flying horses, as well as men with no mouths who lived by inhaling perfume from flowers, and men with feet so large that when the Sun was at its most intense, they lay on their backs and used their feet as sunshades. But he was a great inspiration to others, believing that man was the measure of everything, and everything existed for man's sake. Flowers and plants were either food or drugs; animals were either food or servants – if anything, plant or animal, was of no use to man, then it must exist to offer a cautionary or moral tale. Though many of his ideas were incredible, his zest for life was commendable, particularly as he advised that everyone should seek out new ideas or species – and that we should try to absorb everything we can. What we don't understand now, he claimed, we will understand eventually as our knowledge becomes more complete – all sound advice for young people in education today.

Sadly, however, Pliny's natural and all-consuming curiosity eventually proved the death of him. He was in charge of the Roman home fleet, stationed near the Bay of Naples, when, on 25 August AD 79, nearby Mount Vesuvius suddenly erupted, and began to pour volcanic lava and ash over the cities of Pompeii and Herculaneum. Sensing the historic significance of the event, Pliny sailed into shore to observe the catastrophe at close quarters. His captain advised him that they should not land, but Pliny famously replied, 'Fortune favours the brave!' He was brave indeed, strapping a pillow to his head to protect him from the falling hot ash, and going ashore. While everyone around him fled in panic, Pliny stayed on, taking notes and even bathing in the sea to cool down. But adverse winds delayed the ship, and he collapsed – already an asthma sufferer, the pumice in the air was too much for him. In the confusion, the ship departed, then returned the next day to find the body of the great man lying on the shore.

A True Hero – or Heron? – of Engineering

What's remarkable about our next character is that, despite being classed by many as the greatest Greek engineer and inventor of them all, historical details about him are so obscure that we can't be sure of his date of birth or death, the century in which he lived, or even his name – Hero or Heron. Yet his reputation lives on today, almost certainly because he lived in an age dominated by Rome, which perpetuated his fame.

Heron, as we'll call him, was supposedly born in the first century BC, but writings discovered in the last fifty years seem to suggest that he was alive during a lunar eclipse in AD 62 in Alexandria, putting his birth date at around AD 20. He probably wasn't Greek at all, but Egyptian, and he likely had a more practical nature than most Greeks, who considered mathematics more in an abstract sort of way. But he did make some new discoveries, including one that took Pythagoras' Theorem a giant step further. He devised a system for calculating the area of a triangle of 'any shape', even without knowing its height or the size of any of its angles. All you

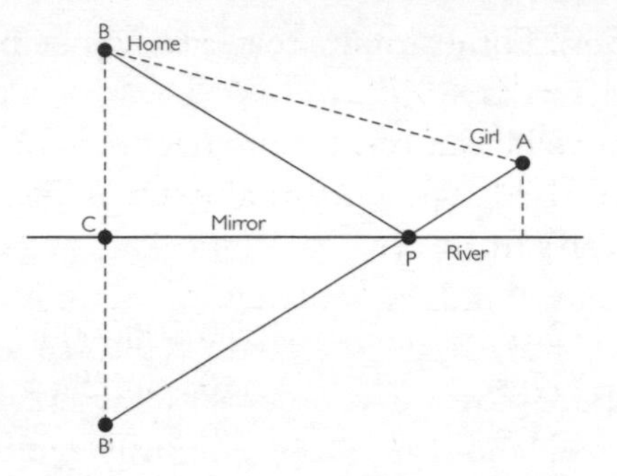

Figure 6.5

need to know is the length of its three sides, and Heron's formula (which we feature and explain in the Wow Factor Mathematical Index).

Heron also explored the mathematics of mirrors, and in doing so discovered the answer to a famous maths puzzle: 'A girl with a bucket is told to come home, but to bring water from the river on the way. The question is, which is her quickest route home?' (see Figure 6.5).

She could go straight to the river, then diagonally home, or first she could go to the point on the river that's closest to her home. But neither of these is the right answer.

If you imagine that the river is a mirror, then the girl will take the shortest path if she heads for where the 'reflection' of her home would be, and then heads straight home from the river. Heron explained that light always follows the shortest path, which is known as the law of least action.

Heron also believed that light comes from the eye at infinite speed – but he was wrong. We now know that light bounces off things and arrives at our eyes. We also now know that light has a finite speed. Nevertheless, for the age in which he lived, Heron's conceptions were quite brilliant.

Heron's explanation of fractions outstripped previous Greek mathematicians (even Archimedes had done little with fractions: he wrote ¾ as ½ + ¼, for example). This hints that Heron may have gained this knowledge from early Egyptian writings: the Egyptians were far more accomplished with fractions than the Greeks. Like them, however, Heron always used 1 as the numerator for all fractions (with the sole exception of $\frac{2}{3}$). For example, to write $\frac{31}{51}$, he broke it down to give $\frac{1}{2} + \frac{1}{17} + \frac{1}{34} + \frac{1}{51}$.

We can prove he was right using decimals and a calculator. First, convert 31/51 into a decimal (0.6078431). Then do the same for Heron's fractions:

1/2 =	0.5000000
1/17 =	0.0588235
1/34 =	0.0294117
1/51 =	0.0196078

When added, this gives 0.6078430 – which is near enough, don't you think?

Heron also used the principles behind the Mighty Five machines (see Chapter 4) and tried to get a better understanding of how they worked – he saw, for instance, that dragging a block of stone up a ramp is more difficult the steeper the ramp. Nevertheless, it's still difficult to drag the block along level ground, because the weight of the stone creates friction. Perhaps, thought Heron, he could explain what was happening if he removed the element of friction

So he imagined moving a wheel or a ball instead of a block. On level ground a wheel is just as easy to roll either way, with very little friction. But a wheel on a slope is more difficult to push uphill. To try to explain mathematically why that should be, Heron drew circles on several slopes, and for each he drew a perpendicular line across the circle from the point of contact with the slope (see Figure 6.6).

On a steeper slope, more of the circle will be on the lower side of the line, indicating (by proportion), why the push is more difficult. Although the maths Heron used to explain all this wasn't perfect, his conception was spot on.

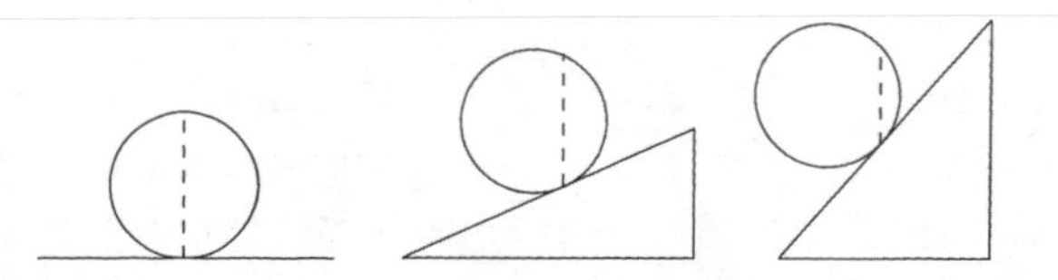

Figure 6.6

Following in Ctesibius' footsteps, Heron created a wind-powered organ, and improved the pumping system on Ctesibius' fire engine. But Heron also used fire to operate other amazing devices, including the world's very first steam engine, which he christened his *aeolipile.* The device consisted of an urn, which was filled with water and placed on a fire. As the water boiled, steam was forced up two hollow arms into a metal ball with two thin spouts turned outwards at 90 degrees near the ends. As the steam escaped, the pressure pushing one way pushed the ball the other way, causing it to spin – an early demonstration of Newton's law that every action has an equal and opposite reaction (see Figure 6.7).

Some historians question whether Heron could have made the joints between the hollow arms and the ball loose enough to turn, and yet still tight enough to prevent steam escaping. There's no doubt that Heron, and others of the same era, were very capable of high-precision engineering, but it might be worth exploring whether the design that has been passed down to us is indeed accurate.

I still have an aeolipile that was made for my stage musical *Let the Force be with You* in the 1990s. Each spout had a fine hole at its end to create jets of steam, but one spout unscrewed so that we could pour water into the ball before starting. It worked extremely well, with no steam leakage. Heron's aeolipile was probably quite small, and produced nowhere

Figure 6.7

Figure 6.8

near the pressure and energy output that steam engineers such as James Watt achieved 1,700 years later. But Heron also produced other devices powered by water and steam.

He mounted dancing figures on spouts with bent ends, hanging below a table, or perhaps an altar, over which was a bowl that held burning material. The heat expanded the air, probably enough to spin the figures round. It would have worked rather like a wind chime works today.

More impressive, though, was Heron's automatic door opener (see Figure 6.8)

Inside a temple, a fire on a plinth expanded the air inside the device, which then forced water through a pipe and into a bucket. As the bucket became heavier, it pulled on door posts that 'magically' opened a pair of doors inwards. However, although a small model would work, it's unlikely the device could have created enough force to open a full-sized pair of heavy wooden doors.

Also for a temple, Heron invented the world's first known slot machine (see Figure 6.9).

A coin in the slot slid to land on the end of a balance arm that tilted and raised a cone shaped stopper to release water. But as the balance arm tilted, the coin slid off it again and the stopper closed off the flow. So each coin bought a small measured amount of water. It was a holy water dispenser.

Figure 6.9 and 6.10

Heron also made a miraculous urn that delivered wine or water, as requested. The urn was divided internally into two sections. One of two hidden holes could be secretly closed off with a finger. Only the uncovered hole would allow in air, releasing the required liquid (see Figure 6.10). The trick is used by magicians to this day.

In explaining how this worked, Heron realised he might produce a book that discussed the science and understanding of the air around us. Observing that air can be compressed, he suggested that it must be made up of tiny particles surrounded by space, so they can be forced together more tightly. He saw that air was a fluid, just like water, and rather like crabs at the bottom of the sea, we are all living at the bottom of an ocean of air that is our atmosphere.

Heron was often involved in theatrical productions. He made a thunder machine, in which heavy stone balls, released on cue, dropped onto tilted resonating boards, cascading from board to board to give the effect of continuous rumbling thunder. He even created scenery that could move automatically during a play. The scenery was mounted in front of a trolley, behind which was a huge hopper full of sand or grain. By releasing a key, the sand or grain would flow through a small hole, creating an effect like an egg timer. A large stone weight on top of the grain would then descend

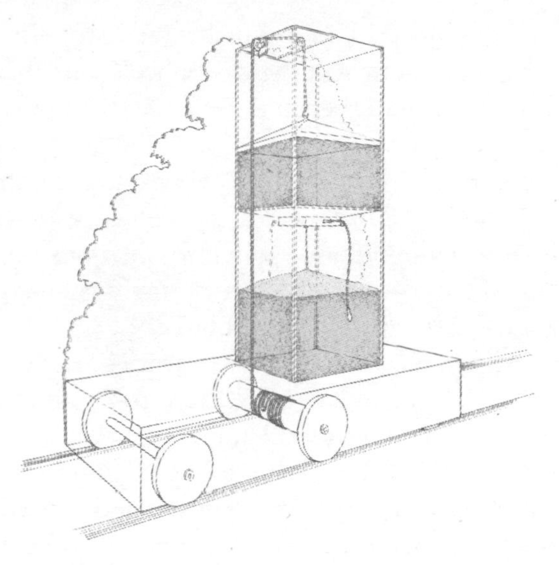

Figure 6.11

at a slow and measured rate. The stone in turn pulled on ropes wound around axle wheels, which slowly propelled the trolley and scenery magically sideways. By placing a peg in the side of the winding drum and looping the rope around the peg, Heron could make the device move a certain distance, then stop for a set time before retracing its path – as required by the play's director (see Figure 6.11).

Heron made use of every kind of device that had been discovered by that time, including siphons, syringes, pulleys and a whole variety of wooden gears, such as worm gears, peg gears and circular toothed gears. He often combined them to change the direction of certain forces and movements.

His most practical surveying invention was the *dioptra* – a vast improvement on the *groma*, used by the Romans in building their vast system of roads (see Figure 6.12).

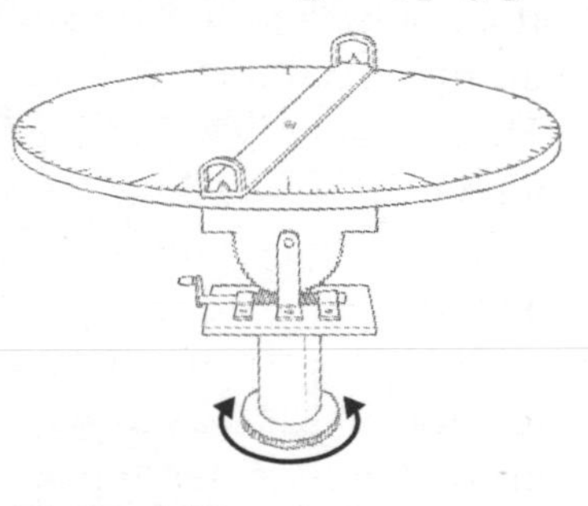

Figure 6.12

The device consisted of a table on a stand set at roughly head height. On the table were two sights that could be lined

up with distant objects like a tree or mountain top. By turning a worm screw, you activated two cog wheels, one that moved up and down and the other in a circle. By manipulating the sights you could view a distant object and read its angular direction on one scale and its angle of elevation on the other. You could also clearly see when the device was absolutely level, which was essential for laying foundations. Today the Dioptra has telescopic lenses and extremely fine adjustment can be made – it is called a Theodolite.

Heron also produced many devices that could move heavy pillars and building materials. One device could move a weight of 1,000 *talents* using the effort of just five – a ratio advantage of 200 to 1. A talent varied in weight over time and from country to country, but it was usually the mass of water required to fill an *amphora* – a standard-size earthenware pot, usually with a sharp, pointed bottom so it could be placed in soil or sand. Greek or Egyptian talents weighed between 26 and 27kg (57–60lb), while a Roman talent was 32.3kg (71lb). So Heron's device could move about ten tonnes by applying 50kg of effort, which is very impressive indeed (see Figure 6.13).

So how did it work? First, an axle with rope wrapped around it bore the load. On the axle was a cogged wheel with a radius five times the axle's width. This turned the next wheel, with a third wheel five times that radius, and so on – producing an advantage of 200 to 1. However, the operator (or slave, as Heron had it) would have to make 200 turns of a handle for the first axle to revolve once and lift the main load only the length of that axle circumference.

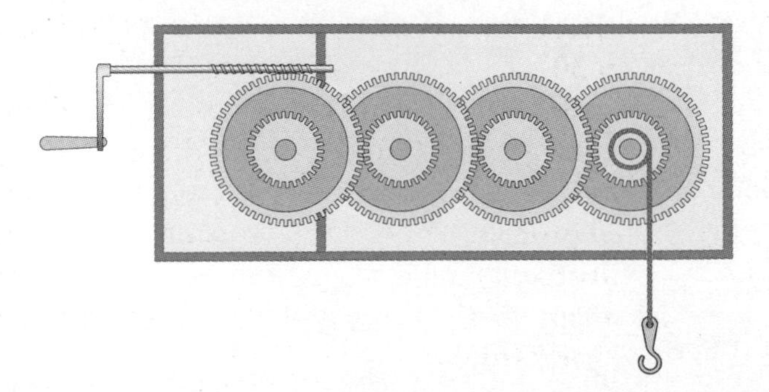

Figure 6.13

What these surviving examples of Greek technology tell us is that we actually know very little of what the Greeks were really capable of, and collectively their skills and knowledge must have been astounding. What is especially sad is that from around the time of Heron, their achievements in the mechanical sciences would hardly be improved upon for about 1,500 years.

Religious Turmoil

As they expanded their empire, the Romans usually allowed those they conquered to continue living as they had always done, so long as they paid their taxes. Problems did arise, however, mostly caused by conflicting religions and beliefs – and, to a small extent, mathematics.

In 37 BC the Romans declared Israel a client state under the puppet King Herod, and carefully assessed how much tax was due from each province. So it is told, some 30 years later, a carpenter and his pregnant wife Mary were forced to take the difficult journey to Joseph's birthplace to be counted.

It's now believed that around the year 6 BC, calculated from the reign of King Herod, Jesus Christ was born. Wise men in those days would have been astronomers and mathematicians, naturally interested in a magical star, rather than kings. There is no convincing local account that there was such a star at that time, although Korean astronomers did record a special celestial event at that time. Little is known of Jesus' youth, but in around AD 27, aged just over 30, he began his public ministry, only to be condemned to death by Pontius Pilate, Roman Procurator of Judea, and crucified in AD 30.

The first recorded council of Christian leaders, the Council of Jerusalem, took place in AD 50, heralding the start of a period of religious struggle, as various faiths fought to promote their beliefs and protect their right to worship as they chose. Gradually Christianity spread, without causing much disruption, perhaps partly because it was often considered trivial in those early times.

The Jewish faith, on the other hand, fighting for its rightful homeland, had far more trouble. In AD 67 the Jews declared war on Rome; Christians caught in the middle in Palestine fled to Pella in Jordan. Three years later the Romans sacked Jerusalem, and in AD 73 Masada fell to Rome, and its Jewish residents committed mass suicide.

In AD 130 Emperor Hadrian rebuilt Jerusalem, renaming it Aelia Capitolina. To spite all faiths, he erected a Pagan temple over the site of Christ's crucifixion and resurrection. Five years later he crushed the second Jewish revolt and expelled the Jews from Palestine. Religious persecution continued, while education under Roman rule went out of the window.

The Roman Contribution to Mathematics and Science: Money

During the years of their empire, the Romans produced little in the way of mathematical or scientific innovation. But their richest and most powerful soldiers and statesmen became philanthropists and benefactors, enabling skilled Grecian engineers (perhaps Heron among them) to produce some very elaborate timepieces.

One such device could be found in the Roman Square or Agora in Athens – an amazing water clock at the top of the 13m Tower of Winds, an octagonal marble structure, the remains of which can still be seen today. It automatically told the time and date, as well as the season and place in the zodiac. It was surmounted by a sundial, while at the very top was an elaborate weathercock to indicate the direction of the wind.

Roman architecture continued to advance as the empire reached its peak during the reign of **Trajan** (AD 53–117). He built a magnificent complex in Rome that consisted of Trajan's Forum and Market (which still stand today), as well as a huge column depicting his early escapades as a general. Trajan tried very hard to endear himself to the people, offering food and support for the poor and hosting a three-month gladiatorial festival witnessed by some five million spectators. Held in the great Colosseum in Rome, thousands of ferocious beasts were slaughtered, along with 11,000 slaves and criminals. Which surely couldn't have been popular with all of the people...?

Historians claim that the Romans of this period used mathematical perspective in their theatrical productions and their buildings, but there seems to be no concrete evidence for this. However, it is probably true that the oldest known use of perspective was devised at that time – and it still stands today for all to see, set firmly in Roman concrete.

The first Pantheon, built by Agrippa in 25 BC, burnt down in AD 80 and again in AD 110, after which the stone and concrete building that still exists today was established. This version was probably started by Trajan, but was finished in the reign of his adopted son Hadrian, who placed Agrippa's name on the frieze above the main entrance. Trajan produced a good number of triumphal arches, but the roof of the Pantheon is an architectural wonder to this day – it is still the largest unreinforced concrete dome in the world.

The building was designed around an imaginary ball 43.3m (142ft) wide. So we have a cylinder surmounted by a hemisphere. The 2m-wide central hole in the top of the dome – the *oculus* – is the only source of light, but amazingly it illuminates the entire building, and the Sun's rays can even be used as a sundial. Rain can and does pour in, but the marble floor slopes slightly, and drains at strategic points let the water out (see Figure 6.14).

The walls supporting the base of the dome are 7m (23ft) thick. The thickness of the hemispherical dome varies from

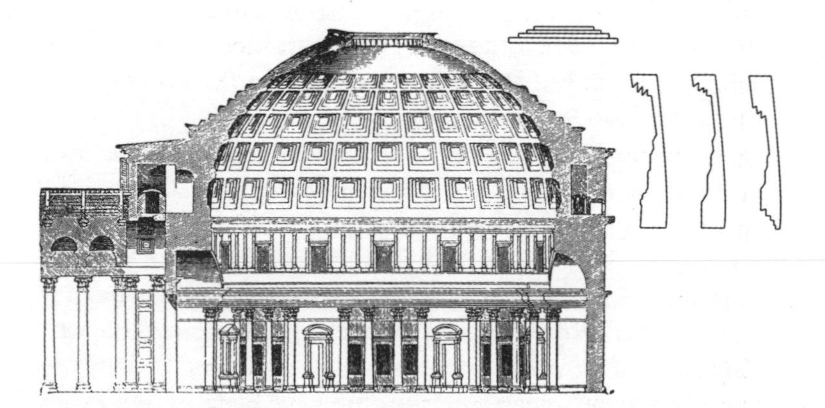

Figure 6.14

6.4m (21ft) at its base to 1.2m (3.9ft) around the oculus. The cement contains granite aggregate to strengthen it, but as the dome gains height this is slowly swapped for pumice, making the dome lighter and slimmer as it rises. The dome's colossal weight is further reduced by five graduated rows of 28 rectangular coffers or sunken panels, each four steps deep. The mould for each coffer would have been supported on a huge wooden structure, and the concrete poured onto it from above. When it set, the wooden support could be removed, leaving the incredible domed roof just as we see it today.

Had the intended focal point to view these concrete coffers been at the centre of the sphere, then each layer of coffers could have been exactly the same shape. But the intended viewing point was the centre of the floor of the building. To make the coffers appear regular from that point, although the left and right steps could be the same, each layer of vertical steps had to be varied at each end so that the coffers would all appear more or less identical from the floor. The varying side elevations are shown in our illustration. It's clear that the building's designer had a strong understanding of perspective.

The amazing achievements of the builders of Imperial Rome were to prove a huge influence on the artists, architects and designers of the Italian Renaissance – as we shall see.

The Last Few Important Mathematicians of Alexandria

Very few prominent and influential mathematicians featured during the period of Roman dominance in Alexandria. But there were some major exceptions.

Menelaus (around AD 70–140) was a minor legend as a mathematician from Alexandria. But he should not be confused with the legendary King Menalaus of Greek mythology, whose beautiful wife Helen, a daughter of Zeus, ran off with Paris, triggering the Trojan Wars. Menalaus tried to end the war by challenging Paris to a personal duel to settle the matter. But when Menalaus was winning, Paris was magically spirited away by Aphrodite. Eventually, as a peace offering, Menalaus presented a gift: a huge wooden horse, which the Trojans dragged through the gates and into the besieged city of Troy.

The willing captive Helen walked around and beneath it that evening, and considered the strange gift from her homeland. She began to imitate the voices of wives of warriors she had known. The soldier husbands hiding inside could hardly maintain their silence. When Helen left and all was quiet, they descended, slaughtered the sleeping guards, won the city and took Helen back to Menalaus – who, amazingly, forgave her, even though she'd caused the war in the first place. But then she did have 'a face that launched a thousand ships'.

In any case, all that is a myth, described in Homer's *Iliad*, which dated the Trojan Horse episode to around 1184 BC. Let's get back to the maths of the first few centuries AD.

We know that Menelaus the mathematician was writing while he was in Rome, around AD 98. An astronomer, he accurately plotted the dates and times of the changing signs of the zodiac, which later helped Ptolemy to calculate exactly how the Equinoxes progressed slightly forwards each year. To do this, Menelaus studied an area of maths that until then had been rather neglected: spherical geometry. He observed that Euclid's geometry changes significantly if you try it on the surface of a sphere. On a flat surface, the shortest distance between two points is always a straight line – but not on the curved surface of the Earth.

A *great circle* is one that divides the surface of a sphere exactly in half – like the Equator, although there are infinitely more examples. Great circles can encircle a sphere in any direction. The shortest distance between any two points on a sphere is always along the line of a great circle; in the opposite direction, the distance between the two points is the longest possible.

All lines of longitude meet each other at the North and south poles, and they cross the Equator at right angles. So a triangle formed by two lines of longitude and the Equator has two right angles at the Equator, but another angle at the Pole that could be anything from a fraction of 1 degree to a shade less than 360 degrees. On a sphere, triangles can have angles adding up to anything from slightly more than 180 to slightly less than 540 degrees, and Euclid's assertion that a triangle contains angles forming 180 degrees or two right angles goes out of the window.

Menelaus saw that all shapes on the surface of a sphere can be described as being projected from the very centre point of the sphere, and that the projections will have the same shape irrespective of the sphere's size. Explaining the geometry involved, he drew up a table of sines to show the trigonometry required to measure the angles. Today we make maps in many shapes and sizes, but to keep distances constant with respect to one another we use projective geometry, much like that used by Menelaus.

Menelaus also discovered an amazing and beautiful plane geometric concept that we now call the Theorem of Menelaus. More amazingly, he discovered that his theorem also works on the surface of a sphere (it features in the Wow Factor Maths Index).

Claudius Ptolemy (about AD 100–170) was almost certainly an Egyptian who gravitated to Alexandria, although he was no relation of the Ptolemaic royal family (his name probably came from his birthplace, Ptolemais Hermii). Like Euclid 450 years before him, he is remembered not so much for his original work, but for bringing together much of the work on mathematics and geography produced by others during the previous few centuries.

He drew mainly from Hipparchus (most of whose work has been lost), although he devised a system of eccentric wheels within wheels to explain the movements of the planets, and to inspire amazingly complex and beautiful clocks. Today, this is known as the Ptolemaic system, and it has the Earth at the centre, with the planets ordered according to their supposed distances from it: the Moon, Mercury, Venus, the Sun, Mars, Jupiter and Saturn.

The system featured in Ptolemy's book, *Megale Mathematicke Syntaxis,* or *Great Mathematical Treatise*. Some called it *Megiste,* which has the same root as the word 'majestic'. Arab scholars, in turn, gave it the name *Almagest* or 'The Greatest', and their version, translated into Latin in 1175, dominated scientific thinking in Europe until the Renaissance.

Apart from placing the Earth at the centre of the Universe, Ptolemy did not always choose the best options as he trawled his many sources of scientific ideas. He accepted Hipparchus' true distance to the Moon, for example, but chose Aristarchus' erroneous distance to the Sun. His detailed map of the known world included lines of latitude and longitude, but his greatest mistake was to use Poseidonius' distance around the Earth rather than Eratosthenes' more accurate figure, making his 'Earth' much smaller than its true size. Nevertheless, his work was lauded by the Romans, and by the Arab scientists who followed, and the Ptolemaic system, wrong though it was, was accepted as fact for another 1,400 years (see plate section).

From the time of Ptolemy, scientific progress continued far more spasmodically, and almost ground to a shuddering halt as the movers and shakers became ever fewer and we entered the Dark Ages. However, Grecian mathematics was not quite finished yet: while Roman lethargy and general religious tunnel vision closed down academic options left, right and centre, a handful of personalities kept the flame burning in Alexandria.

Diophantus (around AD 210–290) is often called the 'father of algebra', and was the first Grecian mathematician to really explore the subject, although the Babylonians had covered many ideas in that area even before Pythagoras. But Diophantus' work on 'higher arithmetic' was revolutionary, so perhaps a better title for him might have been the 'father of number theory'?

His life is so obscure that we don't know when he lived, although we do know that it was in Alexandria. Some scholars have guessed at one birth date as AD 210 and his most important time as around AD 250, but others have suggested that he might have been born as early as AD 50. All we do know is that he bridged a huge gap in Grecian mathematics up until his time.

A strong clue to the length of his life and the age at which he died lies in a problem featured in *The Greek Anthology*, a collection of poems, epigrams and maths puzzles collected in the tenth and fourteenth centuries but dating right back to this period of Grecian scholarship.

The problem states:

> *This tomb holds Diophantus and tells scientifically the measure of his life. God granted him to be a boy for one sixth part of his life and added a twelfth part to this, where his cheeks were cloaked with down. He married after a further seventh part and five years later had a son. Sadly after attaining half his father's life, chill fate took the son. After consoling his grief by the science of numbers for four more years Diophantus ended his life.*

So how old was Diophantus? From the problem we can deduce his age. He spent a sixth of his life as a boy, a twelfth as a youth, a further seventh till he married, then five years until his son was born. The son lived for half of his father's life. Diophantus then lived for another four years. So:

$$\tfrac{1}{6} + \tfrac{1}{12} + \tfrac{1}{7} + 5 \text{ years} + \tfrac{1}{2} \text{ his life} + 4 \text{ years}$$

Looking at the denominators, 6, 12, 7 and 2 all divide into 84, their lowest common multiple. So, converting the fractions into 84ths we get:

$$14 + 7 + 12 + 42 = 75$$

To that we add extra years, 5 + 4, so:

$$75 + 5 + 4 = 84.$$

According to this calculation, then, Diophantus lived to be 84.

Diophantus couldn't have created this puzzle himself, unless it was actually on his deathbed, so we have no idea who originated it, or whether it actually relates to his real life.

Fortunately, however, we do have examples of the maths that Diophantus developed. He wrote a series of 13 books, six of which survive, and another four, in Arabic, that have been discovered in recent times. Collectively they were titled *Arithmetica*, and almost all deal with equations of a type we still call Diophantine equations.

Diophantus began by making simple and beautifully clear statements about numbers and number theory. To quote him, 'It has been proved what was stated by Hypiscles that taking as many numbers as we please beginning with 1, if the difference between the numbers is 1, then the combined group of numbers is a triangular number, if the difference is 2, then the total is a square number and if the difference is 3, then the total is a pentagonal number and so on.'

So we get	1	1	1
Plus	2 total 3	3 total 4	4 total 5
	3 total 6	5 total 9	7 total 12
	4 total 10	7 total 16	10 total 22
	Triangular	Square	Pentagonal numbers

Diophantus liked to set examples where numbers that were 'unknown' had to be found; in doing so, he laid the foundations for algebra. From the start, he explained his new algebraic ideas and rules very clearly, and at first his solutions were often in whole numbers to keep things simple. Slowly, however, he turned the screw, and his algebra became both more complex and more powerful. Remember too that he always worked with the Greek number system, which used letters for digits (1 = alpha, 2 = beta, 3 = gamma and so on).

Diophantus set down and explained the rules of algebra: 'Know that a minus times a minus equals a plus, but a minus times a plus equal a minus' is just one example. He showed for the first time how to write and use equations, and how to swap terms from one side of an equation to the other, which also requires pluses to become minuses and vice versa (for some examples of this early algebra, see the Wow Factor Mathematical Index).

Diophantus covered an enormous amount of mathematical ground, especially when you consider that he had no conception of zero, and had a complete mental block when problems he set gave negative results. Although he used minus quantities in equations, a problem that ended with a negative or minus number didn't make sense to him. As an example, if he saw the equation $3 = 4x + 9$, he would call it meaningless or just plain daft, because the value for 4x would have to be -6, making x equal to -1½.

He only ever looked for one solution in quadratic equations, and there's no evidence that he ever realised there could be two or even more. But among the great mathematicians of the Renaissance his works became required reading, as they pushed back the boundaries of mathematics and opened up the way for Newton and Leibnitz.

Pappus (who was born around AD 260) was the last great mathematician from Alexandria. In AD 320, he wrote a large maths treatise entitled *Synagoge*, or *Mathematical Collection*. It seems to have been an attempt to preserve as much Alexandrian maths as possible before it was lost forever in those changing times, when Greek culture was fading while the Christian religion grew ever stronger and the might of Rome was clearly waning.

There were eight books originally, but the first – and some sections of the others – have been lost. Containing details of the most famous Greek mathematicians, the collection also references more obscure scholars nobody else mentioned – and who, without Pappus' help, would be completely unknown today, making his work doubly important.

Pappus sought answers to all kinds of problems, and celebrated humans – thanks to their ability to think – as special and advanced creatures. But he also demonstrated that we are not the only species with mathematical ability. Take bees, for example…

Bees have a wonderful regimentation, in which every member of a hive works diligently at the task allotted to it, and in service to the queen. Pappus observed how bees learnt to seek out nectar from flowers to make honey – which was highly prized in the days before sugar. But then he studied their incredible mathematical ability, highlighted by the way they construct their honeycombs.

Only three regular straight-edged shapes will cover a plane surface leaving no spaces when butted together: the equilateral triangle, the square and the hexagon. The plane geometric shape that encloses the greatest area is the circle, but circles leave spaces if they're butted together.

Of these three shapes, the nearest to a circle is the hexagon, and this is the shape chosen by bees. The hexagonal spaces are of optimum size, and their walls require the minimum amount of wax for the storage space they produce. It is mathematical perfection.

Pappus' Theorem

Pappus made an amazing discovery in geometry that to this day is called Pappus' Theorem. Draw two lines and add arbitrary points along them: A, B and C on one and A^1, B^1 and C^1 on the other. Now join A to B^1 and C^1, B to A^1 and C^1, and C to A^1 and B^1, to create six lines in total. In all cases the three crossing points will always lie on a straight line (see Figure 6.15). Neat, isn't it?

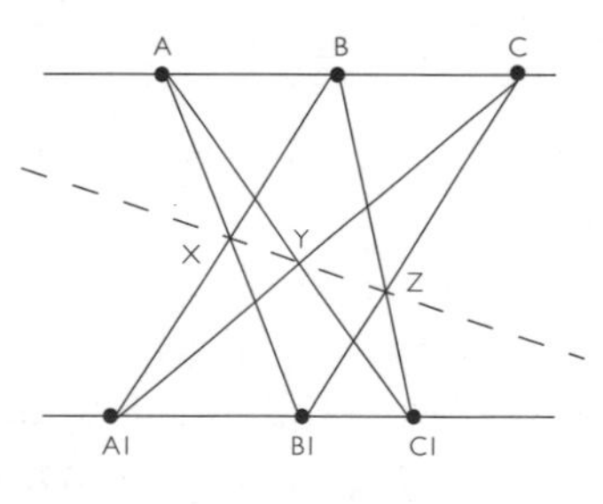

Figure 6.15

In bringing together his historical mathematical findings, Pappus proffered some other ideas that, because we don't know of any other source for them, might have been his very own original work. One of them is the 'complete quadrilateral' (see Figure 6.16).

The Complete Quadrilateral

Draw a quadrilateral, ABCD, with no parallel sides. Extend the opposite sides AB and CD to where they meet, at point F, and the sides BC and AD to where they meet, at G. Join F and G with a straight line. Now draw the two diagonals of the quadrilateral, and call the point where they cross E. Extend them to meet the line FG, and call the two new points H and K. In all cases, the ratio of AE to EC is always the same as that of AC to CH. This also applied to the line BEDK, where the ratio BE to ED equals the ratio of BD to DK.

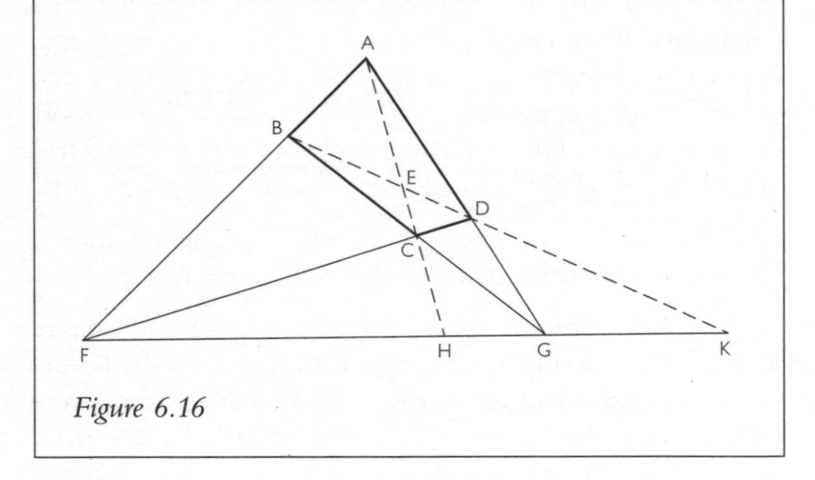

Figure 6.16

This and similar theorems proved useful some 1,300 years later, with the development of projective geometry, as did 'The Problem of Pappus', as more modern mathematicians began to unravel its complexities.

The last two recorded members of the great Museum at Alexandria were **Theon** (around AD 335–405) and his daughter **Hypatia**. Following on from Pappus, and seeing the danger posed by new religions hell-bent on destroying Grecian scholarship, Theon rewrote and edited the works of Euclid for posterity. He was present in AD 391, when Christian emperor Theodosius declared Paganism, the Greeks' ancient religion, as well as their worship of mythical gods, illegal. On his orders, the great Library of Alexandria – and probably the secondary library at Serapeum – were burnt to the ground.

Hypatia must also have witnessed this tragedy. In fact, today Theon is possibly most famous for allowing his daughter to succeed him as a scholar, astronomer and mathematician in such dreadful times. Hypatia was the first known female mathematician. She also worked to the very end of her life.

By this time Christian sects and new monastic orders were becoming ever more powerful, and were in constant conflict with both Jews and the diminishing Pagan community, which still followed the Grecian religion and way of life. Hypatia may have succeeded because, although she was steeped in Grecian traditions, she was happy to teach people of all beliefs. She was certainly a colourful character, and would drive her chariot at great speed through the city. Before long, however, the various religious sects began to cast her as a mystic, opposition to her grew until in 415 she was brutally murdered in a temple. Her death drove the final nail into the coffin of the age of Greek scholarship in Alexandria.

The Birth of Constantinople

But just as the Grecian age was no more, the Roman Empire itself was floundering dramatically. Its frontiers on the Rhine and the Danube had collapsed, and marauding tribes were pushing back the last of the Roman legions. By the third century, barbarians were overrunning the western Roman Empire, which was in danger of utter collapse. But Christianity was also gaining a more powerful hold. In AD 301, Armenia became the first nation to make Christianity its state religion.

The Roman Emperor **Constantine the Great** (AD 285–337), in partnership with **Diocletian** (AD 245–315), took the drastic step of moving the centre of their Empire from the once mighty Rome to the old Byzantine capital overlooking the Bosporus, because it was easier to protect and close to the major army recruiting grounds of Anatolia and the Balkans. There they built a new and impregnable Christian city, far larger than any city that had gone before, and named it Constantinople. They also made Christianity the official religion of the entire Roman Empire – demonstrating how widely its power had spread in less than 300 years – and founded theology, a mix of Grecian logic and Christian revelation. Now, although the mathematics and science of the past weren't widely taught, at least they would be preserved in Constantinople's new and gigantic library. This may have been why Greek replaced Latin as the official language of the Roman Empire; this in itself helped to preserve Grecian maths and science.

Up until then, the Romans had been encumbered by their awful mathematical language, which used the letters I, V, X, L, C, D and M to represent numbers. Mind you, the Greek numerical system wasn't much better: it used the alphabet. In fact, just about all simple calculations and business accounts were performed on counting boards; only the results of maths calculations were written and communicated, and Grecian letters were perfectly adequate for that.

But still the Romans were capable of producing amazing architecture. Their most ambitious achievement, under **Justinian** (482–564 AD), was Constantinople's Hagia Sophia, the largest religious building in the world until St Peter's in Rome was built 1,000 years later (see Figure 6.17).

Completed in just five years, it is a monumental celebration of geometric architecture. Both its famous designers were more mathematicians than architects: Anthemius of Tralles and Isidore of Miletus, an Alexandrian who specialised in solid or three-dimensional geometry. Indeed, the Hagia Sophia's walls and pillars were designed to withstand the earthquakes that

Figure 6.17

occur in this area, and have done on several occasions. Supported by huge arches and many half- and quarter-domes, the huge crowning dome was eventually extended to 31m across, enclosing an internal space larger than ever before imagined. In 1453 it became a mosque, and has been a museum since 1933. Visitors still gasp at the architectural vastness and mathematical splendour of this 1,500-year-old building.

As we shall see in the next chapter, under the Holy Roman Emperor Charlemagne, Constantinople played a pivotal part in the way that mathematical understanding spread throughout Europe. The city became a depository for previous Grecian works, including mathematical treatises that were diligently copied by monks, and which have come down to us through the ages. Nevertheless, Constantinople was never a centre for new mathematical ideas. For those we need to look further afield than Europe.

CHAPTER 7

Maths Origins, Far and Wide

China and Chinese Mathematics

It has been said that between 500 and 1400 AD no new mathematics of note was produced in the entire Christian world. Outside those confines, however, there was constant development. China, in isolation, acquired incredible maths understanding, while the Indus Valley civilisation produced a working number system that was to change the world. In leaning more towards arithmetical skills than geometric ones, the Chinese and the Hindus were more like the Babylonians than the Greeks, although – as we shall see – both peoples realised the importance of π and the Pythagoras' Theorem.

Partly through lack of easy access and partly through the dominance of Greek maths, the history of mathematics in the mighty Chinese Empire was long neglected. Joseph Needham, the world's greatest scholar of all things Chinese, complained in 1959 that modern European maths history simply ignores China despite the existence of more than 1,000 ancient Chinese mathematical texts in Beijing alone.

From around 1500 to 1000 BC, China was ruled by the Shang dynasty. Oracle (or fortune-telling) bones dating from the fourteenth to the eleventh centuries BC, as well as coins, show that the Chinese used an early base-ten numbering system. The numbers 1 to 4 were represented by one, two, three and four horizontal bars, and higher numbers could be formed from straight vertical lines or 'rods' – the number 7, for instance, was the symbol +, just like our modern plus sign.

The early maths used by the Chinese may have been quite advanced. Indications are that by the fifth century BC they understood how to record numbers less than 1 using an exponent system – so $10^{-1} = 0.1$, $10^{-2} = 0.01$ and so on.

In 1027 BC the Shang were conquered by the Chou, who in turn, from around 700 BC, were constantly troubled by insurgents. **Confucius** (550–478 BC) preached unity and stability as being essential to a peaceful and happy life – clearly a cry for some kind of sanity. Sadly, however, few listened, and the years from 480 to 200 BC were known as the time of the Warring States; all of China was involved in conflict in some way. Some of Confucius' sayings still have a modern ring to them. Take this one, for example: 'In a country with good government, poverty is a crime. With bad government, wealth is a crime.'

Throughout these troubled times, the Chinese seem to have had a philosophical belief that they could survive the constant feuding and fighting, as long as the Earth itself did not let them down. They felt they could cope with the hazards of life, but only as long as the Sun rose each morning and the seasons followed one another. Their great fear was that at some point the Sun might move further north or south, becoming ever colder in winter or hotter in summer.

There is perhaps an explanation of the ancient Chinese character here, in the two distinct religions that slowly formed around those times, and which have survived to this day. Taoism, which came first, held that whatever befalls us, nature will prevail, as the Sun will rise again tomorrow and the Earth will recover and always provide. But the writings of Confucius slowly drew many to another belief, which put the onus of survival on man himself, saying that whatever the evils of man or the vagaries of nature, the Earth will prevail, and that collectively we can and will recover, rebuild and survive. These twin sentiments have helped China through a wealth of troubles and disasters over the centuries, underpinning the wonderfully resilient character of the Chinese nation.

Between 221 and 210 BC there was a marked change in Chinese history, with the reign of Emperor Chhin Shih Huang Ti, from whom we get the modern name for China. Upon gaining power, and possibly to stop uprisings and quell civil unrest, he marshalled everyone together to work collectively on huge projects. His accomplishments during his 11-year reign were astounding. First, he ordered the destruction of any walls that divided individual states, at the same time extending the Great Wall, joining together other earlier attempts that had been severely damaged or neglected.

His wall may have finally stretched some 450 miles, but it was not as strongly constructed as the walls visited by tourists today, which were built during the Ming dynasty around 1500 AD. But Chhin's earth-wall building project was massive, and it's pretty certain that tens of thousands of workers died fulfilling his wishes.

Today, Chhin is best known for his life-size terracotta army, which contains more than 8,000 soldiers, along with 700 horses – some drawing chariots and others carrying officers. The entire army – which also included civilian officers, acrobats and musicians – was buried standing upright in long pits. It's an indication of how well Chhin managed people that he allowed the artists making the figures to use their own or friends' faces as models, to help them feel that they too would be part of posterity.

China's terracotta industry was already well established, making tiles, water and drainage pipes; on close inspection many of the soldiers' legs are just pipes with features added. But the soldiers carried real weapons, often coated with special materials to preserve their sheen and cutting edge through an extended burial period of some 2,000 years. Today tourists marvel at the terracotta army, and replicas can be found in museums, galleries and corporate buildings around the world.

Sadly the whole project was really a huge demonstration of egotism on Emperor Chhin's part. What's less well known is that everything he achieved was marred by another of his draconian decisions. He decreed that all books from previous dynasties should be burnt, so that for those who followed, Chinese history would begin with him. As a result, much Chinese history is a matter of conjecture and legend. Nevertheless, legends can give clues about earlier periods – and they indicate that those who governed China long before the Chhin period were profoundly interested in the science of astronomy.

There is the legendary tale of the goddess Hsi-Ho, who was either mother of the Sun or the charioteer who drove the Sun across the heavens. As the Sun's mystical movements were so important for the continuation of life, the legend broadened, turning Hsi and Ho into a pair of brothers, who then became four or even six 'magicians' (see plate section).

The magicians were dispatched by the legendary Emperor Yao to the four cardinal points of the Earth. To the north and south, their task was to ensure that the Sun turned back at the winter and summer solstices. Those sent to the east and west, meanwhile, were to encourage the Sun to continue on its natural course through the equinoxes, and to disappear each evening to the west but return every morning in the east.

Their legends were certainly rich and colourful, but the fact is that Chinese astronomy developed well ahead of those of other cultures. As early as the thirteenth century BC, the Shang people knew that the year was 365¼ days long. Using coordinates, they tracked the path of the Sun throughout the year, relative to the position of the stars at night. From that they deduced that the heavens revolved, and that the whole system was based around circles. But for the Shang, reducing the length of a year to anything other than 365¼ days would have been considered unscientific. Unfortunately, this commendable stance actually harmed their mathematical progress.

The Babylonians first had the idea of simplifying things by reducing the year to 360 days, and using that number to divide up a circle. Only when this idea finally filtered through to them did the Chinese make inroads into circular mathematics. In the second century BC, imperial astronomer Feng Hsiang Shih used the Pole Star to help locate the four cardinal points (north, south, east and west). He also counted in 12-year cycles – 12 years being the time it took Jupiter to complete a full orbit of the Sun – dividing the year into 12 months and the day into 12 'double hours'. He recorded the position of the 28 stars that determine the 'lunar mansions', or the Chinese version of the zodiac.

He then lapsed into mysticism, attaching good or evil traits to the 12 years in Jupiter's cycle. More scientifically, from observing five types of cloud, he also predicted floods or drought, abundance or famine. From this time, Chinese officials established an Astronomical Bureau to keep the

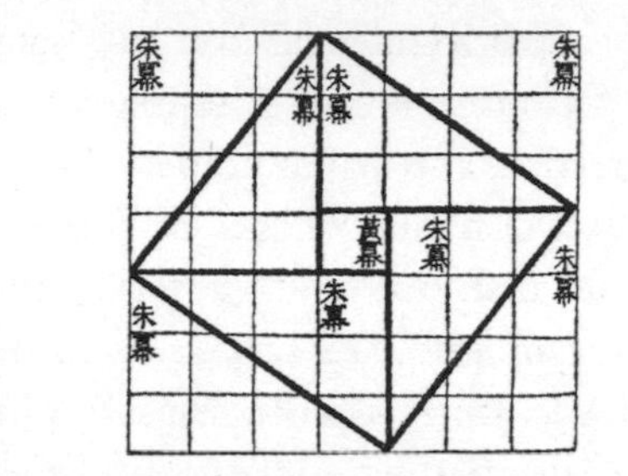

Figure 7.1

Emperor informed of developments – a system that was to last for 2,000 years.

The oldest known Chinese book on mathematics, *Chou Pei Suan Ching*, or *The Arithmetic Classic of the Gnomon and the Circular Paths of Heaven*, dates from the Chhin era. It contains the diagram shown in Figure 7.1, which demonstrates that the Chinese had already considered Pythagoras' Theorem, probably from around the sixth century BC.

The text instructed readers, 'Let us cut a rectangle diagonally and make the width 3 units and the length 4 units. Then the diagonal will be 5 units in length.' This in itself would have been a fascinating discovery, because right triangles that have all sides with whole-number lengths are rare. Like the Ancient Egyptians, the Chinese clearly had a similar fascination with this triangle.

The text also asked readers to 'Form a 5 x 5 square on the hypotenuse and set it at an angle in a 7 x 7 square array of 49 squares.' The four outer triangles together cover 24 squares, as do the four inner triangles, with a 25th square at the centre. This revelation – a mathematical proof of a kind – probably survived more for its fascinating glimpse of mathematical patterns and mysticism than for its geometric insight.

The book also revealed that the Chinese had discovered the magnetic compass by 200 BC, long before it appeared in the West. For them, the magnet pointed south, in the direction of warmer and more attractive climes and away from the cold and inhospitable lands to the north. They also floated compasses on water, a system that worked as well as the balanced needle we're more familiar with today.

The Chinese additionally discovered that if you heat a piece of iron and hammer it into a leaf-shaped mould, it takes on magnetic properties, and when floated always aligns itself with the original direction of the mould in which it was formed. The way a modern ship is constructed can cause it to become magnetised; this magnetism has to be removed so it doesn't interfere with the ship's own compass. The process is called *degaussing*, after the mathematician Karl Friedrich Gauss, who explained it in the early 1800s (see Chapter 15).

From time to time scholars of Chinese history would come across a legendary tale of a 'south-seeking carriage'. For many years it was dismissed as being associated with a particular use of magnetism, but that explanation was wrong. In the third century AD an unknown Chinese engineer understood gears well enough to make a carriage that could be wheeled along by man or horse, and which had the figure of a man on top. The figure held a pointer, and if this was set to face south at the outset of a journey, then no matter which way the carriage travelled or how many turns it took, the figure would always point south.

In fact, the 'magic' was not magnetic, at all. When a modern car turns a corner, the outer wheel covers more ground than the inner wheel, and a differential gear system compensates for this. The south-seeking carriage used the earliest known version of the 'magical' differential gear (see Figure 7.2).

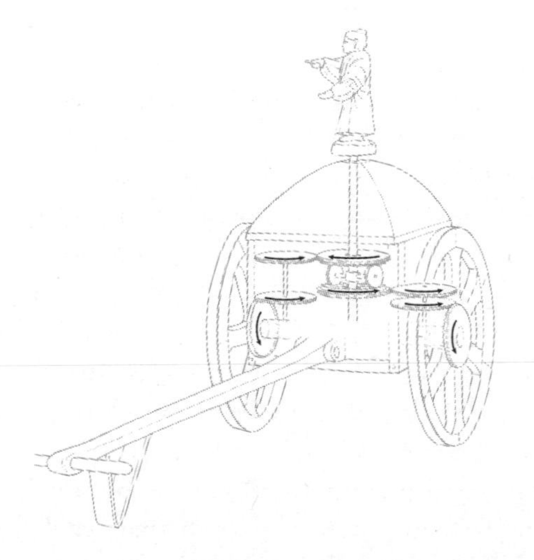

Figure 7.2

Word of the differential gear concept does not seem to have travelled to the Mediterranean, but many Chinese products had been finding their way west via traders for centuries. Indeed, Chinese silk and ironware were on sale in Imperial Rome at the time of Augustus.

But claims that China was often first with new technological ideas and Europe a poor second are supported by the story of paper. It was invented in China in the first century AD, but didn't even reach the Muslim world until the eighth century. It took another 400 years for it to be used in Spain, 200 more to reach Germany and another 100 years before it was used in England.

But because China was closed to the Western world for so long, it's difficult to assess when or even if ideas spread in either direction until the time of the Mongol Empire in the thirteenth century. Under Ghengis Khan, the Mongols were a ruthless people, often annihilating those who dared to resist them. Once they'd taken charge, however, they were remarkably amenable, allowing occupations, religions and traditions to continue. They were also happy to learn from those they conquered, all the way from eastern Europe to the eastern coast of China. When the Mongol Empire was at its height it was said that anyone could travel its length and breadth freely – even a girl with a pot of gold on her head would be completely safe.

The Chinese, rather than the Europeans, may have been first to take advantage of this freedom of movement. Yehlu Chhu-Tshai, followed by others, travelled with Ghengis Khan between 1219 and 1224 AD; Marco Polo's father and uncle travelled to China some 40 years later (between 1260 and 1269), and took the 17-year-old Marco with them between 1271 and 1275. There, under the new Mongol emperor Kublai Khan, Marco became an envoy in the imperial service. On his return to Venice he enthused about the Chinese technologies he'd seen – especially canals, lock gates, bridges and ships – all much more advanced that those in the West.

In around 1090 AD in Honan City, a Chinese engineer by the name of Su Sung produced the first ever mechanical clock with an *escapement* mechanism (see Figure 7.3).

Figure 7.3

A constant water supply fed a ladle until it was full, at which point its weight lifted a key that allowed the wheel to turn slightly, the ladle to pass and the next ladle to take its place. The clock kept perfect time; it also featured a celestial globe and a series of figures that appeared in niches at certain times to ring bells or strike gongs.

Chinese Counting Rods

Around the second century BC, the Chinese adapted the upright or sideways rod patterns that had been used to represent numbers in oracle bones and on coins. They developed a system of counting rods, and were the first to use a decimal place system, just like the one we use today, with units placed to the right, then – moving left – tens, hundreds and so on. So they could quickly spot the number indicated, they used an alternating system in which units were expressed in horizontal bars, but tens, one place to the left, used vertical bars. The hundreds then reverted back to horizontal bars, and so on. The number bars could be written, but more often, and for calculations, actual rods were used.

The use of huge numbers of counting rods led to an early understanding of hexagonal numbers. A bundle of seven rods (or pencils) form a hexagon with one central rod surrounded by six others. If another 12 rods are added, the bundle of 19 rods will still form a hexagonal shape. The Chinese also knew that, according to this progression, 127 counting rods would also form a perfect hexagon, as would 217 and 271 (Figure 7.4 shows the arrangement for a bundle of 127 rods).

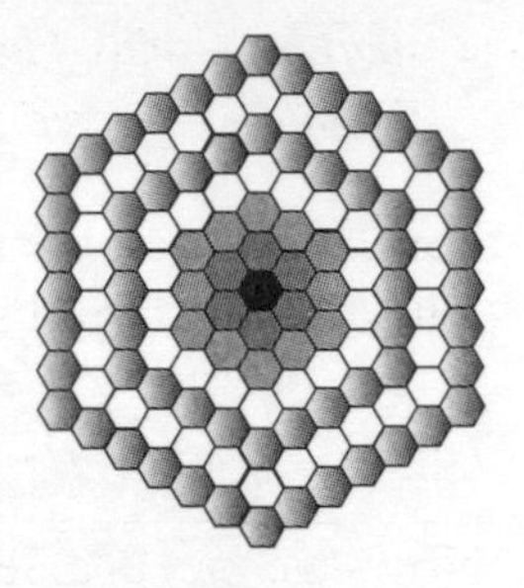

Figure 7.4

Rods added	1	6	12	18	24	30	36	42	48	54
Hexagon Total	1	7	19	37	61	91	127	169	217	271

The 127 bundle comprises consecutive rows of 7, 8, 9, 10, 11, 12, 13, 12, 11, 10, 9, 8, 7 rods.

It is clear that some mathematical manuscripts evaded the Chhin bonfires: shortly after his reign, around 200 BC, the first Chinese mathematician of note, Chiu Chang Suan Shu, produced his *Nine Chapters on the Mathematical Art*. It drew together 246 problems, most of which came from earlier texts, and featured items on surveying and taxation – all essential in managing what was mainly a farming community.

One innovation in the *Nine Chapters* was a maths first: negative numbers. Some 500 years later, Diophantus (see Chapter 6) failed to grasp them, dismissing them as stupid. But the Chinese first listed assets as negative numbers in accounting, and absorbed the whole concept of positive and negative numbers very early on.

Perhaps the most important feature of Chang Suan's book was the first known example of a way to solve maths problems using the 'double false position' method, which would be rediscovered again and again throughout history. Here's a simple example:

An item is purchased jointly by a group of people. They discover that if everyone contributes 8 coins, the total is 3 coins more than the purchase price. If everyone contributes 7 coins, they're short by 4 coins. How many people are in the group, and what's the purchase price?

For the solution we need two sequences – first a series starting with 8 – 3 = 5 and increasing by 8 each time:

$$5, 13, 21, 29, 37, 45, 53, 61, 69 \text{ and so on.}$$

Now we need a sequence starting with 7 + 4 = 11 and increasing by 7 each time:

$$11, 18, 25, 32, 39, 46, 53, 60, 67 \text{ and so on.}$$

As 53 appears in both sequences, the item cost 53 coins. As 53 is also the seventh number in each sequence, there were 7 people in the group. Problem solved.

A Sudden Change of Direction

Around 400 AD there was a remarkable change of direction in the way the Chinese did mathematics. After hundreds of years they suddenly decided that their counting rods had been facing the wrong way. So they switched them all round.

Instead of units being represented by horizontal rods, the rods became vertical, and were now called *hengs*, with a horizontal rod across the bottom to represent six upwards. However, the tens column now had horizontal rods called *tsungs,* with a vertical rod above the bars to represent 60 upwards). The hundreds column now copied the units column, with upright hengs; the thousands column reverted to horizontal *tsungs*, and so on (see Figure 7.5). To make the change clear to everyone, the following verse appeared in

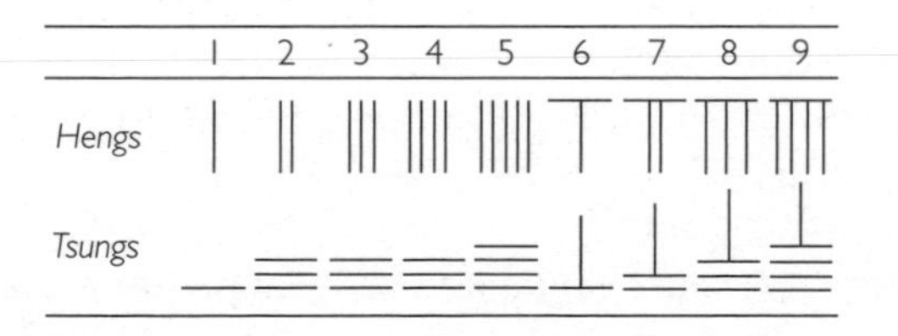

Figure 7.5

Sun Tsu's *Arithmetic Manual* (AD 280). This is a translation, of course:

Units are vertical, tens are horizontal,
Hundreds stand, thousands lie down,
Thus thousands and tens look the same,
Ten thousands and hundreds look alike.

This system proved very versatile. It made decimal fractions easy, for one thing. So the number 738.64, for example, could be represented as shown in Figure 7.6.

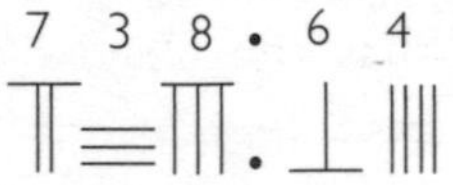

7 3 8 • 6 4

Figure 7.6

The Chinese placed their rods on counting boards divided into squares, not unlike the boards used for the game of Go, which was first mentioned in 548 BC, although it could have been developed much earlier. It's quite likely that mathematical calculations and Go games could have been carried out on the same boards at different times.

But what if a square contained no rods at all? To cope with this the Chinese decimal system needed an understanding of the concept of zero, or 'nothing'. But it wasn't until the ninth century AD that a symbol for zero was used to fill the gap, and that (as we'll see later) was in India. The first written mention of zero in China didn't appear until 1247 AD in the Shu Shu Chin Chang (*Mathematical Treatise in Nine Sections*) – although it's believed to have been used perhaps a hundred years earlier.

Depending on the region where they were being used, counting rods would have been made in their thousands from wood or bone. Those who worked with positive and negative numbers simply changed the colours of the rods, choosing red for positive numbers and black for negative ones. When Europeans first started to work with negative numbers, after Luca Pacioli published his double-entry bookkeeping system around 1500 AD, they went for the opposite combination, using red for negative numbers and black for positive ones – the version we still use today.

Rod handlers must have been very adept at changing digits rapidly by adding or removing rods, and adjusting the numbers accordingly. And although only a maximum of five rods were needed to represent any single digit, it was still a cumbersome way of working. Representing a whole calculation could require a large number of rods, and any rod could easily be accidentally knocked out of line or misplaced.

The Chinese used counting rods right up until the Ming Dynasty (1368–1644 AD), but this seems likely to have delayed their widespread adoption of the abacus. It also seems clear that the abacus did not originate with the Chinese, but eventually it was widely used there, and its use spread to Japan and Korea.

Where Did the Abacus Originate?

In the late 1970s I recall Chinese restaurants in Bristol, and in Acton, London, still using an abacus to tot up diners' bills. It's often assumed that the Chinese – who continued to use the abacus when other countries had set it aside – must have invented it. But the story of the birth of the abacus is far more complex than would first appear.

The first abaci or counting boards may have been used by the Sumerians in Mesopotamia between 2700 and 2300 BC, to record numbers in their sexigesimal number system.

An ancient Arabic word for dust is *abaq*, which suggests that the 'board' could have simply been drawn in the dust or sand, with pebbles used to represent numbers. Even today, our words for tables and counters in shops originated from the time when counting or accounting was done on tables.

The word 'calculate' comes from the Greek *calculi* – a stone or pebble. A play by the Greek writer Alexis from the fourth century BC mentions the use of an abacus and pebbles for accounting, and a little later Diogenes mentions 'men that sometimes stand for more and sometimes for less, like the pebbles on an abacus'.

The Roman abacus usually consisted of at least seven vertical columns along which stones could be slid. The units column was on the right, then the tens, and so on moving left. Quite high up on the board was a dividing line, with four stones in each groove below the line and one stone above – really just a copy of the human hand, with four fingers and the thumb. The lower stones could be slid up to represent 1, 2, 3 and 4. Then, while the first four stones were slid down to their starting positions, the upper stone would be moved up to represent 5. Now the lower stones would be moved one at a time to give 6, 7, 8 and 9. When a further 'one' was added, all the stones in this first right-hand column would return to their original positions, and just one stone in the second column would be moved up to represent 10.

Being able to draw an abacus in sand would have been very convenient, but it's hardly surprising that there's so little early evidence of their use. The Chinese had a wooden version of the abacus between 100 and 200 BC (see Figure 7.7), which was similar to a metal one found in Rome about the same time; a Marble version dated to 300 BC was found in Salamis. The Roman poet Horace, during the time of Emperor Augustus, mentions tablets made of wax on which marks could be made and erased easily.

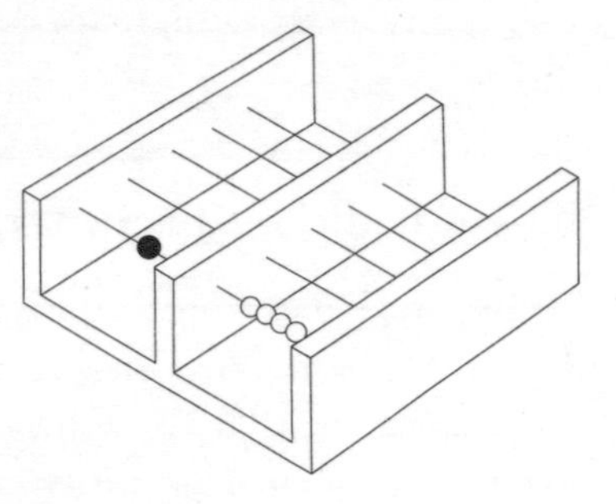

Figure 7.7

However, right up to modern times the Chinese and Japanese became the true experts at using the abacus. If abacus users needed to record the result of a calculation, they could write number symbols under the vertical columns. So it seems likely that when no stones had moved in a column, an extra symbol might be used, or no symbol at all to represent zero. But we'll come to zero a little later...

Chinese Magic Squares

One aspect of mathematics that greatly intrigued the Chinese was completely ignored by both the Babylonians and the Greeks. The Chinese loved magic squares.

The first mention of magic squares in China dates from around the time of the semi-mythical Emperor Yu, in the third millennium BC. The legend goes that he could control the waters, and that water creatures presented him with two charts to help him govern the empire.

First, and more widely known, is a chart that shows the basic 3 x 3 magic square, containing the numbers 1 to 9 with the number 5 at the centre (see Figure 7.8). This was known as the *Lo Shu*, or the *Lo River Writing*, which was traditionally first found inscribed on the back of a turtle.

In around 1261 AD, Chinese mathematician Yang Hui proposed a mystical method for forming this square. First make a diamond of the numbers 1 to 9 (a), then swap the numbers in the opposite corners (b), and finally squeeze the corner numbers together to form a square (c).

(a)

		1		
	4		2	
7		5		3
	8		6	
		9		

(b)

		9		
	4		2	
3		5		7
	8		6	
		1		

(c)

4	9	2
3	5	7
8	1	6

The second mystical chart was the *Ho Thu* diagram or the *River Chart*, which was a gift from the magical dragon-horse that emerged from the Yellow River (Huang Ho). In it, the first 10 numbers are represented in alternating white and black beads, with odd numbers shown as white beads and even numbers as black ones. The number 5 is at the centre as a white cross with a square of 10 black beads around it. Discounting those, the sequences of odd and even numbers in

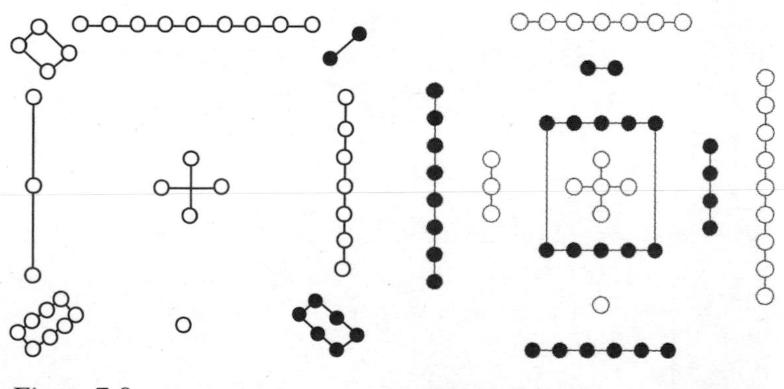

Figure 7.8

Figure 7.9

the diagram each add up to 20 (1 + 3 + 7 + 9 = 20 = 2 + 4 + 6 + 8) (see Figure 7.9).

The 3 x 3 magic square was discussed by Thabit ibn Qurra, who died in 901 AD, and who we meet in the next chapter. But it seems that no one developed the idea beyond the 3 x 3 magic square until Yang Hui in the thirteenth century explained a basic 4 x 4 magic square and how to create it.

Simply write the first 16 numbers in a square (a). Then swap or flip the two diagonal rows (b). Magically, each row, column and diagonal now adds to 34. Not only that – because all opposite pairs add up to 17, then any combination of two opposite pairs will also produce 34.

(a)

1	2	3	4
5	6	7	8
9	10	11	12
13	14	15	16

(b)

16	2	3	13
5	11	10	8
9	7	6	12
4	14	15	1

Yang Hui also produced a beautiful magic square in circular form (see Figure 7.10), with 9 at the centre. Every straight line that includes the 9 adds up to 147, while the five circles that don't include the 9 add up to 138 (9 fewer than 147).

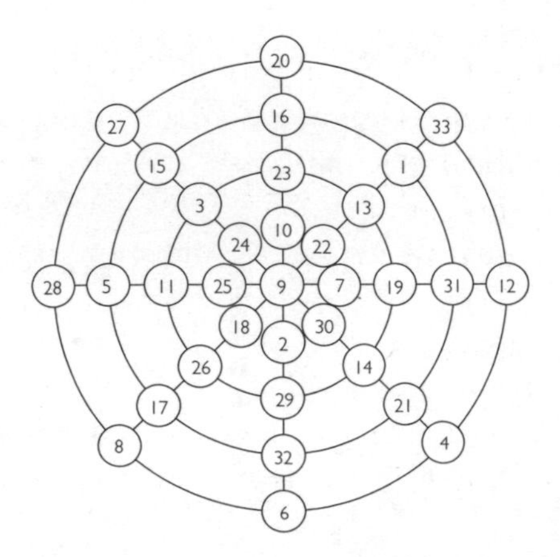

Figure 7.10

The Chinese Love of Pi

Greek mathematicians were fascinated by π, and comparing the distance around and across a circle. But for a few Chinese mathematicians, searching for its true value was almost an obsession. Around 130 AD Chang Heng suggested that π was the square root of 10, or 3.16 – the figure used in Ancient Egypt perhaps as early as 2000 BC.

Then along came Liu Hui. 'Who he?' you might ask. He was a third-century Chinese mathematician who – reminiscent of Thales (who we met in Chapter 2) – showed how to measure the height of a pagoda using similar triangles. There's no evidence that mathematical information travelled either way between east and west in those days, although Liu Hui was active at the same time that Archimedes was placing a '96-agon' inside a circle to calculate pi. Lui Hui, however, advanced the process another five stages to create a 3,072- agon (for an explanation of how he achieved this, see the Wow Factor Mathematical Index).

The Chinese search for the true value of π didn't stop with Liu Hui. Some time before 500 AD, and aided by his son Tsu Keng Chih, **Tsu Chhung Chih** (430–501 AD) took the calculation another three steps further. He worked out the distance around a figure with 24,576 (6×2^{12}) sides inside a circle – a result not accomplished in Europe until around 1600.

Tsu Chhung Chih then brought everything down to earth again, by suggesting $^{22}/_{7}$ as an adequate value for π for most mathematical calculations – the ratio we mostly use today. But he also said that 355/133 was amazingly accurate, and actually equal to his 24,576-side solution, placing the value of π between 3.1415926 and 3.1415927, or correct to six decimal places.

An ancient Chinese unit of measurement (about 23cm or 9 in), the length of a standard house brick, was called a *chih*, and one wonders if this name derived from the two Chihs?

A Problem in Chinese Arithmetic

The Chinese conducted a thorough exploration of arithmetic, always trying to offer simple explanations. Here is one example from the 246 problems in the Chiu Chang – they were probably gathered together before the first century BC, and have been continuously regurgitated by later mathematicians. What's also remarkable is that, unlike the Greeks and the Sumerians, who started their readers off with quite simple problems, the Chiu Chang gets complex from the very start. Try this translation using modern notation…

Divide 119,000 by 182 ⅝

Although the problem seems complex, and involves fractions, the method is simple.

First get rid of the fractions by multiplying both sides by 8:

952,000/1461 (182 x 8 = 1,456 + 5 = 1461)

Now we have a straightforward division sum with the troublesome fraction removed. However, the answer is still quite complex for a beginner: 651.60848 (to five decimal places). (There are more puzzle examples in the Wow Factor Mathematical Index.)

Tradition has it that zero as a numeral, along with the other symbols from 1 to 9, was first developed in the Indus Valley – and that is where we are going next.

Early Indian Mathematics

The earliest evidence of mathematical understanding in India dates to around 3000 BC. Although they'd originated over a period of 500 years, plumb bobs – which, hung on strings, indicated the true vertical – all had a unit weight of about 1oz (27.58g). Subsequent finds of other weights, however, show a progression of 0.05, 0.1, 0.2, 0.5, 1, 2, 5, 10, 20, 50, 100, 200 and 500 units, indicating that the Indians had a well-established, decimal-derived system of weights and measures.

They also seemed to have standard-sized bricks. Harrapan (Indus Valley) towns were built alongside the Indus river and its tributaries which, like the Nile, flooded regularly, helping to fertilise and refresh the land. Unfortunately, these floods also washed away anything that wasn't pretty solid, like buildings made of mud bricks. Before the Harrapan culture declined, around 1750 BC, they had learnt to fire bricks that could withstand the periodic invasion of water.

Bricks in the Indus Valley have been found in 15 sizes, but all of them have sides in the same ratio: 4 units long by 2 units wide by 1 unit deep. A railway builder in the nineteenth century found so many that he used them as a foundation to support a railway track more than 100 miles long between Multan and Lahore. And there were still huge numbers of the ancient but durable bricks to be found.

But we have no real written evidence that the Indians explored mathematics before 500 AD, though there are references to maths in religious works. The Vedic culture in India spanned a period from 1500 to 500 BC, and *Sulbasutras* (or writings) linked to religion and astronomy survive. These include 49 verses of *Jyotisutras*, which give instructions for calculating the positions of the Sun, the Moon and the signs of the zodiac. The Vedics also constructed altars for worship and sacrifices, and their design reveals an amazing fascination for Pythagoras' Theorem and Pythagorean triples.

The Indian Pythagoreans

Baudhayana was a Vedic priest and mathematician, who lived somewhere around 800BC. He is sometimes known as the Indian Pythagoras. A translation of one of his Sanskrit writings says, 'A rope stretched across the diagonal of a square makes an area double the size of the original square.' Katyayana in the third century BC went further, saying, 'A rope stretched along the length of the diagonal of a rectangle makes an area which the vertical and horizontal sides make together.' Neither actually states that you must square the length of the rope or the sides, or offers anything like a proof, but this predates the same idea accredited to Aristotle and perhaps

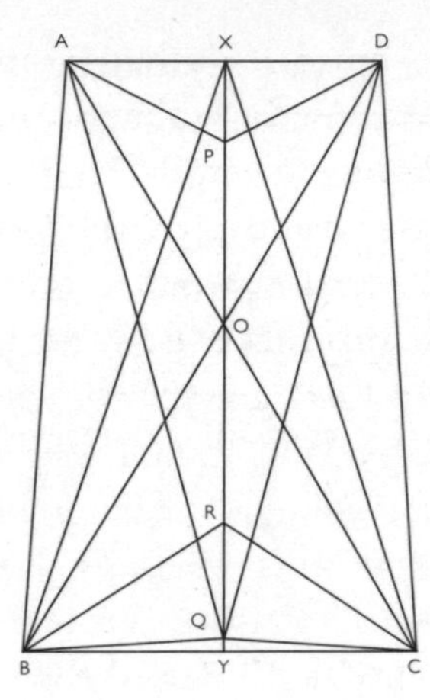

Figure 7.11

even Pythagoras. So it may have come to the Vedics by way of the Babylonians – both seem to have used their maths knowledge in the design of sacrificial altars.

The design for a *smasana* (or cemetery altar), found in Apastamba's Sulbasutra, has clear instructions for marking it out mathematically (see Figure 7.11).

1. Using a rope, mark out the centre line XY to precisely 36 padas. (A pada is approximately 10cm or 4in).
2. From X, mark points P, R and Q at 5, 28 and 35 padas down that line.
3. Draw horizontal lines AXD (24 padas long) through X, and BYC (30 padas long) through Y.
4. Draw lines from A, B, C and D to P, R and Q.
5. As a final check, draw diagonals AC and BD. They should intersect XY at O.

The result is a collection of amazing Pythagorean triangles:

APX and DPX have sides
5, 12, 13

AOX and DOX have sides
12, 16, 20

BRY and CRY have sides
8, 15, 17

BOY and COY have sides
15, 20, 25

AQX and DQX have sides
12, 35, 37

BXY and CXY have sides
15, 36, 39

...and all are whole-number Pythagorean triples.

The Vedics even used Pythagorean triples involving fractions, such as 2½, 6 , 6½ and 7½, 10, 12½, though these were simply some of the above triples halved.

They also found a neat geometric way to turn a rectangle into a square, which features in the Wow Factor Mathematical Index.

Vedic Maths

Although the Vedic culture in India stretched between 1000 and 500 BC, very few mathematical sources have survived from that period. **Swami Bharati Krishna Tirthaji** (1884–1960) wrote 16 volumes on Vedic maths, of which only one survived (it was printed in the USA in 1965, after his death). He claimed that his source for the work was a *parasista* (an appendix from a subsutra), but no one has since traced these sources, so we can't be sure that his story is correct. Nevertheless, the maths he featured in his work is still taught today, largely because it provides wonderful examples of how to make arithmetic simple, and how to empower people by improving their mathematical ability. Here's an example.

Multiply any two numbers that are a little less than 100

To multiply 96 x 94, for example, first subtract each from 100 and multiply the result.

So:

100 – 96 = 4 and 100 – 94 = 6   4 x 6 = 24

That's the back half of the answer.

Now, add the original numbers and subtract 100.

So: 96 + 94 = 190 – 100 = 90.

That's the front half.

Putting both halves together we get 96 x 94 = 9,024. How neat.

It's advisable to always do the two calculations this way round, as with slightly smaller numbers you may get a rear-half solution larger than 100 – in which case, you have to add 1 to the front-half solution.

So multiplying 92 x 87 gives 100 – 92 = 8 and 100 – 87 = 13. 8 x 13 = 104, the back half.

Now 92 + 87 – 100 = 79 – the front half. But you must add 1 to the 79 to get 8,004.

(The Wow Factor Mathematical Index contains more examples of Vedic maths.)

The religions of Buddhism and Jainism were established around 500 BC as the Vedic religion and its ritualistic sacrifices subsided. Jainism in particular was very rich in mathematical ideas.

Like Archimedes, Jainists thought about very large numbers, and even infinity. They discussed squares of squares of squares (or cubes) and square roots of square roots of square roots (or cube roots). They may also have been the first people to think of possible permutations...

If you have three items, *a*, *b* and *c*, for example, there are six ways to pair them up: *ab*, *ac*, *ba*, *ca*, *bc* and *cb*. While still in the Vedic period, a religious writer, Sushruta, calculated that for the six different tastes identified at the time (bitter, sour, salty, stinging, sweet and hot) there were 63 different combinations – which perhaps explains why Indian food has such a variety of flavours.

Taking six symbols – a, b, c, d, e, f – there are six single items, 15 pairs, 20 triples, 15 groups of four, six groups of five and 1 group of six, making 63 in all.

The Origin of Zero in India

Then in the valley of the Indus, someone had a brainwave so tremendous,
Have number symbols 1 to 9 and use those numbers all the time,

But always be aware of space and put them in their proper place,
So if you wrote, say 321, it meant three hundred and twenty one.
But if you take away the tens, it leaves 31 which doesn't make sense,
The three still needs to be three hundred, so keep the digits split asunder-ed,
To remember that there's nothing there – put a ring around it to mark it – there.
This unknown chap became a hero, the day that he invented Zero.

Lyrics from the final verse of *What is it? It's Nothing*
by Johnny Ball, 1999

It's difficult to pinpoint just when the idea of using a circle to represent zero first appeared. Carved Kharosthi numerals dated to between the fourth century BC and the second century AD appear very similar to Roman numerals, with I for 1 and – unusually – X for 4. So the numbers 1 to 8 were written as I, II, III, X, IX, IIX, IIIX and XX, suggesting that they were using an octimal (base-8) system rather than a decimal one. That was unusual, as almost every number system groups numbers in fives or tens most – probably copying the human hand.

Early Brahmi numerals, which definitely followed a base-10 system, did not use the place system that followed. Following the Greeks, they used separate symbols for 1 to 9, then 10, 20, 30 and so on, and then 100, 200 and so on. They used horizontal bars for 1, 2 and 3 – one theory is that our numbers 2 and 3 arose because two or three horizontal bars were joined with diagonal lines so they could be written quickly. Whatever the truth, there is a clear link between these early numerals and our modern ones.

Indian numbers and mathematical ideas were communicated north and to Arab communities via three early Indian astronomers. The first, Aryabhata, was born in 476 AD, which happened to be the year that the Roman Empire finally fell. As the Greeks had done, he referred to a system in which letters (in this case from Sanskrit) were used to represent numbers. The first 25 numbers were represented by consonants; the tens from 30 to 90 by the seven unclassified and lesser-used consonants; and 100, 110 (and so on) by the language's 10 vowels. It's possible that this system was employed purely because using letters enabled scholars to write maths-based poetry. Aryabhata was also a noted astronomer: his name was chosen for India's first artificial space satellite, which launched from Russia in 1975.

It's not until the sixth century AD that we finally find a term for zero: the word *sunya* or 'empty' was used to represent nothing. The idea of using a symbol for zero must have come after this, but it may not have been accepted, or understood immediately: how was it possible to multiply a number by 10 simply by adding 'nothing' to the end of it?

The first known truly decimal place system that used a symbol for zero was the Nagari number system, which dates to around 650 AD. It has been suggested that the mathematician and astronomer **Brahmagupta** (598–665 AD) was responsible for bringing the Indian number system to the Arab world. This may just be a legend, however: the concept of zero seems to have been introduced close to the end of his life, and was widely adopted even later. We do know that Brahmagupta worked at what was then the centre of Indian science at Ujjain in west central India, and was an eminent scientist at the time. It's something of a surprise to learn, then, that the treatise he wrote in 628 AD still denied that the Earth revolved.

Zero was definitely widely used from around 800 AD. An inscription from the city of Gwalior dated to 876 AD – the first carved instance of zero known – shows it as a small circle (see plate section). The numerals found here formed the basis of those passed to Muslim scholars, which even today we know as Arabic numerals. As we shall see, however, the time between zero's acceptance in India and its adoption by Muslim scholars in Baghdad may have been very short indeed.

So, rather than having zero and the Hindu number system brought to them, Arabs may well have acquired both as they absorbed new ideas from the many cultural groups they came into contact with as their empire spread with startling speed.

The Amazing Mathematics of the Central Americas

In considering the origins of maths outside Europe and the Mediterranean, we can't possibly ignore what could be a depiction of the oldest known date. In the Mayan city of Quirigua in Guatemala there is a tall stela (or stone obelisk) covered in hieroglyphs, which shows a calendar that could calculate dates going back 400,000,000 years.

Of course, there were no humans in the world that far back – or indeed in the Americas, until they first crossed the Bering land bridge from Asia between 20,000 and 12,000 years ago. During that time there must have been successive sporadic movements into what we now call the Americas. All Native American tribes originated in Asia, but it's unlikely that sea travel across the Pacific was possible at that time.

The first mathematical activity in the area seems to have been between 2200 BC and 300 AD, a period now known in modern-day Central America, and in Peru, as the Preclassic period. The Peruvians built the world's first known canals between 5700 and 3400 BC, and an observatory, near Lima, not later than 2200 BC. When the first European explorers arrived in Peru in the sixteenth century they found a magnificent road system stretching more than 2,000 miles from north to south. One major road followed the coast, while others ran roughly parallel to it, but through the incredibly difficult terrain of the Andes Mountains. About halfway down this long, thin country is the inland capital Cuzco, set between the renowned and magnificent mountain city of Machu Picchu and the world's highest large freshwater lake, Titicaca. The lake straddles the border between Peru and Bolivia, and today local people on both sides tell tourists the same traditional but rather vulgar story, but in reverse: their side was named for the Titi while the other side represents the caca (see Figure 7.12).

The Peruvians were experts at crafting magnificently complex stonework. While the Egyptians could fit square blocks together precisely, the Peruvians achieved quite incredible precision with oddly shaped stones of different sizes, using no mortar. One 700-year-old stone wall in Cuzco contains a stone with 12 different angles that all fit perfectly with the stones next to it (see plate section).

Figure 7.12

To achieve this precision, stonemasons used cords and sharp pieces of obsidian, the hardest volcanic stone. By holding one end of the cord at a central point they could stretch the cord to cut elaborate curves that were a perfect fit with neighbouring stones. Behind the accurately cut surfaces, fine rubble was used to bed each stone firmly in place, so the masons only needed to apply absolute precision when shaping the outer faces of the stones. Nevertheless, the craftsmanship is astounding.

At the same latitude as Cuzco, but on the coast, is Nazca, the area famous for its mysterious lines carved in the desert floor. These form birds and animals so large that they can only really be observed from high above the ground, as though they were created purely for the enjoyment of the birds – or perhaps even the Gods. The precision and complexity of the patterns, probably produced between 200 BC and 600 AD, suggests that their designers had a strong mathematical ability to measure accurately over long distances. But how and why they did it is a complete mystery.

The Peruvians famously devised a mathematical system based on a collection of knotted wool cords they called a *quipu* (see plate section). Quipu means 'talking knots': the knots were used to keep records (probably for the purposes of taxation) of animals, crops or belongings produced or owned by individuals. Each quipu was an individual's personal record, and some were over 1.5m long. Different coloured cords could have indicated just about anything, and we may never know their meaning, the knots on each cord were part of a number system that was clearly decimal. One coloured cord might represent a goat, for example, and then a series of knots would record the number of goats.

There must have been many thousands of quipus, but fewer than a thousand have survived. One that could be about 5,000 years old was found in an arid cave at 11,500ft – but it's the only one of that age. When the Spanish conquered Peru in around 1533, they destroyed all they found, as a way of expressing their total dominance and control. (For more on the quipu and the knots it used, see the Wow Factor Mathematical Index).

When the first modern computers were being designed in the mid-twentieth century, one of the most difficult problems to solve was how to get the machines to account for sums of sums, and ever more complex numerical arrangements. The Peruvians had already tackled these problems centuries ago: quipus regularly contained sums of sums of sums.

Central America

A series of different civilised cultures rose to prominence in Central America in ancient times. The Olmecs, for example, between 1200 and 900 BC, produced huge stone heads perhaps 3m tall (see plate section), which were found defaced and buried at San Lorenzo Tenochtitlan in central Mexico. The faces, made from volcanic rock, were beautifully carved, but they'd been transported more than 100km to the site, which indicates that the Olmecs must have had a strong system of social organisation.

During this period the Mayan number system emerged. The Mayans used it to record time, study astronomy, and help them understand the best time to plant and harvest.

Oaxaca in the Mexican highlands was home to the Zapotec culture, which traded minerals and blue jade. Monte Alban, an expansive city site high above the Oaxaca valley, has the earliest examples in the Americas (dated to around 500 BC) of a number system carved in stone, which used a disc shape to represent 1 and a bar to represent 5. In the massive main plaza at Monte Alban, an arrow-shaped building 'points' at a different angle to the rest of the city buildings, and seems likely to have been an observatory (see plate section).

Some way to the north-east of Oaxaca lies the site of Teotihuacan, which contains several major pyramids; by 150 AD it covered 20km^2 and had a population of about 200,000. Today's Mexico City is built on the southern part of the site, where the Toltecs and then the Aztecs once ruled. Between 1000 and 1500 AD these cultures traded with the Mayans, whose huge territory covered an area of modern-day Central America that takes in southeastern Mexico, Belize, Honduras, Guatemala, El Salvador and Nicaragua. In around 300 AD they started to erect *stela*, into which they carved symbols that indicate they must have used a calendar of incredible mathematical accuracy. This was the start of the classical period in the history of Central American maths.

The Mayan Calendar and Number System

The Mayans used a base-20 number system, yet it employed just three symbols. Of these, a dot or small disc represented 1, and a bar was used for 5. The bar was long enough to fit four discs neatly above it with minimal space between them – so a number like 19, say, had four discs closely stacked on top of three bars (see Figure 7.13).

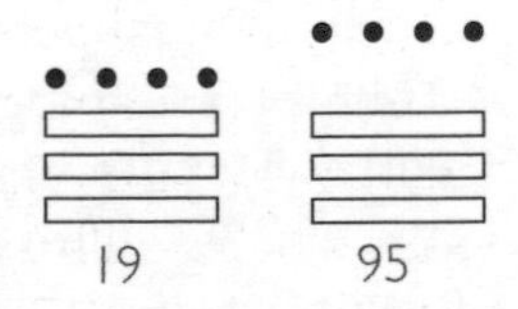

Figure 7.13

But if the same symbol had a bigger space between the discs and the bars, the number represented became greater:

In the lower position, 3 bars of 5 = 15

In the next upper position, 4 dots of 20 = 80, making the total number 95

In the third position up, the units became 400s, and in the fourth 8,000s; in the fifth 160,000s, and in the sixth 3,200,000s. In effect it was an infinite number system.

Most significantly, the Mayans used zero, which was represented by an empty snail shell. This was definitely in use by the third century AD, well before anyone in India – or indeed anywhere else in the known world – had the same idea. The empty snail shell had clear thinking behind it: although it didn't contain anything, it had at one time, and might again in the future. That's as neat an explanation for zero as you could wish to find!

Mayans believed that each day had 'contents' that were used up by the day's end, at which point another content-filled day appeared. Somewhat more mystically, however, they felt that time moved in a cyclical pattern, rather like a huge wheel, so there was a possibility that another day with the same contents might appear some time in the future. Their day was called Kin, which was also their name for the Sun, and the term they used for the passage of time throughout the day. At night, they believed, Kin fought with the Dragon of the Night, but always won to reappear at dawn. The Mayans had a god for everything – rain, death, hunting, corn, the Moon – and the effect that each god, pair of gods or even trio of gods had was complex and varied, and depended on their individual personalities.

The Mayans independently came up with the same idea as the Babylonians regarding days of the year, choosing 360 days instead of 365. This may have been because they had a base-20 number system. Their year, or *Haab*, comprised 18 months (or *uinals*) of 20 days – each of which had its own god – and this made up their Long Count (or full year). The five extra days to bring the year up to 365 days were known as *Uayeb*. Every fourth year the Uayeb – a time full of anguish and uncertainty marked by fasting and abstinence – was six days long to account for the leap year and to keep the calendar accurate.

For the Mayans, however, the Haab was not the calendar that ordinary people used. Just to complicate things, they had another calendar of a totally different length, and the whole thing became an incredibly complex system of dates guided by superstition.

There were 20 separate gods for the 20 days of the Mayan month. These Gods heralded the year, but because of the cyclical relationship between them, only four could mark the first day of the year. They were Kan, then Mulluc, both of which were considered favourable, and then Ix and Cauac, which were unfavourable.

As well as their month gods and day gods, the Mayans also had 13 'gods of numbers' – for them 13 was a number loaded with superstition – as it still is for many of us. The day gods ran alongside the number gods until, after 13 days, the number gods started again, although the day gods continued up to 20, completing a month (or uinal) before they started again.

This cyclical system produced a short year, or *Tzolkin*, of 260 days (20 sets of 13), and it was this sacred calendar that ordinary people followed. Each day always juxtaposed at least two gods, which determined the kind of day it was to be, and the whole system ran for 52 years, after which it all started at the beginning again.

This meant that the whole populace knew and remembered the 13 number gods, the 20 day gods and the 18 month gods; holy men calculated what day it was, and its godly significance, then announced their findings to the people. The calendar contained many special days, such as the day of the dead, or the day your loved ones offered their hands to help you. The short calendar was used in all of Mesoamerica, and survived the Spanish Conquest. It's often used by rural Mayans even today.

The Mayans represented the 260 day Tzolkin year with a wheel with interlocking cogs to run alongside the long Haab year wheel. The complexity of the system throughout its 52 year cycle was staggering (see Figure 7.14).

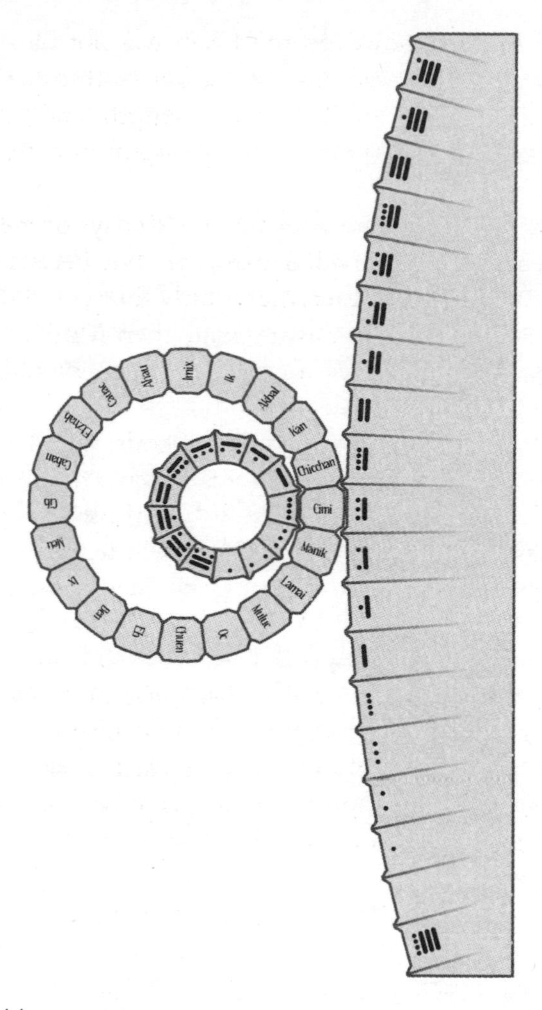

Figure 7.14

On the eastern side of Stela C at Quirigua in Guatemala, there's an inscription that features as part of the Mayan Long Count calendar. This pinpointed the last Creation as 13 August 3114 BC on our Gregorian calendar. The date of the next Creation correlated to 21 or 23 December 2012. As this date approached, some alarmists said it indicated the end of the world – but this was not a Mayan idea at all: to them time was cyclical, and kept going round and round forever.

Added to all this, the Mayans also accurately calculated the phases of the Moon, and could predict solar and lunar eclipses. They also tracked the movements of Mars, Venus and Jupiter with extraordinary accuracy – and all without telescopes.

All the cultures of Central America built vast cities and stone structures – often temple-pyramids with huge flights of steps rising to the summit, and inner labyrinths of passages. They played a game that the Mayans called *pok-ta-pok*, but which the Aztecs knew as *tlachtli*. Every settlement still has an ancient oblong court shaped like a capital 'I', although they vary greatly in size. No one knows the rules of the game or the sizes of the teams that played, but there is evidence that some players might have been beheaded, probably for losing – although it could have been for winning. No one is sure.

What we do know is that these Central American civilisations were incredibly diverse, colourful and successful, and very able to use the mathematics they needed to build their cities, feed their families, lead fulfilling lives and accurately record the passage of time. Eventually, however, everything changed: European adventurers discovered this New World, and chose to plunder it rather than contribute to it. They took everything of value they could find and in exchange brought in dogmatic religion, violence and European diseases.

CHAPTER 8

Mathematics Was Never a Religion

Mathematics could never be a religion – perhaps the one thing we could and should all worship is life itself...? Still, the story of mathematics after the fall of the Roman Empire in Europe (in around 476 AD) to the end of the Dark Ages (in around 1400) is mostly a fragmented tale of two religions.

During this 1,000-year period, in the Christian world, barely a handful of new mathematical ideas saw the light of day. At first there was no other powerful religion of note either. Then, quite suddenly, a new religion did emerge, one that was immediately confrontational and warlike. But it also spawned a fabulous new age of scientific and mathematical learning that greatly influenced – for the better – the way we use mathematics even today. That religion was Islam.

The Rise of Islam

The prophet Mohammad was born in Mecca (in today's Saudi Arabia) in around 570 AD. At the age of about 40, the story goes, he had a vision of the Angel Gabriel, who gave him the first ideas of what was to become the Koran (or 'recitation'); he then formed the ideas that would become Islam. He gathered a few followers around him, but when their numbers swelled they were classed as troublemakers and driven out of Mecca, from where they fled some 200 miles north to Medina in 622.

Just eight years later Mohammed had gained enough followers to form an army, which returned and took Mecca in 630. Although Mohammed died two years later, the Islamic movement was now established, and his successor Abu Bakr was proclaimed Caliph (leader of the Muslims). After his death in 634 AD he was followed by Caliphs Omar and Othman, who together conquered Mesopotamia and entered Persia and Syria. Jerusalem then surrendered to the pair, and Omar built a mosque in honour of himself that later went by the name of Dome of the Rock. Even the once mighty Alexandria fell to them as they entered Egypt.

Nevertheless, rather like the Romans, when the Islamists took over they allowed the local people to continue with their normal lives, including following their own established religions, such as Judaism.

Over the next few decades the Islamists gained strength, realising that if they could take Constantinople they could enter and possibly influence the whole of Europe. As things turned out, Constantinople survived, due largely to the efforts of one man – and one simple, yet vital, military invention.

Callinicus, who was born in around 620 AD, might have been Egyptian (some suggest he may even have been Jewish). In any case, he fled to Constantinople to avoid the advance of Islam. Although he dabbled in chemistry and medicine, he achieved fame with one world-changing idea. He invented 'Greek fire', used to great effect in 670 AD.

Bitumen based, this fiery, tar-like substance could be lit, and with the weapons readily available lobbed at approaching ships. On impact it stuck to whatever it came into contact with: oar, deck, sail or human flesh. Victims' immediate reaction when hit was to jump into the sea to quench the flames. Sadly that was no solution at all, for Greek fire continues to burn in water – if anything even more fiercely. So Constantinople survived, and the Islamists turned west. (Much later it was established that the only thing that could quench Greek fire was vinegar, provided you had copious amounts of it handy.)

The Arabs entry into southern Italy was a little more successful, but their most amazing success was in sweeping west, right across North Africa as far as the Atlantic coast. In 711 they crossed to Gibraltar and swept through Spain. By 732 they were more than halfway into France, although they were repelled at Poitiers by the leader of the Carolian Empire, the Frankish king Charles Martel, known as the Hammer. The Islamists withdrew to Spain, which they were to hold and govern for 700 years (see Figure 8.1).

It is this Islamic period, and the Islamic scholars who rose to prominence during this time, that we must thank for not only preserving the mathematical and scientific learning going right back to Thales, but also for building on it. Arabic teaching was eventually translated into Latin, and it slowly spread throughout Europe, bringing an end to the Dark Ages and ultimately ushering in the Renaissance.

A Few Bright Shining Lights in the Early Dark Ages

From the fall of Rome, mathematical knowledge in Europe was hardly recorded and even more seldom taught for a thousand years. But there were always a few individuals determined to be different and to boldly go their own way, seeking an understanding of the past and attempting to take things one step further.

The Venerable Northumbrian Monk

One early glimmer of light came via an obscure monk in an even more obscure part of the world: Jarrow in north-east England. There lived the **Venerable Bede** (673–735 AD). Bede did not follow the path to becoming a bishop, refusing high office so he could fill his time with study and writing. Although he never travelled more than 50 miles from home in his entire life, he had books brought to him, and his writings and reputation did the travelling for him. He wrote a history of early Anglo-Saxon England, although he got most of this from reading Pliny.

As a mathematician and astronomer he tried to pin down the true dates for Easter. He noticed that the spring (or vernal) equinox fell on the 18th March rather than on the 21st, as the calendar had slipped three days over time. This required a calendar adjustment beyond the ideas of Sosigenes and his leap year. Bede suggested that leap years should still occur in every year divisible by four, except in every hundredth year, when the leap year would be skipped to keep things on track. This adjustment was actually adopted 900 years after Bede suggested it, as part of the new Gregorian calendar. Today, for almost complete accuracy, if the first two digits of a century year are also divisible by four, as in 2000, then it is, or was, a leap year afterall.

Bede is remembered for suggesting that all dates should be measured from Anno Domini or the 'year of our Lord' – an idea first proposed by Dionysius Exiquus some 200 years earlier. In any case the calculation was wrong as Jesus was probably born in around 6 BC or 'Before Christ'.

Bede also observed that the tides are governed by the movements of the Sun and the Moon, and that they are irregular, changing from place to place. This meant that every seaport needed its own almanac to predict the tides. As the Moon follows its 28-day cycle, the tides in any one place lag behind and occur 50 minutes later each day. Being in England, Bede was well placed to observe this – something that had been missed by Mediterranean scholars because their tides are more modest than the widely varying ones that occur in the North Sea.

Although he didn't understand the reason, Bede realised that the main entity governing tidal movement is the Moon, and that high tides occur even when the Moon is on the other side of the Earth. As it turned out, we had to wait some 900 years before Isaac Newton explained all (see Chapter 13).

Bede also noticed that the highest tides – the spring tides – occur when the Moon and the Sun are aligned, around the spring or autumn equinoxes. In midsummer and midwinter, the Sun is at its highest or lowest in the sky, and is most out of alignment with the Moon, which follows a steady ecliptic orbit around the Earth. In the middle of the ocean, tides are barely noticeable, but along shallow coastlines they can be 20ft, or higher. In river estuaries, under certain conditions, they can be as high as 40 or 50ft.

Bede was an exceptional scholar at a time when learning in Europe was at a standstill. But further progress required a kick-start from someone, and it duly arrived. Martel the Hammer's son Pepin in turn fathered **Charlemagne** (AD 742–814), who grew to be 193cm (6ft 4in) tall. He formed an alliance with the Church to reduce strife and conflict, and became Holy Roman Emperor himself (between 800 and 814), uniting most of Western Europe, excluding Islamic

Spain. In Charlemagne we at last had an influential European who, although comparatively unschooled himself, realised the importance of learning.

Alcuin of York (732–804 AD) was inspired by his teacher, Bede, and traveled to Rome, where he caught the eye of the emperor Charlemagne, whom he taught to read. Sadly, although he tried again and again, the emperor never mastered writing; some of his childlike efforts have survived. Charlemagne made Alcuin responsible for the copying of ancient texts, using an army of scribes in monasteries all over Europe. As vellum, or light leather, was expensive, Alcuin had a brilliant idea for how to get far more information on each page. The first letter of a page would be enlarged and decorated (or 'illuminated'), but everything else would be in lower case. So the text you're reading now, apart from capitals, was Alcuin's invention.

It's also thanks to Alcuin that we have the famous wolf/goat/cabbage problem:

> *A man has a wolf, a goat and a cabbage. He needs to cross a river, but his small boat will only hold himself and one other thing. Left alone, the wolf will eat the goat and the goat will eat the cabbage. How does he get everything across the river safely?*

Easy: he crosses with the goat, leaves it on the other side and returns. He then takes either the wolf or the cabbage across, but brings the goat back. Now he leaves the goat behind, and crosses with the cabbage or the wolf. Finally he comes back for the goat and crosses for the last time.

From Charlemagne's time onwards, there was more teaching in Europe, although it was still all done by monks for monks, and for the most part featured religious texts only. Worse still, they often erased Greek writings to recycle the vellum, losing thousands of valuable records from the past. As far as maths is concerned, the Dark Ages in the Christian world were a complete blackout.

The Islamic empire, on the other hand, was rather different. By 755 AD it stretched from Spain and the borders of France in the west, all along the coast of North Africa (as well as

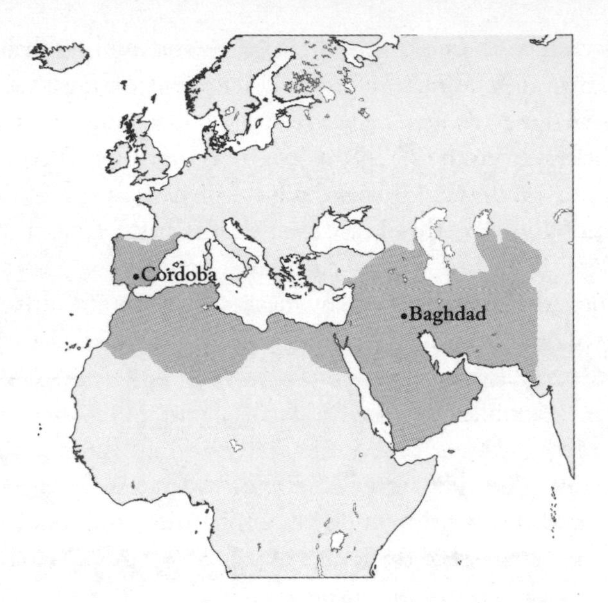

Figure 8.1

parts of Italy), to Isfahan in the east. So that its leaders could manage it better, the empire was split into the Western Kingdom (with its capital at Cordoba) and the Eastern Kingdom, centred in Baghdad.

Although their religious ideals were confined, the Arabs constantly absorbed intellectual ideas from the Greek, Jewish, Persian, Egyptian and Indian worlds. Indeed, Baghdad became a fabulous new centre of learning, where the ninth-century Golden Age of Islamic mathematics began. Huge advances were made by a succession of brilliant scholars, of whom the most famous was **Al-Khwarizmi** (AD 780–850).

Al-Khwarizmi and the Fabulous House of Wisdom

Having seen the demise of Alexandria as a teaching centre, the Arabs were determined to surpass those achievements. They built a huge library and centre for astronomy and learning in Baghdad, which they called Bait al-Hikma – The House of Wisdom. There they faithfully translated all the Greek, Babylonian and Hindu works they could find. The long-time chief librarian, and the greatest mathematician to work there, was Al-Khwarizmi.

He wrote two books, the first (in Arabic and published in around 830 AD) was *Hisab al-jabr w'al-muqabala*, or *Transposition by Restoration and Reduction*. In the title is the first instance of a new word: *algebra* from the Arabic *al-jabr* or 'transposition'. Although Diophantus and the Hindus had dabbled in using letters to represent numbers, this branch of maths had been unnamed until now. Even our word 'arithmetic' comes from shortening his name: Al-Khwarizmi – Khwarizmi – arizmi – Arithmetic.

His second book, of which only a Latin translation survives, was *Algoritmi de numero indorum*, or *The Algorithms of Indian Numerals*. It includes the Hindus' decimal number system, and its use of zero. This book set the standard number system the world uses to this day. Compared with the unwieldy Greek or Roman number systems, it is far simpler – and far more powerful.

For multiplication, Al-Khwarizmi used a grid system later used widely in Tudor England. It became known as Elizabethan maths, but it was clearly Arabic in origin. The grid system reduces complex multiplication to a series of simple single-digit sums (see Figure 8.2). Corresponding single digits are multiplied and the results placed in the relevant squares, each divided by a diagonal line. If a diagonal adds to more than 10, one is added to the next diagonal. So 367 x 193 = 70,831

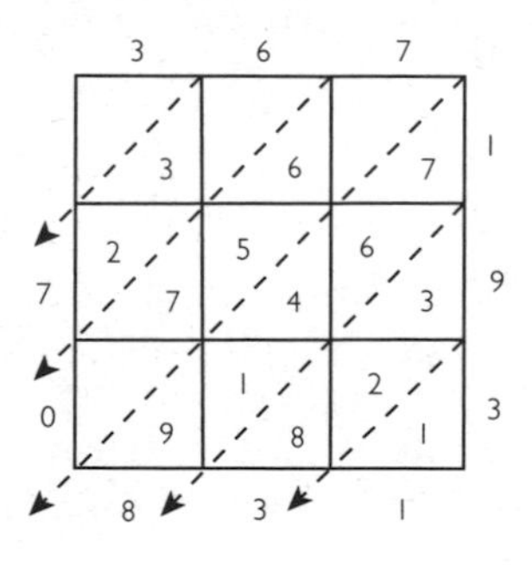

Figure 8.2

A First Step in Algebra

From the title of Al Khwarizmi's first book, *al-jabr* means 'to restore' and *w'al-muquabala* means 'to reduce' or 'simplify'. Al-Khwarizmi showed that by grouping unknown numbers (represented by x, for example) and known ones, we can simplify and solve simple algebraic calculations. Let's solve the following equation:

$2x + 5 = 8 - 3x$	First group the x's together and separate them from the numbers.
$2x + 3x = 8 - 5$	Note - on swapping sides + and - functions are reversed. Simplify -
$5x = 3$	So $x = 3/5$ths

Al-Khwarizmi chose to call an unknown a *root*; today we use the term 'x'. He didn't always use algebra to solve problems, however – sometimes he used geometry (for an example, see the Wow Factor Mathematical Index, which compares his maths to the way we do it today).

Al-Khwarizmi also drew up a *zij*, or collection of astronomical tables, which would be the master work for 500 years. These works show that Indian rather than Greek mathematics was now the main driver – via Leonardo of Pisa (who we shall meet soon) it spread into Europe, and became a major cause of the grand Renaissance.

The actual link between India and the Arab world was possibly a man called Karaka, who came to Baghdad in AD 773, on a diplomatic mission from Sind in northern India. With him he brought the works of Brahmagupta and others – some years later, Caliph al-Mansur ordered that they be translated into Arabic, and Al-Khwarizmi started work.

Al-Khwarizmi was also part of a geographical team that surveyed the earth's surface in their area, and accurately measured one degree of longitude at the latitude of Baghdad at 91km. They also talked in depth to travellers about the terrain they had covered, and their journey times, and mapped the accurate locations of 1,200 important places on Earth, including lakes, mountains, rivers, towns and major cities. Al-Khwarizmi named this work *The Image of the Earth.* He even corrected Ptolemy's erroneous estimate of the west-to-east length of the Mediterranean Sea, shrinking it by a third, although Ptolemy's number was the one that percolated into Europe 700 years later.

Around this time the Arabs began using spherical trigonometry and the astrolabe to work out the direction of Mecca from wherever they were. By knowing the positions of both places, they could calculate the distance between them – and, more importantly, the angular direction to Mecca for their prayer sessions.

A Small Selection of the Many Eminent Arab Mathematicians of this Era

From the setting up of the House of Wisdom and the early work of Al-Khwarizmi, more and more Arabic scholars picked up the baton, working to improve their general understanding of mathematics and all things scientific.

An early example was **Thabit ibn Qurra** (836–901 AD), an Arab mathematician who discovered rare pairs of numbers called 'amicable numbers'. So what are they? Let's take 220 and 284.

The factors of 220 are 1, 2, 4, 5, 10, 11, 20, 22, 44, 55 and 110, which, if added together, give 284.

The factors of 284 are 1, 2, 4, 71 and 142, which together add up to 220.

So 220 and 284 are said to be amicable and they are also the smallest pair of such numbers.

The next two numbers he found with amicable factors were 17,296 and 18,416. In the seventeenth and eighteenth centuries Descartes and Euler were to discover more, and an Italian schoolboy in 1866 found the second smallest pair: 1,184 and 1,210.

Thabit ibn Qurra translated into Arabic (from the original Greek) Apollonius' conics and also the work of Archimedes, Euclid and Ptolemy. Hunayn ibn Ishaq was a Nestorian (or Christian scholar) working in Baghdad, and Thabit also improved on the maths in his translations of Euclid's *Elements* and Ptolemy's *Almagest*, and he translated Ptolemy's *Geography*. His translation of one of Archimedes' works – which detailed how to construct a regular heptagon – was only rediscovered in the twentieth century, the original having been lost. (I've included it in the Wow Factor Mathematical Index, because Archimedes' method really is quite unique.)

Thabit ibn Qurra also wrote about the theory of numbers, and was the first scholar to mention the famous 'chessboard puzzle':

> *The inventor of chess offered the game to a king, who was enthralled by it – especially when he saw that everyone else can be slaughtered, but the King is never harmed, just hemmed in. As a sign of his gratitude, the king told the game's inventor that he could have anything he asked for.*

The canny inventor had a simple request: all he wanted was one grain of rice on the first square of the chessboard, then two grains on the next square, then four, then eight, and so on, doubling the amount until all 64 squares contained grains of rice. Thinking the man a fool, the king readily agreed. But the king's courtiers soon realised that this gift would bankrupt him. In fact, to fulfill the promise the king would have to hand over more rice than was produced in an entire year – more, in fact, than is produced in the world in a year even today.

The answer, as Thabit explained, is 2^{64} grains of rice, or simply a matter of adding the sequence 1 + 2 + 4 + 8 and so on for 64 stages. The total number of grains is 18,446,744,073,709,551,615, much higher than most people would intuitively expect. Eighteen and a half million million million grains of rice would make a whole lot of pudding.

Thinking scientifically, Thabit also proposed a theory of motion in which both upwards and downwards motions are caused by weight. He suggested that there were two competing attractions: one between heavenly bodies (meaning the Solar System, the planets and the Earth), and the other between all parts of each and every element separately. This was a remarkable insight, and wasn't fully explained until Isaac Newton, some 800 years later, discovered the laws of universal gravitation. Thabit was a true milestone on the long scientific road to understanding.

Abu al Wafa (940–998 AD), a Persian who came to Baghdad aged 19, eventually became the greatest Arab astronomer. He built the first curved quadrant (or quarter-circle) wall to observe the heavens. It would have been set on a north/south line, and by looking through the central focal point of the curve, it would have been possible to easily read and record the angles of heavenly bodies as they passed that line.

Abu al Wafa also produced the most accurate set of sine tables so far. He established several trigonometric identities, such as sin (a ± b), in their modern form, in contrast to ancient Greek mathematicians, who had expressed the equivalent identities in terms of chords. He also discovered the law of sines for spherical triangles.

Al-Karaji (953–1029) flourished around 200 years after Al-Khwarizmi. In his book *Al-Fakhri* he explored the sums of powers (or polynomials), for which a certain number needs to be written several times.

For instance, 2^1 means 2 written once, whereas 2^2 means 2 x 2 = 4, and 2^3 is 2 x 2 x 2 = 8.

It's a sort of mathematical shorthand. To multiply such numbers you simply add the powers. So $2^2 \times 2^3 = 2^5$, or 4 x 8 = 32.

Al-Karaji also explained how to form a table of binomials into the mathematical triangle we know today as Pascal's triangle. This seems to have been much earlier than the Chinese version featured in the previous chapter, which was dated to around 1303. That said, no one can say for sure when or where this triangle was first discovered: what we do know is that it came to Europe via a revolutionary book by Leonardo Fibonacci of Pisa, who we shall meet very soon.

Ibn-Sina (around 980–1037) – known as **Avicenna** in Europe – was a Persian polymath, and one of the most significant writers of the Islamic Golden Age. Of his 450 works, around 240 have survived, including 150 on philosophy and 40 on medicine. Avicennian logic influenced several early European scholars, and in modern Iran he is regarded as one of the greatest Persians who ever lived. A monument to him was erected outside the Bukhara museum; in Hamadan in Iran the Bu-Ali Sina University houses the Avicenna Mausoleum and Museum, which was built in 1952.

Ibn-Sina's book on physics has sections on arithmetic that positively glow with his love of the subject, and his obvious joy in playing with numbers. He discusses odd and even, perfect, deficient and abundant numbers, and presents some beautiful discoveries…

6 is a perfect number because it is divisible by 1, 2 and 3, and 1 x 2 x 3 = 6.

8 is deficient because its divisors 1, 2 and 4 add up to 7, or less than 8.

12 is abundant because its divisors 1, 2, 3, 4 and 6 add up to 16, or more than 12.

Avicenna's Beautifully Odd Magic Square

Remember they wrote from right to left, so

9	7	5	3	1
19	17	15	13	11
29	27	25	23	21
39	37	35	33	31
49	47	45	43	41

This is a 5 x 5 array that places the first 25 odd numbers in order. Surprisingly, each of its diagonals adds up to 125, which is 5 x 5 x 5.

For any such square array, irrespective of the unit number chosen, the two diagonals will always add up to the unit number cubed. So the 6 x 6 square compiled of consecutive odd numbers starting with 1 will have diagonals that add up to 6 x 6 x 6 = 216.

Avicenna's Even More Magical Triangle

This triangular array is, I think, one of the most beautiful of mathematical revelations.

The triangle is formed from consecutive odd numbers starting with 1.

1	Total 1 = 1 x 1 x 1
3 5	8 = 2 x 2 x 2
7 9 11	27 = 3 x 3 x 3
13 15 17 19	64 = 4 x 4 x 4
21 23 25 27 29	125 = 5 x 5 x 5
31 33 35 37 39 41	216 = 6 x 6 x 6

Amazingly, the sum of each row is always the cube of the number of digits in that row. It is also easy to predict the start number of any row without having to write the triangle out in full. The row starting numbers progress in ascending even numbers. So to find the 8th row, the left sloping side sequence goes 1 (+2=) 3 (+4=) 7 (+6=) 13, and on to 21, 31, 43, 57 in eighth place. So the eight row starts with 57 that, with the next 7 odd numbers, would add to 512 which is 8 x 8 x 8. This triangle could be continued to infinity. It is truly a mathematical wonder.

Alhazen or **ibn Al Haythen** (965–1039) was the most important physicist of the Middle Ages, although he was lucky to survive to the age when he completed his most important work – the work that made him famous. Being an over-confident young man was nearly his undoing. He suggested to the Caliph Al-Hakim that he could build a machine that would control the Nile – still flooding every year – at all times. The Caliph thought this was a wonderful idea and hired Alhazen to do the job. This would have been a fabulous opportunity for the lad, except for one slight problem – actually achieving this was impossible.

Al Hakim's reputation was known in Europe from tales told by returning Crusaders. He was arguably the most violent ruler of ancient times, with a nasty habit of having even leading officials executed at the slightest whim. Most thought him mad, so Alhazen thought, 'If you can't beat them, join them!' and began to feign lunacy. Luckily the Caliph believed it was a religious crime to harm the afflicted, and spared

Alhazen's life, although he was forced to keep up the pretence until the Caliph died.

Alhazen's work on optics included correcting the claims of Greeks like Heron and Ptolemy, by stating that light from the Sun and other luminous objects is reflected from surfaces and travels to the eye, rather than the other way round. He also had a pretty good explanation for the mathematics of rainbows. He constructed a pinhole camera and observed that the projected image appears upside down, because light rays cross over as they pass through the hole. He made parabolic mirrors, and explained how lenses work by reflecting and refracting light at specific angles on their curved surfaces.

He thought the atmosphere had a finite depth, of about 10 miles, which was no bad guess: as the air is too thin for modern aircraft to fly at much above 8 miles. His works in Latin were to influence Kepler and Newton as they further advanced the theory of optics – although not for another 600 years.

Omar Khayyam (1048–1131) was born not long after the death of Ibn-Sina, who had lived in the same area. The family name implies that his father was a tent maker, but Khayyam was a poet as well as a scholar: his collection *The Rubaiyat* was translated by Edward Fitzgerald in 1859. Many of the 100 or so *quatrains* (or four-line verses) are often quoted, or used as titles for modern works. Two of the more famous are:

A Book of Verses underneath the bough,
A Jug of Wine, a Loaf of Bread, and Thou
Beside me singing in the wilderness.
Oh, wilderness were paradise enow!

And

The Moving Finger writes; and, having writ,
Moves on; nor all your piety nor wit
Shall lure it back to cancel half a line,
Nor all your tears wash out a word of it.

Of course, it's difficult to say how many of these translations were the work of Omar, and how much was embellished by Fitzgerald, but Omar's verses were clearly inspirational. Yet he was primarily a mathematician and astronomer. In 1070 he wrote probably the finest Arabic book on algebra, in which he explained basic quadratic equations, although he was less successful with cubic equations. And, like Al-Karaji, he explored the array of binomial coefficients we now call Pascal's triangle. In 1074, as one of eight scientists and astronomers employed by Sultan Malik-Shah, he produced a revised and improved calendar; to keep it on track he suggested having eight leap years every 33 years. This was slightly more accurate than the Gregorian calendar, which wasn't to be introduced into Europe for another five centuries.

The Beauty of Islamic Art

From its inception, the Islamic faith forbade anyone from creating an image of the human body, especially in religious buildings. So the driving force behind architectural decorative design was geometry. As the Islamic empire spread, the walls, floors and ceilings of mosques and temples were adorned with inspirational mosaics, carvings and tile decorations. Examples in Cordoba and the Alhambra in Granada are still in amazing condition to this day.

Islamic geometry has a spiritual name – 'Abrahamic Pythagoreanism'. The Islamists believe that the wonder and beauty of the entire Universe can be observed through geometric maths, and that numbers and figures are the key to the structure of the cosmos – and indeed everything. This concept alone is breathtaking. (There is an explanation of how they applied the mathematics in the Wow Factor Mathematical Index.)

The great mosque at Cordoba contains double arches, all raised on pillars to create huge spaces with mystical views in all directions. The effect is pure magic. In Islamic mosques you will find arches of many different shapes, all achieved with compasses and by subtly moving the centre of various curves. The above arches show the geometric centres and curves used to produce them.

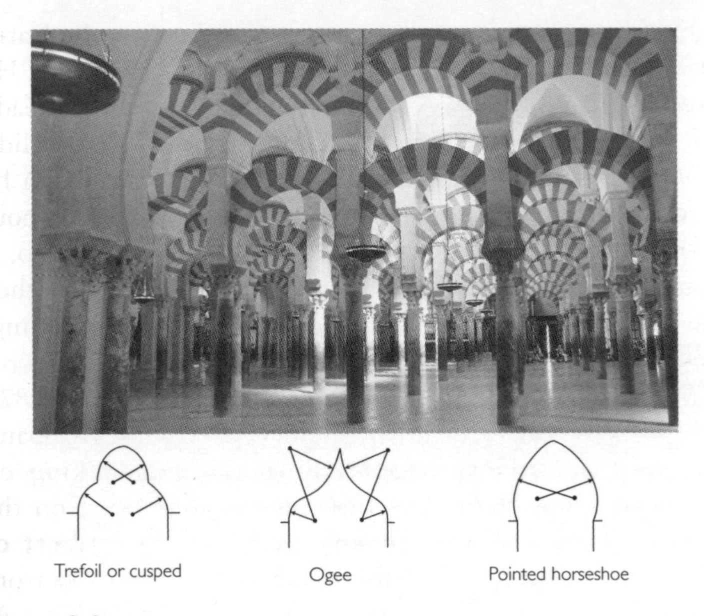

Figure 8.3

The ancient Greeks and Romans were adept at producing mosaic pavements based on mathematical designs. Even today some major cities still pride themselves on their often unique mathematical pavement designs, though the origins date back to Islamic times (see plate section).

In those early days, extending Greek geometry became an essential part of the Islamic psyche. Even today, Islamic worshippers are always surrounded by beautiful geometric designs and images every time they enter a mosque.

Europe Discovers Arabic Mathematics

After the Arabs ceased their advances into Europe, there was a long period of negligible interaction between Islam and Christianity. But there were a few European scholars who, having heard of a mysterious new and easier to use maths system, were desperate to know more. One was the English monk **Adelard of Bath** (*c.* 1080–1152), who eventually taught the future king Henry II. While still a youth he left England for tours in France; from there he travelled on through Greece, Asia Minor and North Africa.

He found access to Arab universities difficult, so he learnt Arabic, shed his monastic robes and dressed as an Arab. He learnt of Al-Khwarizmi's explorations into algebra (already 300 years earlier), and brought back an Arabic copy of Euclid's *Elements*, which he then translated into Latin. Although he tried desperately to enlighten his fellow English monks about Al-Khwarizmi's new number system and the use of zero, it seems that they mostly ignored him. Perhaps, when they asked what he had discovered and he told them 'Nothing' they totally missed the point…

Another monk, **Gerard of Cremona** (1114–1187), journeyed to Toledo, the then vibrant capital of Islamic Spain. He was the greatest of the early translators, working on Thabit's version of Euclid's *Elements* and improving on the work of Adelard. About the same time, in 1145, **Robert of Chester** translated several historically important books from Arabic into Latin, by authors such as Al-Khwarizmi and Al Haytham (965–1039).

Working in Segovia, Robert first translated Al-Khwarizmi's *Algebra* into Latin, but made an error, reading the major trigonometry ratio as *sinus*, the Arabic word for bay or inlet, which we still call the 'inlet' to the nose. So 'sine' is the maths term we have used ever since. Robert could not always find Latin equivalents for Arabic words, so he repeated them in their Arabic forms; we still use many of them today, such as *zenith*, *nadir*, *cipher* and *zero*.

Ultimately, however, rather than a religious figure, it was a European trader who made other Europeans aware of the strength and value of Arabic maths – and in particular Al-Khwarizmi's algebra. His name was **Leonardo of Pisa** (*c.* 1170–1240), and he produced a monumental and quite revolutionary work on mathematics.

The Book of the Abacus

Leonardo spent his youth working for his father, an importer in Pisa who traded with the Arabs in North Africa. The lad stayed in Algeria for long periods, where he learnt Arabic and absorbed their base-10 number system. Seeing how easy it was to use, compared with the Babylonian, Greek and Roman systems, he determined to spread the word to the world. He was only 32 when he produced *Liber Abaci*, the book that would eventually shape the path of mathematics in Europe via the Renaissance, some 200 years later.

His opus begins, 'These are the nine figures of the Indians: 9 8 7 6 5 4 3 2 and 1. With these nine figures and with the sign 0, which in Arabic is called *zephirum*, any number can be written as below will be demonstrated.'

Breeding in Rabbit Succession

Of the mathematical problems Fibonacci listed, the most famous asks:

> *How many pairs of rabbits will be produced in 12 months if you begin with a single pair, and if each pair becomes productive after one month and then bears a new pair every month?*

The first pair is too young to reproduce, so by the second month there will still only be one pair. But from the second month onwards they will reproduce. Meanwhile, the second pair will pause for a month to mature and then it also will produce a new pair of rabbits every month. As each new pair will progress along the same lines, the sequence will be:

Months	1	2	3	4	5	6	7	8	9	10	11	12 –	13	14	15	16
Total pairs of rabbits	1	1	2	3	5	8	13	21	34	55	89	144 –	233	377	610	987

and so on.

So in 12 months, starting with one pair, you would end up with 144 pairs – an explosion of rabbits.

No one told the rabbits about this, though…Rabbits can mate fairly soon after birth, and their gestation period is about 30 days, which is probably where Leonardo got his idea of 'not the first month but every month thereafter'. In fact, rabbit litters usually contain many more than two babies, so his estimates of population growth were very conservative. Rabbits breed, well, like rabbits.

The Fibonacci Series and Phi, the Golden Ratio

The resulting series from Fibonacci's rabbit breeding experiment is known as the Fibonacci Series, which goes 1 1 2 3 5 8 13 21 34 55 89 144 233 377 610 987 etc. If you explore this series you will find it has direct links to the Golden Ratio.

The series builds naturally by taking two consecutive numbers and adding them up to get the next number in the sequence. If we take say 1 and 2 and divide them. 1 over 2 is ½ while the other way around – 2 over 1 = 2 – has no connection. But if you take two consecutive numbers further along the series – say 89 and 144 – you get:

$$89/144 = 0.618 \text{ whereas } 144/89 = 1.618$$

These two numbers and all consecutive pairs in the sequence from now on are more and more accurately in Golden Ratio proportion to each other as they get larger.

Was Mother Nature a Mathematician?

It seems that Mother Nature herself may have discovered the Fibonacci series long before man stumbled upon it, because so many plants follow the series in the way they grow. Cut an apple across the line of the stalk, and each half reveals five seed pods – a Fibonacci number found in every single apple. Rubber plants always have five leaf positions around the stem, and the sixth leaf is always exactly above the first. A pineapple's scales are arranged in shallow spirals going one way and steeper spirals going the other way. In every pineapple (and pine cone) that ever existed, there are eight shallow spirals one way and 13 steeper spirals the other way; both consecutive terms in the series.

As it grows, how does a rubber plant or pine cone 'know' where to set its next bud or sprout? It relies on the Golden Ratio, splitting a circle of 360 degrees into 222.5 and 137.5 degree 'segments'. After one bud has emerged, plants seem to select the starting point for the next bud at about 222.5 degrees around the growing circle. Why? Because 360/222.5 = 1.618,

222.5/137.5 = 1.618 and 360/137.5 = 2.618. Built into their genes, plants seem to have an understanding of phi as a suitable ratio on which to base their growth. No doubt about it: Mother Nature is a consummate mathematician.

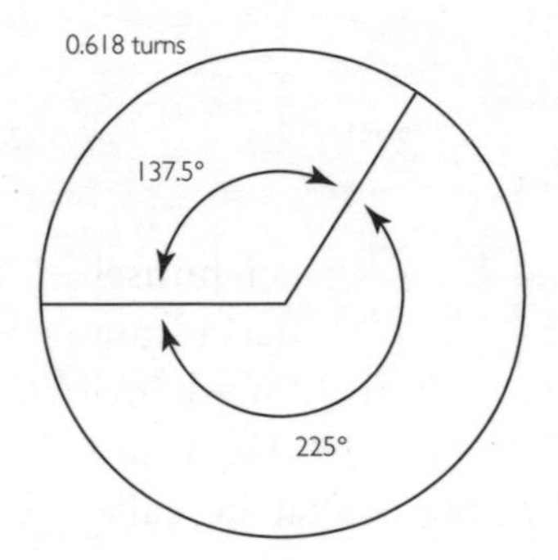

Figure 8.4

Cards of Golden Proportions.

Using Fibonacci numbers we can construct a three-dimensional dodecahedron and icosahedron using 'golden ratio cards' (see Figure 8.5). Make three cards (perhaps use consecutive Fibonacci series numbers 233 x 144mm). Note two of them will require a central slot of 145mm wide, while for the third card the slot needs to continue to the side of the card.

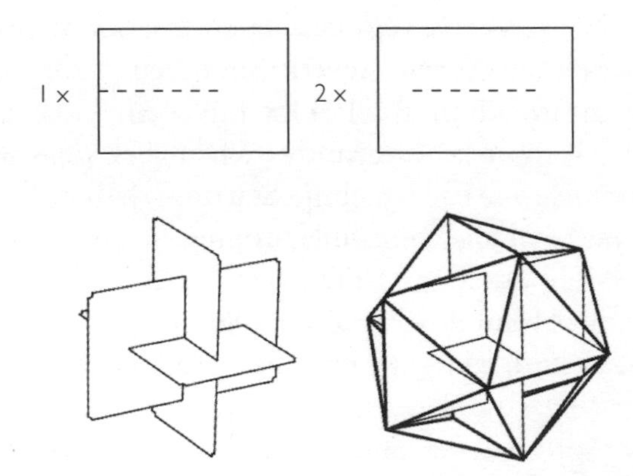

Figure 8.5

Slot them together and then make a tiny notch at each corner. You can then wind string around the corners to form an icosahedron of 20 triangles. Each of the 12 corners are the centre point of each pentagon that forms a dodecahedron, and the link between these figures and the golden ratio becomes clear.

Lucas Numbers

As far as we know, Fibonacci himself never used the term 'Fibonacci series' or 'Fibonacci numbers'. In fact it was nineteenth-century French mathematician Edouard Lucas who gave the series its name. He went even further, exploring variations on the theme that he called 'Lucas Numbers'. A Lucas series starts with any two numbers that, as with Fibonacci numbers, are added before the series continues.

A simple maths trick will explain it all. Ask someone to select any two numbers (it's best to start small, because we'll soon be into larger numbers). Let's say they choose 4 and 3. Ask them to write one under the other – the order doesn't matter, although different orders produce different results. They then add the two numbers to get a third, which they write under the previous two. They then add the last two to get a fourth, then a fifth and so on, until they have a list of 10 numbers. For 4 and 3, the sequence would be: 4, 3, 7, 10, 17, 27, $\underline{44}$, 71, 115, 186.

Now, could you add up that list of 10 numbers in, say, less than five seconds? Yes, you could! To get the answer every time, simply multiply the seventh number in the sequence, which I've underlined, by 11: $44 \times 11 = 484$.

This trick comes in handy when helping with the understanding of basic algebra. Starting with unknowns *a* and *b* (instead of 4 and 3), form a progressive column and note how both sides proceed as Fibonacci series. When you have 10 terms, their total will be $55a + 88b$.

But the seventh in the series is $5a + 8b$, which sure enough when multiplied by 11 gives the same result.

$$a, b, a + b, a + 2b, 2a + 3b, 3a + 5b, \underline{5a + 8b}, 8a + 13b,$$
$$13a + 21b, 21a + 34b; \text{total } 55a + 88b$$

In a Lucas series (as with Fibonacci numbers), as the series becomes larger, each consecutive pair of numbers, when divided into one another, gives a result ever closer to the value for phi. Here's a Lucas series with the first two numbers in reverse order: 2, 1, 3, 4, 7, 11, 18, 29, 47, 76, 123, 199, 322 and so on. Once again, as you get to triple digits, consecutive numbers, when divided, get closer to phi (or the Golden Ratio) – as with $^{123}/_{76} = 1.618421$. Even Lucas series that start with numbers that are further apart home in on phi. So if we take 7, 1, 8, 9, 17, 26, 43, 69, 112, 181, 293, 474 and 767 … and take two three-digit numbers from the sequence, we get $^{767}/_{474} = 1.618143$.

In his book, Fibonacci repeated much of the maths laid down by Al-Khwarizmi 300 years earlier. Slowly but surely, the base-ten Hindu number system was adopted by more and more people, along with a symbol for zero. This changed the way maths was taught and used, as Europe slowly dragged itself out of the lethargy and ignorance of the Dark Ages, and began the golden age of the Renaissance. But this didn't happen overnight – it took another few hundred years.

In God's Name

The huge and successful Arab empire was a thorn in the sides of the Christian European states – one that was becoming ever more painful. The Christians' main grievance was that Islamists controlled the holy places that were at the heart of Christian teaching. So, in the year 1095 there began what was to be 150 years of Crusades to the Holy Land, with the sole aim of reclaiming those biblical centres of early Christianity. Throughout this long period, it was said that there were just three types of European: those who fought, those who worked and those who prayed.

Urged on by the Church and those who prayed, many kings, princes and landowners chose to fight in the Crusades, commanding those they governed – including most of those who worked – to follow them. The first Peasants Crusade in 1095–1096 was a shambles, but in 1096 the Pope instigated the First Crusade, a much better organised expedition. After travelling overland from France and Germany, their numbers swelling along the way, four armies assembled in Constantinople. They then surged forwards, with great success, capturing most of the eastern end of the Mediterranean before taking Jerusalem itself in 1099. It's pretty certain that the Arabs had not been fully prepared for such an onslaught.

The Crusaders' progress was slow: most of them were on foot, led by knights on war horses that proved very effective in battle. A knight's armour, however, weighed about a third of his body weight; added to the horse's armour the combined weight put a huge strain on the very animal that was vital for the knight's survival. Once off his horse, a knight could neither fight the lighter and quicker opposition soldiers, nor run away. Even if he survived, there were no horses of similar size and strength in the east, and he was reduced to being a foot soldier. The Arabs' smaller horses, however, were fitted with iron shoes – as a result of the conflict, use of the horseshoe soon spread to Europe.

For the Crusaders, defending their gains so far from home was difficult, and the chance of reinforcements was slim. To protect themselves, they hastily built huge fortresses, but under constant pressure from sieges these tended to collapse before help arrived. So the First Crusade was followed by the Second 50 years later, in 1147–1149. From Constantinople, the Crusaders went by sea to Acre, and successfully held much of the Holy Land. But then Saladin, the Sultan of Egypt, moved around the Crusaders' east side, and conquered most of Syria. He began driving the Christians into the sea, retaking Jerusalem in 1187.

The Crusaders responded two years later with the Third Crusade of 1189–1191, led by the English King Richard the Lionheart (who did not actually speak English) and Philip II of France. They sailed from southern France and retook Acre. Confronted by the now-legendary Saladin, however, they failed to retake Jerusalem, and so it went on through the Fourth, Fifth, Sixth and Seventh Crusades, between 1202 and 1250.

As was always the case when far-off wars were fought, soldiers returning from their exploits brought home customs and materials not yet known in Europe. New textiles like damask, muslin and even carpets were introduced, along with new foodstuffs like rice, sugar, lemons, spices and new dyes, as well as cane sugar, which had originated in India and, like their maths, then spread to the Islamic world. Besides bringing back samples of them all, the Crusaders more importantly brought back examples of the new Arabic number system. This meant that all the ultimately worthless warfare actually helped to start the Renaissance – the age of new learning – although that was still some time off.

Gothic Art (1140–1200)

While those who fought and those who prayed did just that, those who worked still had to do the bidding of the great and powerful. Although builders were not particularly numerate, the mathematical skills they used emerged into a new religious architectural age, which eventually came to be called *Gothic*. This name was actually coined some 200 years later, as a criticism of the stark barbaric brutality of the style, which suggested it could only have been created by the heathen Goths who had savaged Rome 500 years earlier. In fact it was created by master masons with a deep understanding of geometry, who found new ways to enclose much greater areas of space and create far taller buildings.

Kings and clergy always demanded that their new buildings should push the boundaries of what was possible in terms of size and height, but using the same materials to make them larger than had previously been thought possible created problems. Forming semicircular arcade roofs is straightforward enough, for example, but if you want them to cross each other, the diagonal cross-arches need to be very wide, making them weaker. So to solve this problem they began to use two centres from which arch curves were formed. Pointed Gothic arcades became popular after which most religious buildings used this style (see Figure 8.6).

In the plate section you'll find the geometric pattern on the end of the Church of Saint Denis, north of Paris (where many French kings are buried). It's a highly intricate pattern,

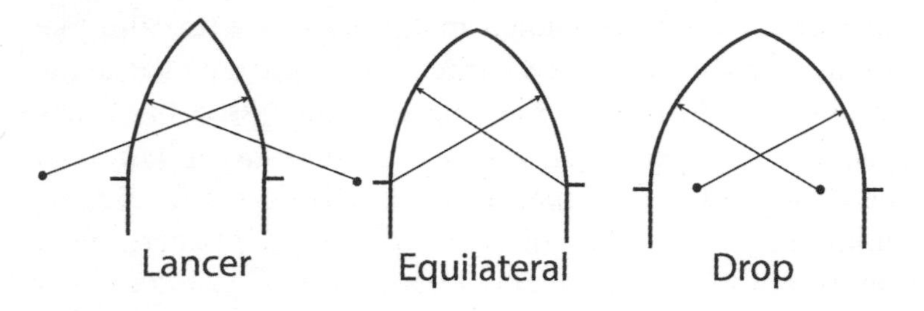

Figure 8.6

with arcades wrapped around the church itself, in which private chapels could be placed. This complex geometric roof system of short limbs supporting each other is called *tracery*, and it was used in almost all cathedrals from that point on.

Architects then attempted to create wider central naves by using flying buttresses sited outside the main building. They first built strong pillars, each topped with an arch pushing in towards the building, to stop the weight of the roof forcing everything outwards. At the time this was cutting edge technology. Gothic cathedrals became rather like ornamental birdcages made of pillars, ribs and arches that bore the colossal weight, took the strain, and permanently supported everything in an explosion of geometric design.

Until that point, because architects had believed that brickwork was necessary to support the whole edifice, church windows had been relatively small. But, again using basic geometry, architects now created complex windows in a series of arched and circular rose shapes, all held together by stone tracery that appeared light and delicate, but which was carefully designed to bear the weight from above and make the whole structure immensely strong (see plate section).

Most of these ancient Gothic cathedrals still stand solidly today, and all were created using state-of-the-art structural technology, to safely produce more space and a greater feeling of grandeur. Gothic architecture was one area where wonderful new mathematics flourished, even in the Dark Ages.

Today we know the names of many of the great church-building masons, who would travel long distances and live on site while their creations became a reality. We also know the names of just a few monks who stood out from the crowd in those dark days, by continually promoting mathematical teaching.

Following Science Religiously

Robert Grosseteste – no sniggering at the back (yes, it does mean 'big balls') – of Lincoln (1175–1253), rose from a poor background to enter the very young Oxford University. He then taught at the University in Paris, and later became Chancellor at Oxford and Bishop of Lincoln. But he never shirked from saying what he truly felt, which could be dangerous. As European monasteries had been the only source of education for many years, Grosseteste suggested that the Church was superior to the State, an opinion that King Henry III didn't share, to put it mildly. Then, as universities started to gain strength, first in Bologna, then in Paris, Montpelier, Naples, Toulouse and Oxford, he started to criticise the Church for not spreading its teaching widely enough. So now he had upset the Pope as well! Undeterred, however, he pressed on regardless.

He made great efforts to include more science in education, and introduced Aristotle's scientific thought to Europe. But he also noticed that the Latin versions of the Arab translations of the Greek originals had errors all over the place. So he sought out Greek scholars who had fled west to escape Turkish dominance during the Crusades, and set them to work making improved translations directly from the original Greek. He then realised that Aristotle was not always right in the first place, and that scientific ideas must constantly develop and improve to stay relevant. This crucial link was another major trigger in starting the revolution that was to become the Renaissance.

Recent evidence, including *The Ordered Universe*, a medieval history project by Durham University, suggests that pre-Renaissance science, beginning with the key figure of Grosseteste, was far more advanced than previously thought. In his book *De Luce*, or *On Light*, for example, Grosseteste made the first known attempt to devise one set of physical laws. He saw that light reflected from objects and came from the Sun, and that the heavens were filled with suns or stars. He also observed that light, like sound, travels in straight lines, but much faster, and he postulated that the birth of the Universe may have involved an explosion and a crystallisation of matter to form stars and planets, making it possibly the earliest Big Bang theory. All that was missing from his unified theory was the idea that light itself might be energy.

Grosseteste read Alhazen's work on optics, and explored lenses, mirrors and rainbows (which, he maintained, were formed when clouds act like lenses to bend light). In his treatise on optics, *De Iride*, Grosseteste said, 'Optics shows us how we may make things a very long distance off appear as if very close, and large near things appear very small, and how we may make small things placed at a distance appear any size we want, so that it may be possible for us to read the smallest letters at incredible distances, or to count sand, or seed, or any sort of minute objects.' Clearly he could see both telescopes and microscopes being used in the future, although they didn't become established for another 400 years. I think he must surely also have considered lenses as a way to improve failing eyesight – but that discovery is usually credited to someone who came just after him.

The Franciscan friar **Roger Bacon** (1220–1292) studied maths, astronomy and alchemy at Oxford, but probably arrived there too late to study under Grosseteste. Bacon was arguably the world's first true scientist (a title usually given to Galileo), attempting to learn by experiment and jumping on anyone who made scientific claims based on philosophical thought rather than experimentation. Like the Italian cleric **Thomas Aquinas** (1225–1274), he disliked Aristotle, once remarking, 'If I could, I would collect all his works and burn the lot!'

Bacon first encountered gunpowder in Chinese firecrackers 'no bigger than my thumb', which may have been brought back from Mongolia by the missionary William of Rubruck. To Bacon's ears and eyes they exploded with a deafening crack much louder than thunder and brighter than lightning. Determining that gunpowder was made from saltpetre, sulphur and carbon, he recorded the correct proportions in a code that was only translated in the early twentieth century. But the translation seems to have introduced errors, because the mix it revealed wouldn't explode, producing only smoke.

Like Grosseteste, Bacon devoured the works of Alhazen on optics, and is credited with being the first to use spectacles. A drawing he made seems to indicate that he experienced the illusion of an object seen through a round bottle filled with water, appearing to face and move completely the opposite way (see Figure 8.7), although Theodoric of Frieberg was to improve on this interpretation in 1311.

Bacon gained a powerful supporter in Guy de Gros (the big guy) who, after becoming Pope Clement IV, commissioned Bacon to write a universal encyclopedia, *Opus Majus*, to try to bring together science and religious thinking. Men of religion worried that scientific discovery was liable to go further than the Bible in explaining the world, but Bacon and others regarded what they were doing as – if you like – peeling God's onion to reveal ever more of the wonders of his works. Which meant that there should be no conflict between Church and science: one could 'follow science religiously'. Pope Clement applauded Bacon for his work, but after his patron's death in 1268 Bacon suddenly found himself without support, and his Franciscan enemies began to move in.

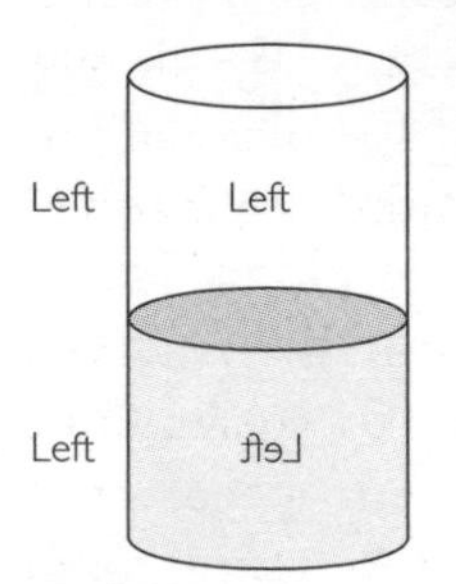

Figure 8.7

Bacon prophesied that man would make land vehicles, larger ships and even submarines. He argued that the Church should adapt to new discoveries, even if they showed the Bible to be wrong – for which he was accused of heresy. It was said he made a brazen head that could speak and answer any question it was asked – although as this implied witchcraft surely this was just a tale made up by his enemies...?

He continued to promote a glorious future for science and mankind, and even suggested that man would one day fly. He cited the example of Eilmer of Malmesbury who, some 200 years earlier, and inspired by the Greek fable of Daedelus, fitted home-made wings to his hands and feet in an attempt to glide, or maybe even fly (see plate section). In the event, Eilmer broke his legs and never walked again, although he claimed that his mistake had been not providing himself with a tail.

There is no record of Bacon's activities from 1279 to 1289, so it's likely that he was imprisoned in Oxford for about 10 years; he died shortly afterwards at the age of about 72. But his fame outlived his detractors; subsequent generations named him Roger Bacon, Doctor Mirabilis (or 'wonderful teacher').

Peregrinus was a French military engineer, born in around 1240, who dabbled in mechanics. He attempted to build a motor to drive a model of Archimedes' planetarium using magnets, despite scepticism among scientists of the times that a magnet – or indeed anything else – could influence another object at a distance. So Peregrinus explored and explained much about magnetism: such as like poles repel and unlike poles attract, and that if you break a magnet in two you end up with two magnets.

Probably inspired by Chinese science, Peregrinus also designed a compass with a needle (pointing north) floating on cork, or – his preferred method – suspended by a thread. He believed it pointed towards the zenith of Ptolemy's spherical model, which had the Earth at its centre, and which William Gilbert corrected 400 years later. The device revolutionised direction finding, especially at sea, and it aided the first wave of oceanic explorers – although that had to wait another 200 years.

Like Bacon, **William of Ockham** (1288–1349) believed that scientific truth should be sought through experiments. Once facts are proven, he said, all other concepts not supported by them should be declared abstractions, and should be 'lopped off', or removed from the discussion. This idea became known as 'Ockham's razor', and is often cited as an important yardstick in the development of science. Ockham died young, succumbing in his fifties to the first wave of the Black Death, which ravaged Europe right up until 1400.

The Mean Speed Theorem

In the plague-ridden years of the fourteenth century, however, scholars of Merton College, Oxford seemed to have kept healthy, and also produced an original bit of mathematics – the mean speed theorem. News of this spread to Europe, where Frenchman **Nicole Oresme** (around 1325–1382) turned the maths into pictures with an early graph. Time is the latitudinal line and velocity the longitudinal line, and the journey is represented by the sloping diagonal (see Figure 8.8). The theorem states that a uniformly accelerating body travels the

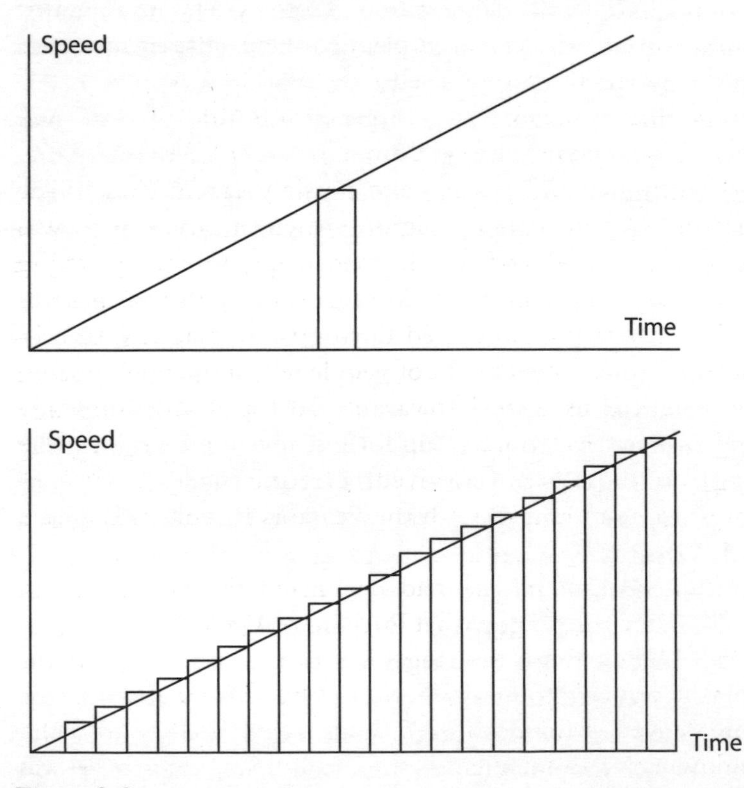

Figure 8.8

same distance in a given time as a body of uniform speed that is moving at just 'half' the final velocity.

If the journey is divided into small rectangles, and the diagonal passes through the centre of the top edge of one – creating a smaller copy of the main diagram – then the theorem still stands for that rectangle, even if it were to become infinitely thin. This is because the excess triangle (in the top left) is cancelled out by the shortfall triangle (in the top right).

This was an opening shot in the battle to produce a mathematics that could measure things that are constantly changing: the calculus. It was 300 years ahead of its time.

Strangely, after this initial and clearly understandable idea, as calculus developed over the next 300 years, mathematicians never seemed to use this simple system again. Instead they always placed their rectangles with the top-right or top-left corner reaching the sloping line. This mystery has plagued me for years. Whereas what plagued them in the fourteenth century was – the plague itself.

The first wave of the Black Death (or bubonic and pneumonic plague) struck Europe between 1348 and 1350, carried by rats and people, and possibly originating in the Crimea. Whatever its origins, when it arrived it was immediately fatal. It was said that people stopped grieving for the dead, because they all expected to go soon. Half of all Europeans probably died during those plague years; in some provinces 80 per cent of people fell, and whole villages and even towns were left vacant. To top it all, kings and landowners tried to make up for lost revenue by raising the taxes on those who survived, creating widespread civil unrest came to a head with the Peasants Revolt in England in 1381.

After years of plague had decimated the population of London, the legend goes that Richard or **Dick Whittington** (1354–1423) arrived, seeking his fortune. According to the story he was able to marry into a wealthy family and become Lord Mayor of London three times purely because so many landowners' and merchants' sons had died. In truth he was actually also from a wealthy family, although he did become a great benefactor, helping those who were suffering because their families had lost breadwinners to the disease.

Even more importantly, as opportunities finally increased, so did the desire for education, and at long last the Dark Ages came to an end, and the dawn of the Renaissance commenced.

CHAPTER 9

Discovering the Unknown World

To Boldly Go

One of the main reasons humans emerged from the Dark Ages with a growing thirst for knowledge was the desire for a better understanding of the world they lived in. The concept of flight was really beyond scientific imagination and reserved only for birds and biblical angels. So the early 1400s saw the beginning of great age of oceanic exploration under sail – this was their final frontier. And, of course, navigation and mapping oceans and new lands required mathematical skills.

Though a few people still held onto the belief of a flat Earth, with a danger of sailing over the edge, scholars and mariners were well aware that the Earth was round, and men already made rich by foreign trade believed that further exploration might produce unimagined new wealth.

The Mediterranean was a buzzing hive of commerce, but almost all trade there was controlled by the Venetians or Muslims, forcing countries like Christian Spain and Portugal to look elsewhere if they wished to expand. Their exploration of the Atlantic began with Henry the Navigator, the son of Portugal's first King John, who had already been knighted for his heroism in North African conflicts. Having fallen in love with this mostly unexplored continent, and determined to discover more, in 1418 he set up a school of navigation and an observatory at Cape St Vincent, Portugal's most southerly point. Within a year his first ship had sailed, heading south with the aim of travelling right round Africa as, according to erroneous legend, Hanno had done 2,000 years earlier. By now his navigators would have had magnetic compasses, the Arab astrolabe (complete with star map and the latest refinements), and the cross-staff – a primitive quadrant – to help them calculate latitude (see Figure 9.1).

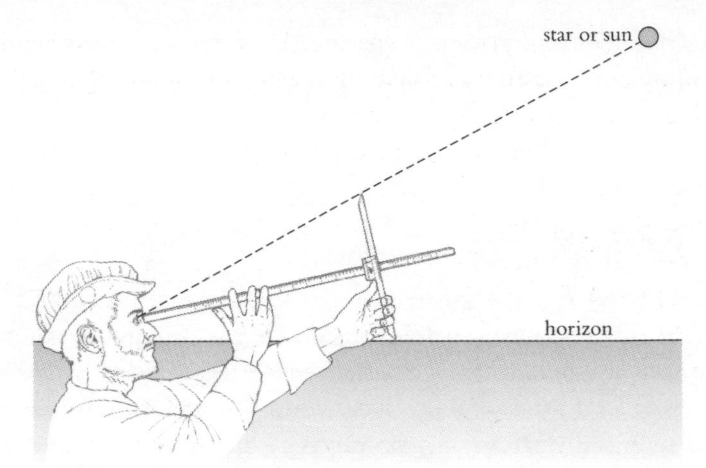

Figure 9.1

Although Henry was celebrated in Portugal as the country's first great sea explorer, he actually suffered the same problems that had faced Hanno. The best of his several voyages only reached the extreme bulge of Africa – around today's Dakar – due to adverse currents. Little did the Portuguese know that these exploits were almost totally insignificant when compared to what the Chinese had achieved just a few years earlier.

Zheng He or Zheng Ho? And Who Was He, or Ho?

In 1405, at the command of the Chinese Emperor Cheng Zu, his eunuch admiral Zheng He 'went down into the West'. This sounds strange, as we would say he went 'down south'. But Chinese maps showed the East (or Orient) and the direction of the rising Sun at the very top of the world, and the West at the bottom – so going 'down to the West' was normal to them.

Zheng He left China on the first of seven expeditions to discover and trade with new countries and provinces. The Chinese clearly had great confidence in their shipbuilding skills: for the journey they created a huge armada of some 300 ships. These included 62 vessels that were about 135m long and 40m wide, displacing 1,500 tons of water each. Inside they had 13 compartments separated by waterproof bulkheads; in some craft the first and last compartments could be flooded to keep the ships stable in heavy seas.

The whole fleet carried about 37,000 men and women through the South China Sea and into the Indian Ocean. Some ships carried just water, some livestock or rice and vegetables, and the entire fleet had everything a city on land would require.

The Chinese were well practised in using both the magnetic compass and their own sextant to judge latitude by the angle of the Sun or the Pole Star. But they also had some success in measuring longitude by recording directions and sailing speeds and times, using sand clocks that recorded 144 minutes, or one-tenth of each day. Studious use of these clocks gave them some idea of the time at their original port, so they could make fair estimates of their longitude. They also understood sea currents far better than other European sailors of the time.

They were not afraid to sail across open seas. In fact they found this safer than skirting coastlines in sight of land, as the greatest danger for their heavier ships was the risk of running aground. They were also adept at using their sails and huge rudders to maintain a course by steering in any direction – except straight into a wind. When confronted with a headwind, they simply dropped anchor and waited for it to subside.

These 'treasure fleets' carried vast amounts of gold, porcelain, jade and other precious items, and at every new landfall they made proclamations to the local people on behalf of their emperor, 'The Son of Heaven', bestowing gifts on tribal kings and rulers. They met little trouble: the sight of such a large fleet moored offshore usually subdued any thoughts of armed opposition, and once the gifts had been presented, friendly relations were pretty much guaranteed.

On returning, some ships went ahead, carrying high-ranking envoys from the new-found countries, to show them the wonders of Chinese civilisation, and to encourage trade and lasting friendship. They were then taken home on the next voyage. As we shall see, this was in marked contrast to the behaviour of European explorers.

The first treasure fleet seems to have sailed just beyond today's Singapore, into the Indian Ocean. But new voyages set sail every second year, and by the fourth, ports in India, Iran, Arabia and Egypt had been visited. The fifth voyage sailed down the eastern side of Africa and brought home – to the Emperor's delight – a giraffe.

The sixth voyage, in 1421, was probably the greatest of them all, composed of four separate units, each seeking out new lands. It's pretty certain that one of these units reached Australia; Joseph Needham has collected evidence that one even crossed the Pacific, visiting what we now know as Peru in South America. There is evidence that indigenous Peruvians from that time used jade, drums, musical instruments, games and dragon motifs that all bore a striking resemblance to similar Chinese items, lending further weight to this theory.

One section of the mighty sixth Chinese armada may even have sailed right around Greenland. On his visit there some 60 years later, a young Christopher Columbus claimed to have seen evidence of a Chinese presence in the country. They must have chosen the time of year for their visit very well, for thanks to relatively ice-free seas, they sailed across the top of Asia to return via the Bering Strait and Japan to home. This is perhaps more likely than one might think: similar trips were carried out by German ships to aid Japan in the Second World War, and all without satellite navigation to help them avoid ice-bound seas. There is also evidence that some Chinese actually settled up the Sacramento River, north of what is now San Francisco. Sure enough, once they had negotiated the path into San Francisco Bay, the prevailing winds and currents would have taken visitors in sailing ships naturally in that direction, along the Bay's north coast.

Gavin Menzies, a British ex-submarine skipper, has made several claims concerning early Chinese exploration, supported by his vast knowledge of the sea currents of the world. In one of his books he features a Chinese map that shows Africa, but on which the northwestern bulge is much smaller than it is in reality.

However, the strong prevailing currents that run up the coast of Africa and then west under the bulge would have given ships travelling with them a false impression of their speed, which could be why they mapped a smaller bulge. Indeed, Hanno (2,000 years earlier) – and the Portuguese – found it very difficult to travel the opposite way, and eventually turned back, to be swept home on those very currents.

The final voyage of the Chinese treasure fleet left in 1431 with 317 ships and 27,000 personnel. The following year a separate group of Chinese envoys travelled overland to talk trade options and possibilities in the commercial capital of the Mediterranean, Venice. On their return, however, the new Ming emperor, beginning the process of purposefully isolating China from the rest of the world, banned all future sea explorations and left the great ships to rot.

European Aggression

The exploits of the Chinese, largely because of the way they behaved, were far different in tone from those of the Portuguese explorers who followed Henry. In 1488 the explorer Bartholomew Diaz tried to sail right down the west coast of Africa. At each place they landed, Diaz's men often mutilated the indigenous people – cutting off men's hands and women's noses – for no other reason than to induce fear and subservience. After Diaz came Vasco da Gama, a similarly violent explorer, who in 1497 succeeded in rounding the Cape of Good Hope and crossing the Indian Ocean. Only when da Gama reached Calicut on the west coast of India did his attitude change. The far more diplomatic Chinese had already set up peaceful trade relations here, and da Gama now followed suit, becoming a European envoy for Calicut shortly before he died there in 1524. By then, however, the Europeans had made a far more monumental sailing achievement – and all because of a mathematical mistake.

When East Became West

By then, the whole of Europe – and especially the Church – still emphatically believed (even after 1,300 years) in Ptolemy's estimate of the size of the Earth, gleaned from the work of Poseidonius. Copies of Ptolemy's map of the known world (see plate section) were republished, and a Florentine by the name of **Toscanelli** (1397–1482) got hold of one. This map suggested that the Earth was at most 18,000 miles around, and possibly only 16,000. Toscanelli drew his own map (see Figure 9.2), and maintained that, because Asia was so very wide, its eastern shores might not be more than 3,000 miles west of Spain.

In 1481 Christopher Columbus was so inspired by this notion that he set off on a round of fundraising visits to the crowned heads of Europe. Having already visited Greenland, Columbus was an experienced sailor, but he now pleaded with Italian nobles and the English court for their financial

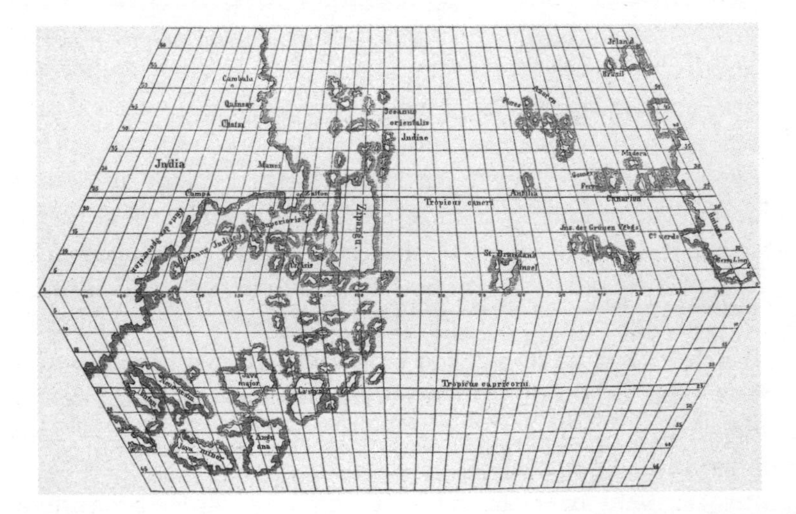

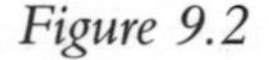
Figure 9.2

support to enable him to sail west and discover a new, quicker and safer route to the East, and thus be party to all the trade riches this would make possible. It was an attractive proposition, and most of the people he approached were interested enough to listen. After all, they had heard Marco Polo's stories of the fabulous Orient – perhaps Columbus's idea was worth the gamble?

Unfortunately, the Italians and the British, unimpressed by Toscanelli's map, turned Columbus down, as did the Portuguese, who had already put their money on sailing around Africa. The Spanish rulers Ferdinand and Isabella, however, flush from their 1491 success in taking back Granada and at last ridding Spain of Muslim occupation, agreed to take a gamble, and financed the trip. So with three small ships and 120 men, mostly from local prisons, Columbus sailed west across the ocean blue, in 1492.

En route, he noted that his magnetic compass seemed to dip, as though seeking a point below the sea, or even the Earth's surface. He told no one about this, however, in case they became afraid of mystical forces they did not understand. An explanation for this phenomenon would arrive about 60 years later, courtesy of English scientist William Gilbert.

With the crew growing fearful of the unknown and pleading to turn back, they at last sighted land. Columbus was overjoyed, certain he had arrived in the exotic Orient. But he was actually in the Bahamas, on the very eastern fringe of the Caribbean. In fact, he'd struck land at a point almost diametrically opposite the Philippines, and half a world away from where he thought he was. To reach Asia he would have had to travel another 12,000 miles – or four times as far as he'd already come, even assuming Central America hadn't been in the way. Had Columbus known from the start how large the Earth really was, he would never have suggested making the trip in the first place, and history would have been very different.

As it was, over the next 10 years he made three more trips across the Atlantic. Always hemmed in by the

Caribbean, however, he never realised that the vast Pacific Ocean even existed, or that it stretched halfway across the world.

Sadly, although he was a fine seaman, Columbus was a poor administrator: he died in poverty, still believing he had found Asia. But the lasting impact of his discovery of new land ensured his fame: King Ferdinand, at the last moment, decided to honour him with a stupendous funeral and a huge monumental tomb in Seville Cathedral. Columbus's body, however, lies in a grave in the Dominican Republic.

Go West Young Men

When news of Columbus's landfall in the New World arrived in Europe, the Portuguese insisted on their right to share any new discoveries with Spain. In 1494, ignoring all other countries, Pope Alexander VI drew up the Treaty of Tordesillas (which rhymes with 'silly ass') to divide the world between the two nations, simply by drawing on a very incomplete map a preposterous line of longitude 100 leagues west of the Azores. A league was about 3 miles, and the line turned out to be approximately 45 degrees west. But as they could not measure longitude, and had little idea just what had been or might be discovered anyway, the line was absolutely meaningless. As it turned out, all of the Caribbean and the as yet undiscovered North America went to Spain – along with the whole of South America, other than the bulge of Brazil (see Figure 9.3).

The Spanish and Portuguese did not even consider the Dutch and the British. So when news of this outrageous decision – and rumours of the area's possible vast wealth – filtered back to the rest of Europe, their seafarers were positively encouraged to become pirates and simply get out there and take their share.

Meanwhile, in 1497 an Italian father and son, John and Sebastian Cabot, chose to sail west from England, as simple maths told them that the further you are from the Equator,

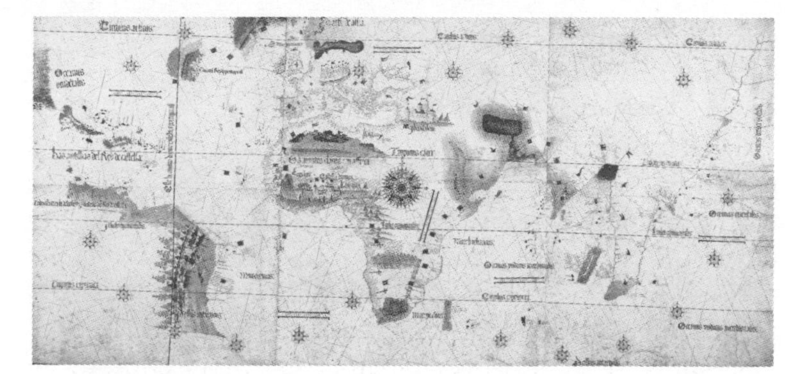

Figure 9.3

the shorter the distance around the spherical Earth, which surely offered a shorter route to Eastern Asia? They managed to reach eastern Canada and possibly land that is now the United States – but they knew this wasn't Asia. So they tried to find a way past by going further north, but to no avail. On Sebastian's subsequent voyage in 1509, however, he did find Hudson's Bay.

But the finest navigator and explorer of this era was Amerigo Vespucci of Florence. He sailed slightly south of the route taken by Columbus and, after navigating the tricky currents around the Equator, landed on the easternmost tip of South America in 1499. After that he sailed north-west up the coast towards the West Indies.

From an almanac he carried he knew when the Moon and Mars would be in alignment above Nuremberg, Europe. So on the same day, he calculated his longitude at 82.6 degrees west of that, placing him somewhere around what is now Santiago in the present-day Dominican Republic (see Figure 9.4).

Vespucci's second voyage took him to the same most easterly point on the coast of South America. But this time he turned south – hugging the coast on favourable currents – to today's Argentina, before eventually turning for home. Having heard of the Cabots' discoveries much further north, he clearly understood that this was not Asia but a vast new

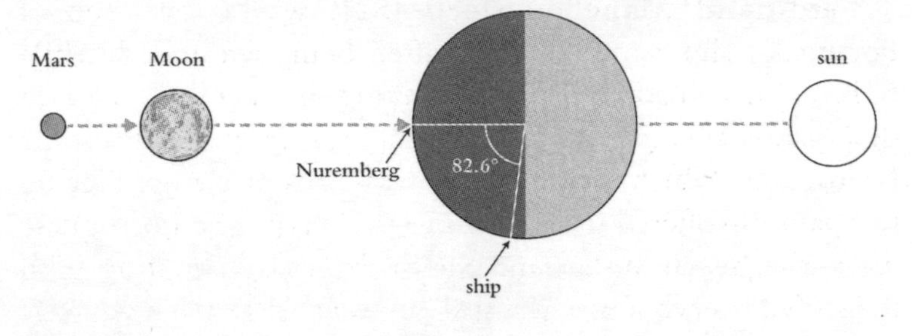

Figure 9.4

continent stretching an extraordinary distance from north to south. Today, quite rightly, the continent bears his Christian name, Amerigo.

He seems to have clearly understood the true size of the Earth, and prophesied that to the west of this New World there would be another huge ocean yet to be discovered, which would have to be crossed to reach the mystic East. If, as he suggested, going around this huge lump of land was going to be incredibly difficult, could someone possibly go across it?

Vasco Núñez de Balboa was a Spaniard so troubled by misfortune and debt that in 1500 he sailed for the New World and a fresh start. He failed as a planter in Haiti and, once again in debt, and to avoid criminal charges, he stowed away in a barrel on a ship, reaching the north coast of South America before moving on to Panama. Surely, he thought, if I could find gold my troubles would be over?

What he didn't realise was just how narrow Panama really is. The present-day Panama Canal, opened in 1914, is only 82km (51 miles) long and, surprisingly, runs south-east to north-west. Crossing from the north coast, Balboa became the first European to see a huge stretch of water that he called the South Sea. This was the new ocean that Vespucci had predicted, and it actually stretches over 10,000 miles to the west. Sadly Balboa did not find gold or clear his debts – or have much luck at all: he was eventually framed by enemies on trumped-up charges and executed.

Ferdinand Magellan (1480–1521) was a loyal son of Portugal, who was left lame after being wounded while fighting in North Africa. Receiving no pension, and in desperate straits, he made the mistake of trading with some Moroccans, which in those days was a major crime. Fleeing to Spain, he offered his services to the navy. The Portuguese were already sailing around South Africa and trading with Asia – but everyone else was still hoping to discover a westerly route. They now knew that the New World was an incredibly long land mass, but it couldn't reach from Pole to Pole, surely? If a way around it could be found, it would give Spain great advantages over its arch rivals. Magellan was given five ships and, setting sail on 10 August 1519, he followed the route used by Vespucci across the Atlantic to South America, before sailing further south on the natural ocean currents until at last he reached what we now know as the Straits of Magellan.

Opposing currents at the entrances to the straits made passage through them incredibly rough and violent, so that when Magellan eventually emerged into a new and calm sea, he christened it the 'peaceful ocean' – Oceanus Pacificus. In fact, it is no more peaceful that the Atlantic, but it is much bigger, as they were to discover. For 98 long and mostly calm days, they sailed north and west on the natural anti-clockwise currents, without sighting any land. When they tried to estimate the depth of the ocean, they did not have lines long enough and it appeared bottomless.

The stillness and vastness of the Pacific, coupled with growing starvation and dehydration, reduced the crew to despair – and no wonder. If, from a position over Christmas Island, you look at a globe (or indeed the Earth from space), apart from the east side of Australia and the west coast of North America, all you see is sea or small islands. The Pacific Ocean, till then unknown in the west, covers almost half the entire globe (see plate section).

Just in time, and now 21 months into their voyage, Magellan and his crew sighted Guam, where they found food and water. Now they could island-hop west, but in what is now the Philippines they got into a squabble with

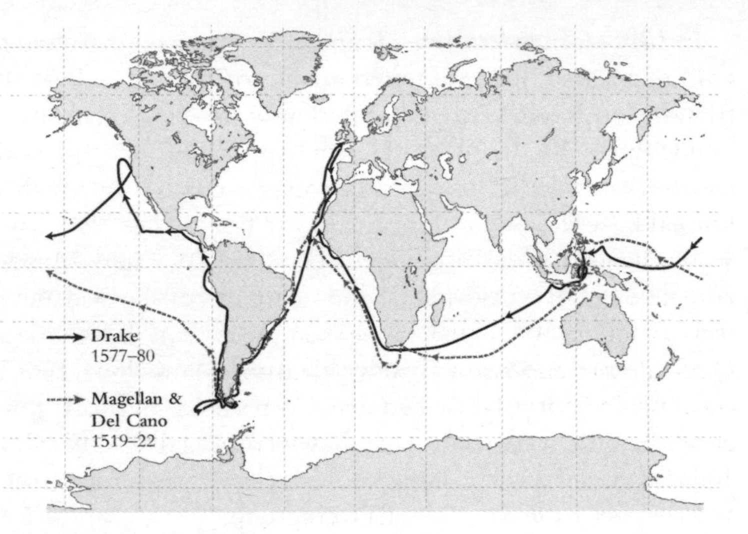

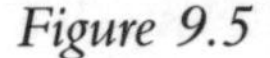
Figure 9.5

natives and Magellan was killed. One of his officers, Cano, took control and brought their last remaining ship, the *Victoria*, back across the Indian Ocean, round the tip of South Africa and home on favourable currents, reaching Spain on 8 September 1522. They had sailed right around the world (see Figure 9.5).

At long last the true size of the Earth was known. Poseidonius and Ptolemy, by estimating heavenly measurements, had been chronically wrong, while Eratosthenes' simple but accurate geometry had been proved right. Now was the time to put the whole picture into perspective.

Greed, Wealth and Destruction

Unfortunately, not a lot of good came from the discovery of the New World. From the start, Columbus had set in motion a series of events that would lead to the eventual destruction of the Mayan and Peruvian civilisations. His stories of beautiful lush islands peopled by very primitive tribes, and tales of gold and unknown riches for the taking, encouraged a flood of adventurers.

In 1519 Hernando Cortes founded Veracruz on the western Gulf of Mexico. He progressed inland to Tlaxcala, where the friendly locals told him of the fabulously wealthy Aztec city of Tenochtitlan to the West. He then persuaded the Tlaxcala people that together they might conquer the Aztecs for their mutual benefit.

The legend goes that when Tenochtitlan's ruler, Montezuma, saw Cortes riding a horse – an animal he had never seen before – he believed he had been sent by the god Quetzalcoatl and immediately fell into submission. This is certainly a story fabricated by Cortes and his followers, because on their first approach they were actually repulsed on 30 June 1520 – what became known as the 'sorrowful night' – and they retreated to Tlaxcala to regroup.

Tenochtitlan was a series of islands built on a massive group of lakes. So Cortes designed boats that he and his allies hauled back to the city and assembled on the shores of the lakes for a second attack. Now, with the aid of huge numbers of Tlaxcala warriors, they overcame Montezuma. One part of the legend that might be true is that the highly superstitious Aztecs may well have accredited their loss to the 'will of the gods' and put up no further resistance. In just a few years, Spanish influence swept away much of the traditional way of life, and when the locals began to fall victim to European diseases like common flu, for which they had no immunity, the collapse became total.

In 1530, another Spaniard, Francisco Pizarro, who had travelled with Balboa, arrived in Peru. The Inca people, in the middle of their own civil war, were totally unprepared and unable to put up much of a fight. Pizarro moved inland to Cajamarca, where he captured the Inca king Atahualpa, who pleaded that if he was released, as a ransom, he would fill a huge room with gold. Pizarro allowed the gold to be brought until it did indeed fill a room, after which he accepted the ransom but killed Atahualpa anyway. He had conquered Peru with just 180 followers.

In the nineteenth century, it was said that so much gold had come from the New World in the early sixteenth century that just about every gold ring or piece of jewellery in Europe contained some gold from Central America. However, an even greater source of wealth for Europe came from the silver discovered and easily mined in Central America up until the seventeenth century, when the mines became worked out.

Tragically, the age of these civilised, highly complex yet successful societies, previously unknown in Europe, had ended. Much of its culture, especially its incredible mathematics, was all but lost, although today much effort is going into recreating it for its cultural heritage.

Mapping the Known World

As more and more information filtered back from explorers about the size of the planet and the new lands that were being discovered, the need for someone to make new and ever more accurate maps became essential.

Gerardus Mercator (1512–1594) was born in the Netherlands, where his childhood was filled with stories of adventurers discovering new lands. Inspired by these tales, on graduating from the University of Louvain, he set up a map-making business.

He had considerable mathematical knowledge, and produced original map-making tools – even giving copies of them to Emperor Charles V. He made an adjustable 'proportional' three-legged compass: the gap between one pair of legs could be adjusted so it was always, say, half or a quarter larger than the gap between the other legs. He also refined triangulation techniques for land measurement, and introduced *centering*, a technique in which places were located with more than two – and sometimes six – other reference points.

In 1538 he made his first map of the world, which he painstakingly improved over 30 years, producing a much better version in 1569 (see plate section). Nevertheless, his understanding of Eastern Asia and the Americas was still sketchy, so much so that even the Atlantic coast looked far from accurate. But his maps were superior to other efforts of

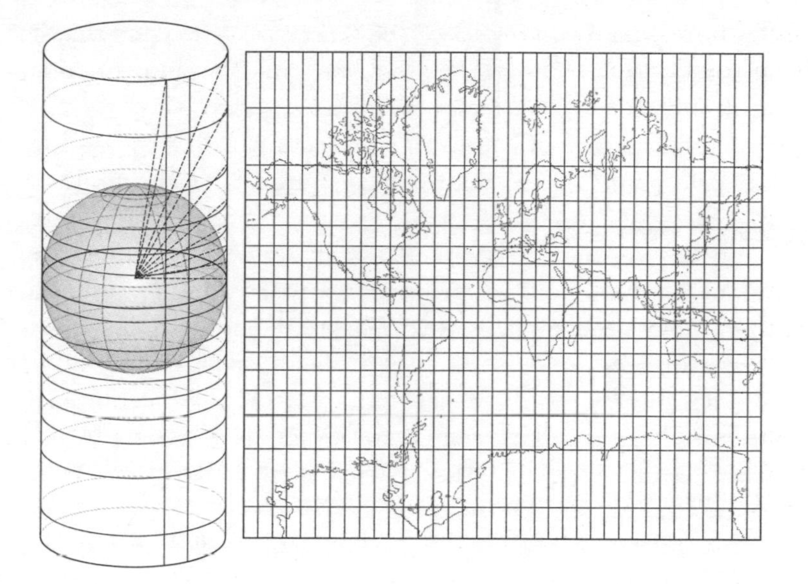

Figure 9.6

the time largely because of his own innovation: the Mercator projection.

Just as artists had done to capture perspective, Mercator used a grid system to help establish correct locations. But instead of making the eye the focal point, he used the very centre of the Earth. Imagine a cylinder wrapped around a globe. Rays from the centre of the Earth passing through any point on the surface of the globe would then arrive at a point on the cylinder, which could finally be unrolled into a flat map (see Figure 9.6). For points along the globe's 'Equator', a straight line is projected around the cylinder. Other lines of latitude are formed in the same way, but on the cylinder the gap between latitude lines gets progressively bigger, and the further away from the Equator you go, the greater the distortion.

So while the North Pole is a single point on the globe, on a cylinder projection it ends up as the entire top edge of the map, and just as long as the Equator. So Mercator's map ended at about 70 degrees north (around the top edge of

Russia) or south (the edge of the Antarctic) – beyond that it was far too inaccurate. Which was fine, because no one ventured closer to the Poles than that anyway.

The advantage of Mercator's projection is that by joining any two points on the map with a straight line, the angle of that line gives you the compass bearing you need to complete a journey on the surface of the real Earth. These lines are called *rhumb lines*, and his projection is still used today, though there have been many refinements and hundreds of newer projections.

The distortion of land sizes on a Mercator projection can be quite startling. Greenland, for example, looks about the same size as Africa. In fact, Africa is 14 times bigger than Greenland, which is only as big as Algeria.

I wonder if, in making his maps, Mercator was ever aware of this amazing topological puzzle:

> *Take two maps. Screw one up into a ball and place it on top of the other one. What are the chances that a point on both maps will be exactly in line vertically?*

The answer is simple: there always must be one point that is vertically aligned on both maps. It doesn't even matter if the two maps have different scales. Take a map that includes the place where you happen to be, screw it up and throw it onto the floor. The ball and the point on the floor must be vertically in line (see Figure 9.7).

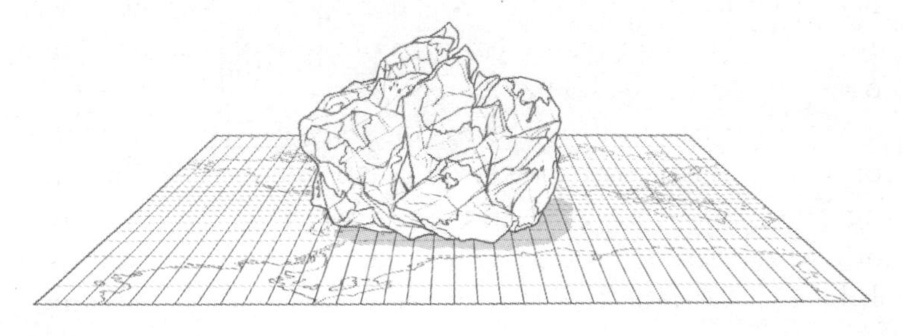

Figure 9.7

But the Earth Is Round

Mercator's major aim was always to create a spherical globe, on which he could accurately copy the Earth without distortion. He made his first globe (which had a diameter of 370mm [14.5in]) in around 1536–1537, then improved on it in 1541 with a model that had a diameter of 420mm (16.5in). But making his globes turned out to be very complex and costly. Firstly, creating a light, round ball out of wood isn't easy. Then it needs a base and an axis to revolve around, all before you can even start the actual mapping.

But he doubled his market by producing celestial globes of the heavens in the same size, and offering them as a pair. The 1541 paired globes were a huge success, but then a sudden explosion of religious conflict all but wiped out his entire business.

Martin Luther, a German Catholic priest, had defied the Pope in 1517 and refused to pay indulgencies or taxes to Rome, for which he was excommunicated in 1520. This marked the birth of Protestantism: from that time onwards conflict between the Catholic and Protestant faiths was to rumble on for many years.

In 1544 the Catholic Queen Maria of Hungary decided to purge the country of all Lutherans. Mercator was a Protestant living peacefully in a Catholic area that included the University of Louvain, which he attended. By order of the Queen a handful of local people were summoned by the Inquisition on charges of heresy and Mercator was among them – all because his map of the Holy Land seemed to vary from biblical details.

Mercator had returned to his birthplace, Rupelmonde, to sort out the affairs of a recently deceased uncle. Declared a fugitive, he was arrested and imprisoned in Rupelmonde's forbidding castle, around whose walls he had played as a child. He was soon told that of the others arrested, two men were hanged and one beheaded, while two women were buried alive. This displays the Inquisition's incredibly cruel logic: men had to be seen to be destroyed, while women were just put out of sight. For Mercator, things were looking grim.

With his wife and friends pleading his innocence, after seven long months he was suddenly released. Although this was a great relief, he was now in debt: his business had closed and, besides supporting his wife and six young children, he also owed for his upkeep in prison. As soon as he could, he moved his family and business to the small German Protestant town of Duisburg, between Antwerp and Frankfurt (which was already – as it is today – the site of the world's largest annual book fair). Here he forged his second career, which was more successful than the first (he was lucky enough to live to the age of 82).

As an introduction to his new book of maps in 1578, he copied Ptolemy's original map – including its errors: the vagueness of Africa and Asia, and a Mediterranean 200 miles too long – to illustrate how far he, and mapmaking, had progressed.

Mercator spent his last few years producing a book of accurate maps of various parts of Europe, which appeared a year after his death. His accurate triangulation and land surveying revealed remarkable discrepancies in land sizes. Since the Romans had departed 1,200 years earlier, no one had measured any distances accurately, and most landowners estimated their land to be much bigger than it actually was.

The cover of Mercator's book displayed Atlas the Greek Titan, holding the world on his shoulders. Every map book since has been called an atlas. From that point on, modern mapmaking replaced forever the earlier work of the Greeks, and navigation at sea became much more reliable.

William Gilbert (1544–1603), an English physician and physicist, helped explorers by clearing up any doubt over what makes a compass needle behave the way it does: the Earth itself is a giant bar magnet. Actually, it's more accurate to say that the Earth is wrapped around a giant bar magnet that points roughly, but not quite, north to south, as magnetism is stronger near the poles and weaker around the Equator (see Figure 9.8).

Figure 9.8

Gilbert managed to magnetise a cannonball, giving it north and south magnetic poles. He then hung a magnetic needle horizontally on a thread, so that any tug from a magnetic source would indicate the direction of pull. When it was dangled near the 'equator' of the ball, the needle pointed towards the north pole, but definitely dipped as though it was pointing to somewhere inside the ball. As Gilbert moved the needle north, the dip increased until the needle was over the pole, when it pointed straight down, aligning itself with the ball as though it too were a bar magnet. Gilbert realised that what we call the north pole of a magnet or compass needle is really the *north-seeking* pole – in actual fact, its south pole. He also discovered that the magnetic North Pole is not actually at the North Pole. As I write these words, the magnetic North Pole is roughly 4 degrees south of the pole itself, and about 160 degrees west. But each year it moves between 34 and 37 miles, heading at present from northern Canada towards northern Russia. Over time, however, that direction seems to change.

Gilbert also experimented with amber or tree resin. Thales had observed that when wool spinners held rubbed amber close to wool, the individual threads sprung apart, allowing them to select one easily. Gilbert found that amber's attraction and repulsion were different from the magnetism of iron, because amber attracts light objects of many materials, whereas magnetic iron attracts only iron. He also discovered that some rock crystals and gemstones have the same property as amber.

Gilbert was court physician to Queen Elizabeth I and he amused her with tricks using magnets and amber. As the Greek for amber is *electrum*, he called his tricks 'electrics' and the term seems to have stuck. In his book *De Magnete* (*Of Magnets*), he showed that by careful experimentation he could prove the facts about magnetism, and refute the wild 'magical' claims of others. This scientific approach won the approval of Galileo, who suggested that Gilbert was the founder of scientific discovery by experimentation – although that accolade could just as well have gone to Roger Bacon.

Gilbert always believed that the Sun and not the Earth was at the centre of things, even suggesting that heavenly bodies might move through the heavens under some kind of magnetic attraction. He would also have known the greatest adventurer of the age – Francis Drake.

With Mercator's maps and Gilbert's magnetic compass explanations, **Francis Drake** (1545–1594) had gained fame by plundering Spanish galleons between Europe and the West Indies. On his return, he persuaded a friend by the name of Doughty to seek permission and funding from Queen Elizabeth for a voyage around the world, during which he would hopefully claim new lands for England.

They set sail in December 1577, but after experiencing difficulty crossing the Equator, Doughty lost his nerve and pleaded each day that they turn back. After a few weeks they sighted land (probably the coast of what is now Brazil), and that evening Drake threw a celebration party for Doughty, although he also informed him that he was guilty of treason and mutiny, for which he had three options. He could be held prisoner until he returned to England to be hanged, or he could be cast ashore at the next landfall. Doughty chose the third option, and next morning they went ashore, where Drake dropped Doughty to his knees, explained that at sea there can be only one captain and commander, drew his sword and beheaded him.

Three of Drake's four ships were lost rounding the Southern Cape, but once in the Pacific, in his remaining ship the *Golden Hind*, Drake sailed north and plundered Spanish ships off Peru. He then landed in today's California, which he claimed for the Queen, calling it New Albion. The commemorative plaque made by Drake's men was found at the site many years later; Drake also recorded that the natives were so that friendly he and his crew found it hard to leave.

Drake and his crew arrived home in September 1580, and on the deck of his treasure-laden ship a delighted queen dubbed him Sir Francis. Undoubtedly, having Mercator's maps, and being able to navigate using rhumb lines, had made the journey far easier than Magellan's 58 years earlier.

It would be a long time until 1761, when 68-year-old John Harrison finally solved the longitude problem for good. He had produced five amazingly accurate clocks, all of which are still working today at the Royal Observatory in Greenwich. 'H4' was a silver watch 13cm in diameter (see plate section) that, in trials on a voyage to Kingston, Jamaica, lost only five seconds, placing Kingston's longitude to within 1 nautical mile.

At last, by knowing the time at your port of origin and the local time according to the Sun or the stars, you could accurately work out longitude. This enabled Captain James Cooke to map much of the vast Pacific by sailing in any direction he chose with confidence.

When the crew of the *Bounty* mutinied in 1789, Captain Bligh was allowed to keep a similar watch, and he successfully navigated a 3,618-mile journey in a longboat with 18 men. Meanwhile, the mutineers, under the command of Fletcher Christian, headed for Pitcairn Island. With no watch to help them, they resorted to the old methods, sailing south until they arrived at the correct latitude and then 'guessing' whether to turn left or right. Fortunately for them they guessed correctly; they found Pitcairn Island, and their descendants still live there to this day.

CHAPTER 10

The Huge Awakening and a New Age of Learning

After the seemingly interminable Dark Ages, the Renaissance dawned purely because the human spirit, and people's desire to learn, cannot be suppressed forever. Yet amid this great new clamour for knowledge, very little new mathematics emerged during the early years. Much of what happened was more concerned with rediscovering and exploring the work carried out by the Greeks, Hindus and Arabs hundreds of years earlier.

The Mathematical Art That Drove the Renaissance

Beginning in Italy – or more precisely Florence – the Renaissance centred round art and architecture, but it was all driven by early mathematical ideas. A major influence was the work of Alhazen (see Chapter 8), news of which arrived in Europe along with the new Arabic number system. His conception of sight as a cone originating in the eye of the beholder helped explain how we see things in real life. As a person approaches, the cone that frames them gets ever wider. Or as the comedian Chic Murray used to say, 'I knew this fella was coming towards me. He was getting bigger all the time.'

Figure 10.1

To understand how perspective changed Renaissance art, check out Giotto's 'non-perspective' painting *The Arena Chapel at Padua*, a detail from his *Last Judgement*, 1306 (see Figure 10.1). All the figures in the picture are the same size, and the three angels seem to be inside the model of the church, rather than behind it and further away.

Brunelleschi's Cupola

The man widely regarded as the father of the Renaissance is **Filippo Brunelleschi** (1377–1446), who suggested the rules for drawing an accurate representation of a square-tiled floor. In mathematics we know that parallel lines never meet, but lines retreating from a viewer appear to get closer, eventually meeting at a point in the distance. Brunelleschi gave his drawing a horizon where all the lines meet at point B, which he called his 'first vanishing point' (see Figure 10.2).

The transverse left-to-right lines of his floor are also parallel and equidistant from one another, though they seem to get closer the further away they are. But in drawing such a floor, how does the artist decide the rate at which the lines get closer? Brunelleschi needed another vanishing point, still on the horizon, but well off to one side by some way. Our drawing features two such points. This point was not where the transverse lines met, but where all the lines running diagonally through the corners of the square tiles met.

Sure enough, with the diagonals lightly sketched in, he now had a clear guide as to where to put his parallel transverse

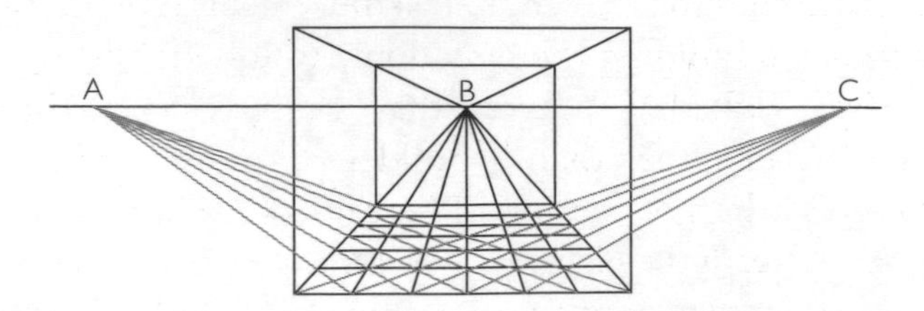

Figure 10.2

lines so they got steadily closer towards the horizon. Now he could create a true representation of a square-tiled floor, just as we see it in real life. The next step was to repeat this three more times, with a left, right and top elevation, creating a box he could use as a guide to perspective, to help him produce a very realistic picture.

From this time onwards, artists began to first sketch a perspective plan of the floors and walls of buildings, onto which they would later introduce the final picture using a main vanishing point. It was then easier to adjust the sizes of human figures depending on their perspective depth in the picture. Piero della Francesca's *Flagellation* illustrates this perfectly (see plate section).

Brunelleschi used a box-shaped pinhole camera to examine the perspective of the images he was trying to capture. This did not record an image, as a modern camera does, but the upside down image remained in place on a paper screen while the artist analysed and copied it. Then, when attempting to capture the true perspective of the Baptistery of St John, which sits in front of the magnificent cathedral in Florence, Brunelleschi devised a brilliantly simple mathematical tool that worked in the same way.

He made a first attempt at a drawing in perspective, then punched a small hole in the centre of it. He then held the back of the drawing right up to his face and looked at the building itself through the pinhole. Then he brought a hand mirror up at arm's length, to reflect the drawing back to his eye. By comparing drawing and building, he could see the errors in perspective he had made. After several attempts he produced a drawing of the building in perfect perspective.

Brunelleschi's most celebrated achievement, though, was in designing and building the new dome to top Florence's as yet unfinished cathedral. Given that he was an artist and mathematician rather than a builder or an architect, it was amazing that he got the job. The cathedral's huge nave and roof were already in place, as was a roofless octagonal tower with 6m-thick walls, awaiting a dome of some kind. At 45m wide, this tower was 2m wider than the Pantheon, making the Florence dome the widest ever attempted. It wasn't going to be an easy job, because there weren't enough wooden beams in the entire province to create a suitable support structure. So the dome would have to be self-supporting throughout the building process. But what would be its actual shape?

The tower already extended above the rest of the building with an 8m drum of masonry that featured eight huge round windows, and walkways around the inside at the bottom and the top. It seems that Brunelleschi first placed beams from the top edge of this drum that reached across from the centre line of alternate windows to support a floor he could work on. He chose the centre of these beams as the compass centre of the curve of his huge arched dome, making it uniquely tall and far more elegant than a simple semicircular dome (see Figure 10.3).

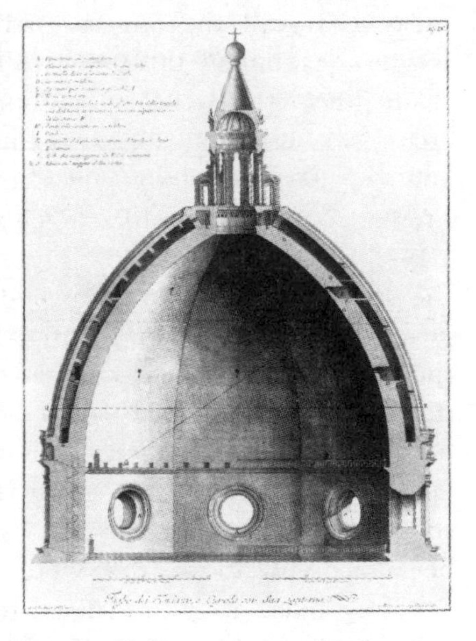

Figure 10.3

His dome was to be twin-skinned; the inner dome would be decorated, while the outer dome would bear the tiles and withstand the elements. But both would be supported by eight 4m-deep ribs of stone that followed his chosen line – curving up and inwards. Five horizontal chain and stone hoops were hidden within the two domes at regular intervals, to stop it collapsing inwards, and to prevent the weight at the top making it bulge outwards. The chains would support both the huge inner and thinner outer domes built of nothing but bricks, many layers thick. The gap between the two domes was there not only to reduce the weight but also to provide access to the lantern Brunelleschi planned for the very top. Visitors in their thousands still climb through this narrow gap to the lantern today – a total of 463 steps.

To strengthen the brickwork, every fifth brick was placed upright, creating a strong herringbone shape. Just a few layers were laid each week so the mortar could set, making progress very slow, until all the bricks were finally in place – all four million of them.

The magnificent dome took 16 years to complete, but even today it is still the largest masonry dome in the world, weighing about 29,000 tons. With no formal experience, Brunelleschi the artist, by applying geometry, had become architect, master mason and finally a supreme structural engineer. He even designed theatrical shows in which, amid fireworks, people flew like angels under his fabulous dome.

Although Brunelleschi designed the lantern that topped the structure, it wasn't finished until 25 years after his death. The problem of getting the materials into placc at such a great height was solved by **Mariano di Jacopo detto il Taccola** (1382–1453) – 'The Jackdaw' – an Italian polymath, artist and engineer of the early Renaissance. Jackdaws are known for their habit of stealing items that shine or glisten, so maybe Taccola got his nickname because he used ideas that weren't necessarily his own...

Using oxen at ground level to move capstans and pull ropes that acted on several different cranes, Taccola lifted and lowered pieces of the lantern carefully into place. He had read all he could on engineering and construction, including the works of Heron and Vitruvius, and wrote two books, *On Engines* and *On Machines*. His drawings demonstrate that he had a huge range of new ideas, though not necessarily for things that would work in practice.

Taccola also attempted a perpetual motion machine (see Figure 10.4). From the drawing, it's clear it couldn't revolve clockwise, as the top bent arm would have to get to just past 3 o'clock before it would straighten. Going anticlockwise, the top arm would bend and perhaps cause a shock movement downwards, but the right-hand arms are further from the centre of the machine, and this added weight and leverage would stop any movement.

Taccola additionally had an idea for a sort of automatic ferry: a boat that moved downstream propelled by the natural

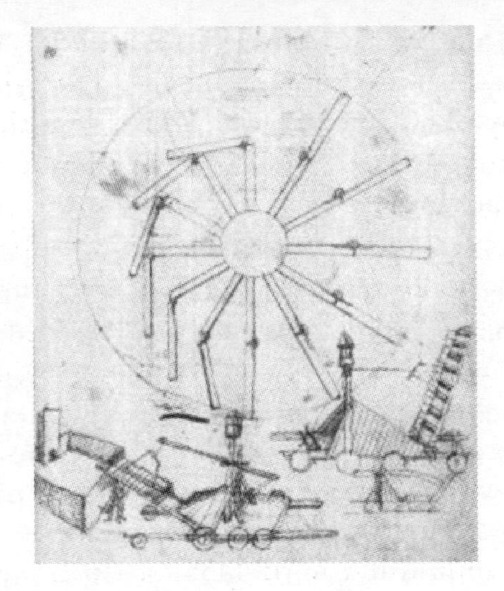

Figure 10.4

current of the water. For the boat to travel in the opposite direction two large paddle wheels on a long axle, and connected to a rope tethered upstream, would be dropped into the water on each side of the boat. Pushed by the current, they would turn and slowly wind the rope in and pull the boat back upstream. Once again it seems like a good idea, but the push on the paddles would not be enough to move the weight of the boat unless they were far bigger than the boat itself. But Taccola's many ideas clearly influenced others – possibly even Leonardo da Vinci...

Oriental Influence

In 1432 a diplomatic party from China arrived in Italy, visiting Venice and Rome. All who the Chinese met were left agog at the many concepts and ideas still unique to China: among them canal locks, mechanical clocks, shipbuilding and paper making. But very little actually came of the meetings, and the envoys returned to China, which very soon afterwards closed its doors to the West for about 400 years.

Some people had taken note of the Chinese achievements, however. Paper production in particular grew rapidly, creating a great innovation in 1439 when a German goldsmith, **Johann Gutenberg** (1398–1468), developed a printing press that used moveable type. The type was made from an alloy of 60 per cent lead, 15 per cent tin and 25 per cent antimony, and at quite low temperatures individual pieces of type could be quickly moulded by hand. Once cooled, however, they were hard enough to be used many times. The poisonous nature of the lead and antimony wasn't even considered and a moulder, while perhaps moulding 3000 pieces in a day, definitely didn't go in for regular hand washing.

Gutenberg (which was Johann's mother's name; his father's name was Ganzfleisch, or 'goose meat') was often hampered by financial problems. His invention did not start to make an impression until 1455, when he produced his first Bible printed in two columns with 42 lines per page. As the Pope was advised, it 'could be read even without spectacles'. Printing houses began to spring up all over Europe, especially in Italy. In England, **William Caxton** (1422–1492) opened up his printing press in Westminster in 1476. By the end of the fifteenth century possibly eight million books had been published, but in the sixteenth century the number rose to more than 200 million and a learning explosion occurred across all subjects, including mathematics.

The Perspective Explosion

Following Brunelleschi's lead, artists now realised the importance of perspective. **Paolo Uccello** (1397–1475) demonstrated this in his 1430s vase, a *Study of a Chalice* (see plate section). This three-dimensional network of lines is astonishingly beautiful, detailed and accurate. Amazingly, it looks remarkably like a computer-generated image we might see today, some 600 years later.

Leon Battista Alberti (1404–1472) joined the Papal court in 1432, the same year that the diplomatic party from China arrived. But Alberti was most influenced by earlier Roman architecture, earning him the title 'finest architect since Vitruvius'. He designed or added to several new buildings, of which Santa Maria Novella in Florence is perhaps the most striking (see plate section).

Built between 1448 and 1470, the lower facade was already constructed, but Alberti added the upper sections and decorated the whole in a flurry of mathematically inspired ideas. To solve the problem of bridging the different levels of the high central nave and the much lower side aisles, he added two large scrolls to the facade. This purely geometric idea was to become a standard feature of churches in the later Renaissance, Baroque and Classical periods.

For Alberti, a picture was 'a transparent window through which we look out into a section of the visible world'. To make that work, however, the perspective had to be right. Alberti may have been the first person to use a grid of black threads to produce perspective, although a similar system had been used by astronomers to map the heavens.

Alberti found a disciple in **Luca Pacioli** (1445–1517), whose knowledge of mathematics influenced a huge number of people. Luca was first employed by a rich Venetian merchant who taught him the tricks of his trade – and Arabic maths. Realising that this knowledge was far too important to leave solely to merchants, Pacioli began to travel, working as an itinerant maths teacher for universities in Perugia, Naples and Rome, where Alberti persuaded him to become a friar.

Pacioli borrowed freely from Alberti and others, but his most important legacy was probably to pass on his business skills: he introduced what is still the standard accounting method to this day – double-entry bookkeeping. The system is simple and effective: in essence you have an income page and an expenditure page, but every amount you earn and every expense are entered on both pages as a plus on one side and a minus on the other. If everything is correct, the books balance.

Around 1494, Pacioli wrote a book, *The Collected Knowledge of Arithmetic, Geometry, Proportion and Proportionality*, which among its contents included exponents. It became hugely popular, overshadowing even Nicolas Chuquet's work (more on him later). But perhaps the greatest event in Pacioli's life was meeting Leonardo da Vinci.

An illegitimate child, **Leonardo da Vinci** (1452–1519) was raised by his father. Despite having no classical education, he grew up to become a truly unique individual who left an indelible mark as a – or perhaps the – Renaissance man. Quite an achievement for someone who could not read Greek, and who learnt neither Latin nor mathematics until later in his life, during his time working with Pacioli.

Leonardo wrote in his local language and always backwards, from right to left, possibly because he was left-handed: it was as natural for him as it was for right-handed people to write from left to right. Some feel he meant his writing to act as a code, but very few people actually read his work in his lifetime, so in effect it was pretty secret anyway.

Personally, I feel that he may have originally written backwards just to annoy his teachers and to say, 'I am an individual – I am unique.' In essence, although we teach children in groups of up to 30 and expect them to work together with their peers, surely the object of all education is to discover, reveal and develop each person's individuality: it is our uniqueness that should be celebrated, whether in the world of the arts or the sciences.

Leonardo was certainly a one-off, and his dedication to learning how to do everything he attempted exceptionally well was the mark of the man. He went to great lengths to capture the human qualities of the figures he painted, and before settling on their outward appearance always thought carefully about the inner structures of their muscles, bones and sinews. He made a detailed study of every part of the human body, and was known to have dissected at least 30 cadavers, something frowned upon by the Church – although Leonardo only revealed this in his old age, when his name was well established. In fact, he was a strict vegetarian, and did not condone the killing of any living thing.

Ironically, however, a major source of income in his middle life came from designing military arms that could be used to kill thousands. He often mesmerised rich and powerful men with his ideas, then took their commissions and their money with graceful thanks. In fact, many of his inventions would not have worked with the technology of the day, and not one of his military ideas was actually produced in his lifetime.

Leonardo sketched designs for underwater breathing apparatus, which used pumps to get the air down to the diver, and even inflated shoes to enable someone to walk on water. But these ideas would not become reality for another 500 years. In a more practical vein, he designed machines to beat gold, twist ropes and dig canals. One design that did work was an early hand-operated lift, installed in Milan Cathedral.

Many of his ideas were for elaborate water-handling devices, and were installed in stately homes or even assisted city water systems. His fascination for the mathematical laws that govern movement and levers is evident in his designs for clock mechanisms, steering gears and a whole range of architectural ideas, including double staircases that wrapped around each other. As he matured his overriding love became mathematics – and especially geometry, which he regarded as essential to becoming an accomplished artist, firmly stating: 'perspective is the rudder and guide post of painting'. In fact, mathematical concepts often made him wild with excitement. When he discovered Hippocrates' lunes, for example, he became obsessed with exploring them again and again.

Aged 30, and brimming with confidence, Leonardo joined the court in Milan of Ludovico Sforza (known as 'Il Moro'), where he served for 17 years as painter, engineer and musician, even inventing a lyre shaped like a horse's skull. It was also during this time that he drew *Vitruvian Man* (see Figure 10.5), perhaps his most famous work after the *Mona Lisa*.

To my mind, *Vitruvian Man* captures the essence of education: 'all education is theft!' This concept should not be at all shocking: no genius ever wanted to take his discoveries to the grave, but left them to be picked up and carried forward by those who came after him. For me, that's the overriding message of this book – all of human endeavour is a perpetual relay race, in which each new generation is ready to pick up the baton and carry it that little bit further.

This is what universities are really about. It is their job to ensure that the students within their walls 'steal from the best', by exposing them to the works of the great and not so great who preceded them, and by helping them to build on that experience by creating new and original ideas.

In naming his picture *Vitruvian Man*, da Vinci gave honest credit to its originator, who had come up with the idea some 1,500 years earlier. But because he had a better understanding of the proportions of the human body, he added a great deal more. To begin with, he placed his figure in an 8 x 8 unit square, rather than the 10 x 10 square Vitruvius had used.

From ancient times, people had used their own body dimensions to measure the world around them. Sadly this practice has all but died out today – even in schools. A person's height and the reach of their outstretched arms is roughly the same. The Roman's called it a *stature* – the word 'statue' comes from the same root. Try measuring your stature. Standing in bare feet, mark your height on a wall; then, with one middle finger on the floor, see if you can reach the height mark with your other middle finger. You won't be too far out.

For the Romans (and other Italians), a stature was 6ft – as a *fathom* it was also used to measure the depth of water. To measure 30ft of rope, for example, you need five statures – and using outstretched arms the task can be completed in less than 10 seconds.

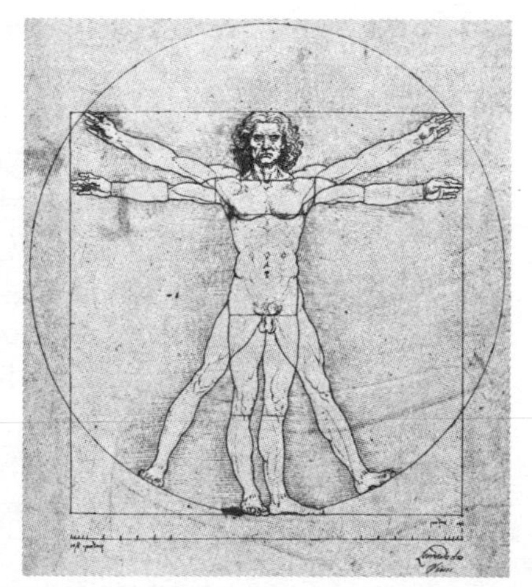

Figure 10.5

A yard was half a stature, and the distance from an adult's nose to the ends of their outstretched fingers. Bring your middle finger to your nose and you'll see that your elbow is exactly halfway along this line, dividing the yard into 2 cubits – a measurement the Egyptians used widely. Half of a cubit is a span – the length of the outstretched hand from the thumb to the end of the little finger. It's about 9in, and is the measurement that the organ keyboard is designed around.

In Da Vinci's drawing of Vitruvian Man, the square is 8 spans wide and 8 spans high. Ask yourself, what point is halfway up the human body? Most people will say the waist or the navel (the belly button). But is that so? Halfway up da Vinci's drawing – or four spans – coincides with the hip bone or pubic bone: the point at which we naturally bend forwards. The navel is a further span higher and denotes the centre of a circle with a radius of 5 spans. Sure enough, when the 8-span tall man raises one arm, he becomes a cubit or 2 spans taller and reaches 10 spans across the circle. Note that da Vinci added two pairs of legs and arms to show that man is dynamic - and long before moving pictures were ever contemplated.

It's often said that da Vinci observed the Golden Ratio in human body proportions. It is true that if you measure Vitruvian Man's outstretched feet from mid-foot to mid-foot, and multiply by 1.618, you get the distance from the middle of his foot to the top of his head. It links to the regular pentagon, but we don't know whether da Vinci intended this.

However, the actual human body does not quite live up to the idea that limbs grow in golden proportions. I have always known that the distance from my elbow to the end of my fingers is 18in. But for the Golden Ratio to be involved, the length of my forearm needs to be 11.124in, and the distance from wrist to fingertips needs to be 6.875in. Sadly, the distances are 10.5 and 7.5in.

Leonardo and Pacioli met while both were members of the court of Il Moro. It may even have been Leonardo who suggested to Il Moro that he hire Pacioli: Leonardo was not particularly gifted in arithmetic, and he was eager to learn all he could. They became close friends, collaborating on a book about geometric shapes, in which Pacioli made models of Archimedean, Platonic and semi-Platonic shapes, as both solids and net constructions. He also tried fitting the Platonic solids inside one another and exploring their similarities. From there it was only a small step to stellating them by adding pyramids of three, four or five sides to each face. How fortunate, then, to have Leonardo on hand. He made quite modest drawings of them, adding delicate shading to give the impression of three dimensions.

Pacioli also worked on the formal production of letters of the alphabet for printing, using quite complex geometry, and basing each curve on a circle of particular size to give each letter the best image possible. This work began the art of typeface design, which today has given us thousands of digital fonts available at the click of a mouse.

Leonardo left Il Moro's court after 17 years, a changed man. A society lady of the day, Isabella d'Este, sent a messenger to Florence to commission a Leonardo portrait of herself, but in reply she received this message: 'Leonardo's mathematical experiments have so distracted him from painting that he cannot bear to see a brush.' It was the ultimate brush off as Leonardo now realised that maths was more important than painting, because it underpinned every aspect of his ideas and his work. He is even quoted as saying, 'He who does not know the supreme certainty of mathematics, is wallowing in confusion.'

In 1515, aged 70, Leonardo was commissioned to design a gift from the city of Florence to the King Francis of France: a mechanical lion that moved forwards and backwards and reared on its hind legs, upon which its chest opened to reveal lilies, the emblems of both France and Florence. King Francis was absolutely thrilled, and employed da Vinci for the last two years of the artist's life.

Thanks to more recent studies of Leonardo's work, we now know that he figured out how a ball accelerates as it rolls down a slope, which Galileo discovered for himself later. As we shall see, this was a most important factor in advancing mathematical and scientific ideas, culminating in Newton's discovery of gravity.

Da Vinci also followed Alberti regarding perspective. To achieve true perspective when drawing a human figure, he said, 'Set a frame with a network of threads between your eye and the nude model and copy the same squares on the paper.' But his work in this area was to be surpassed by an even more influential perspective artist, and a quite remarkable engraver…

The Great Engraver

One of 18 children, **Albrecht Dürer** (1471–1528) was born in Nuremberg, the son of a Hungarian goldsmith. His father's favourite son, Albrecht would surely have followed him into the family business, had he not shown, by the early age of 13, a prodigious talent as an artist. He was apprenticed to a painter who also specialised in engraving and woodcut design, skills that Dürer was later to excel in.

Aged 19 Dürer had a Wanderyahre or gap year, visiting cities in the Netherlands, as well as Basel in Switzerland. In 1494 he returned and married, but when the plague hit Nuremberg he made his first trip to Italy, alone. There he met Bellini and Pacioli in Bologna; he later communicated with Leonardo and others, becoming perhaps the greatest exponent of perspective, which he called 'the crown and keystone of the edifice we call geometry'.

Dürer used perspective to create a series of engravings that are often seen even today. He also used a rather cumbersome method that employed a weighted string hanging from a hook on the wall, which could pass through the frame to any point on the subject, and that point could then be recorded (see Figure 10.6).

In 1514, now a publisher in his own right, Dürer produced two of his most important mathematically inspired engravings. Of the two, *Saint Jerome in His Study* is unusual, because the single vanishing point is very close to the right-hand side of the picture, creating the effect for the viewer of almost being in the room with him (see plate section).

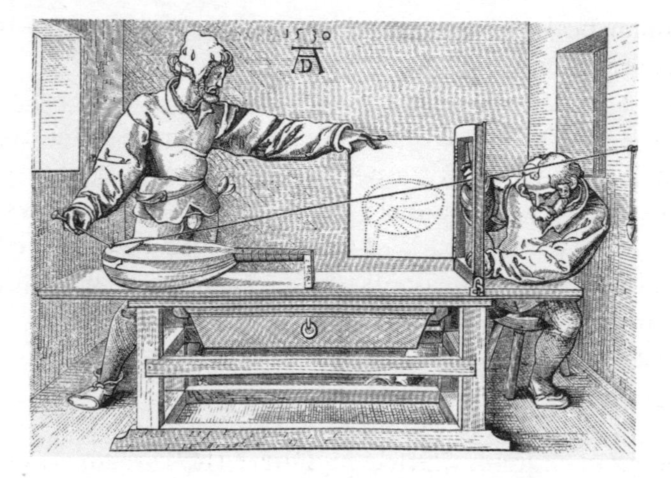

Figure 10.6

But the most famous of all his engravings is *Melancholia I*. The 'I' stands for 'imagination', the first of three forms of melancholia brought on by contemplating one's mortality and brooding on the subject matter in the picture (see plate section). The tools of geometry, which can strike fear into the heart of non-mathematicians, lie on the floor. Also a magic square that suggests luck – good or bad – has a 15 and a 14 at the centre of the bottom row, indicating the year the picture was produced.

In the background there is a truncated rhombohedron with a faint human skull on it – an image now known as 'Dürer's solid'. But other items in the picture suggest the problems of life itself and of the passage of time: the hour glass, the scales, the purse and keys (perhaps implying that you can't take it with you?), and the ladder – perhaps leading to Heaven. And there's the small despondent figure of 'genius', looking forlorn and exhausted.

The engraving may have represented Dürer's own personal melancholia – he was known to have remarked around that time, rather despondently, 'What is beautiful, I do not know?' But we all get the blues once in a while, and for Dürer it was probably a passing phase, because otherwise he led a very successful life. Perhaps the engraving was meant as a warning: don't let all things mathematical get you down – instead embrace them, because they will surely enrich your life...

This was definitely his aim when he wrote a book on mathematics in around 1525. It covered the construction of geometric shapes using compasses and a straight edge, and it proved a great help to artists looking for realism in their pictures through geometry. Although it was written in German, it was very quickly translated into Latin, and it could be said to be the first ever book on applied mathematics.

Dürer went to great lengths to study geometric shapes, especially the ellipse that's formed when you slice a cone. At this point he devised the method – now standard practice for all architects and designers – of conveying the overall intention of a drawing by showing it from three vantage points: plan (overhead) and two elevations (front and side) (see Figure 10.7).

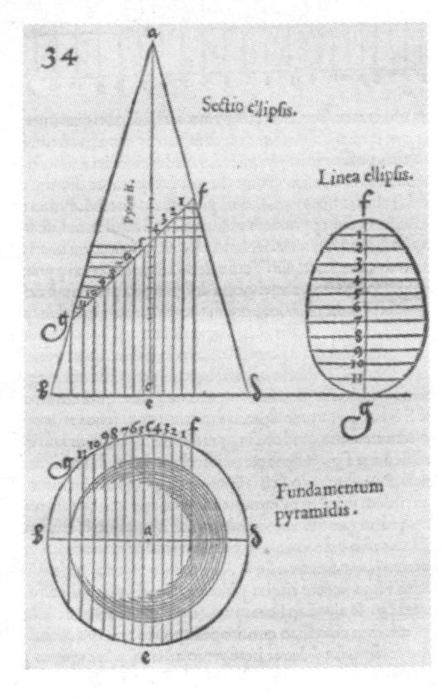

Figure 10.7

In producing the ellipse, however, Dürer refused to believe his eyes, and as a result got his maths wrong. He was convinced that an ellipse sliced from a cone would be sharper at the end nearer the peak of the cone and wider at the end nearer the base. As you can see in Figure 10.7, his ellipse looks slightly wider at the lower end, like a hen's egg. But after slicing cones at many different angles he realised he was wrong, and that the shape produced is always a uniform ellipse, as Apollonius had discovered (see Chapter 5).

Dürer's work on geometry features in his four books on measurement. In the second, he discusses how to produce regular polygons using a ruler and a pair of 'rusty compasses' – a pair of compasses that's fixed, so its span can't be changed. Regular polygons were a feature of both Islamic and Gothic architecture, and featured in the building of large fortresses. Dürer discovered that masons and architects already understood much of the maths involved better than mathematicians did, and very soon he had both sides swapping ideas, to their mutual benefit.

Thanks to Dürer, 'workshop' geometry – geometry that uses a rusty compass – became a very popular pastime for contemporary mathematicians like Tartaglia, Cardano and Benedetti, as well as Kepler and Galileo (all of whom we meet soon). Like Pacioli, Dürer also used geometry to produce letters of the alphabet.

Figure 10.8

Meanwhile, a chap called Segismundus de Fantis introduced very elaborate Gothic script. For this letter 'A' he uses a square and six different-sized circles, though for some letters 13 circles were used. Gothic script really caught on in Germany, but Dürer simplified it, using paper strips of squares cleverly folded to produce the entire alphabet (see Figure 10.9). As well as being a great and accomplished artist, Albrecht Dürer was certainly the most important early publisher of mathematical ideas.

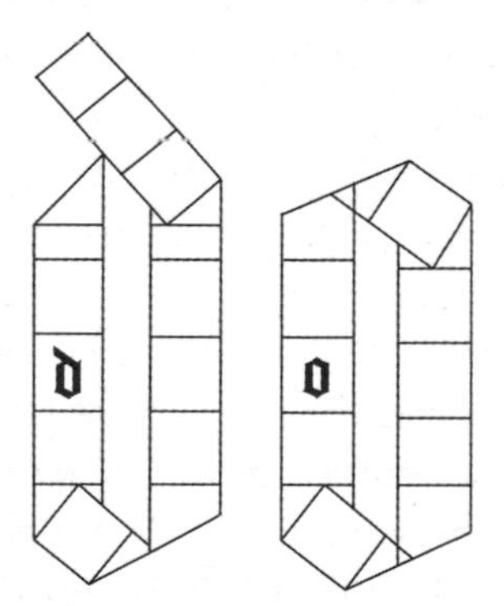

Figure 10.9

Little New Maths But Great Stories

The early part of the Renaissance was all about learning anew the mathematics of the ancient Greeks, Hindus and

Arabs, much of which came via Fibonacci of Pisa 200 years previously. However, some mathematicians explored avenues not previously considered. One revolutionary mathematician of the time was Parisian doctor **Nicolas Chuquet** (1445–1488), who in 1484 produced a manuscript entitled *Triparty en la science des nombres*. Although this treatise wasn't actually published until the nineteenth century, it did feature in a 1520 textbook by fellow French mathematician Estienne de la Roche, and was widely read and very influential.

The first part covered Hindu and Arabic maths, while the second dealt with square and cube (and higher) roots. In part three, however, which he titled *Rule of the unknown thing*, he covered algebra, and was the first to use *exponents* as we use them today, writing 6x as $.6.^1$, for example, or $12x^3$ as $.12.^3$ He also showed how exponents can make multiplication easy by using addition.

As an example, take the two series of numbers shown below. The top row of numbers contains the exponents of 2, which are natural numbers starting with zero. The bottom row shows the corresponding binary sequence (the numbers in base-2).

0	1	2	3	4	5	6	7	8	9	10	11	12
1	2	4	8	16	32	64	128	256	512	1,024	2,048	4,096

We can easily multiply any two numbers on the bottom row by adding the corresponding exponents. Take 16 x 32. The exponent above 16 is 4, and the one above 32 is 5.

$$\text{So } 2^4\,(16) \times 2^5\,(32) = 2^9\,(512).$$

A little later, **Michael Stifel** (1487–1567) became a Lutheran minister, at the suggestion of Martin Luther himself – and immediately married his predecessor's wife. Highly religious, in 1532 he published (anonymously) a work in German entitled *A Book of Arithmetic about the Antichrist: A Revelation in the Revelation*. In it, he felt compelled to warn that Judgement

Day and the end of the world would occur at 8 a.m. on 19 October 1533. When the day passed without incident, he decided to stop making prophesies.

But he added to Chuquet's work on exponents, by using fractional exponents for negative numbers.

$$\text{So, } 2^0 = 1, 2^{1/2} = -1, 2^{1/4} = -2, 2^{1/8} = -3 \text{ and so on.}$$

Chuquet also coined the terms 'million', 'billion', 'trillion', 'quadrillion', 'quintillion' and so on, and explained the system of very large numbers: 'a million is worth a thousand thousand units, a billion is worth a thousand thousand millions, a tryillion is worth a thousand thousand byllions, a quadrillion is worth a thousand thousand tryllions', and so on.

He then described how you would actually say a huge number. His example went,

'Seven hundred forty-five thousand three hundred and twenty-four tryllions, (745324), 804300 byllions 700023 millions 654321' – although, as shown in the copy of the manuscript in Figure 10.10, he mistakenly added an extra 0 to the byllions.

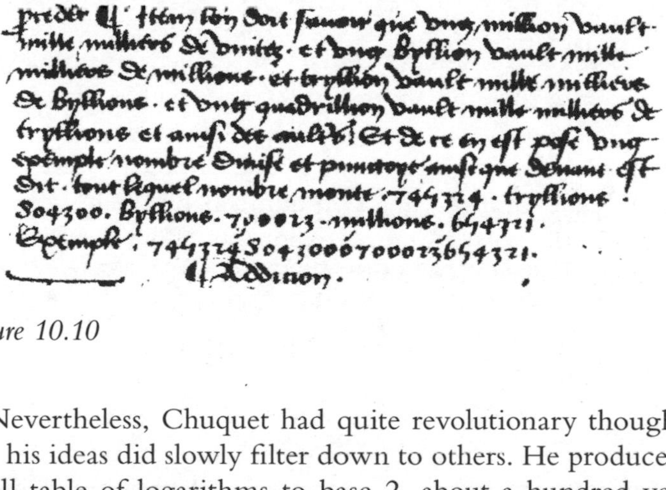
precedent ¶ Item lon doit savoir que ung million vault
mille milliers de unitez. et ung byllion vault mille
milliers de millions. et tryllion vault mille milliers
de byllions. et ung quadrillion vault mille milliers de
tryllions et ainsi des aultres. Et de ce en est pose ung
exemple nombre divise et punctoye ainsi que devant est
dit. tout lequel nombre monte 745324. tryllions.
804300. byllions. 700023. millions. 654321.
Exemple. 7453248043000700023654321.
¶ Addicion.

Figure 10.10

Nevertheless, Chuquet had quite revolutionary thoughts, and his ideas did slowly filter down to others. He produced a small table of logarithms to base 2, about a hundred years before John Napier, who is usually credited with discovering logs. Chuquet saw that by using base 10 for his exponents he could abbreviate the system as follows:

$$10^0 = 1, 10^1 = 10, 10^2 = 100, 10^3 = 1{,}000, 10^4 = 10{,}000,$$
$$10^5 = 100{,}000, 10^6 = 1{,}000{,}000 \text{ and so on.}$$

Modern notation	10^6	10^9	10^{12}	10^{15}	10^{18}	10^{21}
Modern short-scale system	Million	Billion	Trillion	Quadrillion	Quintillion	Sextillion
Chuquet's long-scale system	Million	Thousand Million	Billion	Thousand Billion	Trillion	Thousand Trillion

In 1974 Great Britain joined the USA in using the short-scale system, in which a billion is a thousand million in the old system and a trillion is a billion (and so on), and the system is used by English-speaking and Arabic countries. France stayed with Germany in using the old long-scale system, although in the sciences today the short-scale system is invariably used worldwide.

Italian Equation Juggling

Al-Khwarizmi (see Chapter 8) had already shown that it's possible to solve quadratic equations by turning arithmetic into geometry, and the Renaissance mathematicians picked up on that. But soon more ancient documents were discovered with clues to solving more complex equations. Then a sort of game developed, in which mainly Italian mathematicians challenged each other to solve equations only they knew the answer to.

Squaring and Cubing Unknown Values

In essence, squaring and then cubing an unknown value and a number is quite straightforward. Let's take x + 5.

Firstly, $(x + 5)^2 = (x + 5)(x + 5) = x^2 + 10x + 25$.

So $(x + 5)^3 = (x^2 + 10x + 25)(x + 5) = x^3 + 15x^2 + 75x + 125$

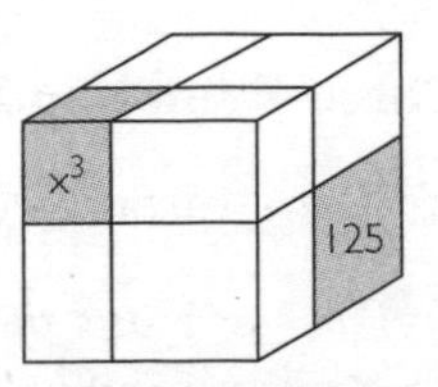

Figure 10.11

From the diagram, we can see that this calculation is fairly easy.

Doing this in reverse, however – factoring or finding the root, or starting point, of a cubic equation, especially one that has been deviously concocted – is a much more complex task.

Following an original challenge from Pacioli, Scipio – or **Scipione del Ferro** (1465–1526) – professor of maths at Bologna University, solved cubic equations using algebra. He showed his findings to his pupil Antonio Maria Fiore, who saw immediately that he could make money from this knowledge, even though it was basically one mathematical trick.

In those days one-upmanship was very much in the Italian character, and in 1535 Fiore arranged a public joust or contest, at which he challenged one particular man to 'open combat' in problem solving. Sadly, the man he chose to challenge was Niccolo Fontana. Big mistake.

Fontana, known as **Tartaglia** – or the Stutterer – (1499–1557) was just 12 when French cavalry invaded his village, and three slashes from a soldier's sword almost severed his head. His father was killed in the attack, and his mother, now penniless and with no possibility of outside help, copied dogs, which lick their wounds: slowly, and quite amazingly, she licked him back to health. Sadly she could not cure his stutter – caused by the shock – which stayed with him for life. His brain, however, was unaffected, and may even have been stimulated; in any case it was to make him very famous. But he and his mother were desperately poor, and as a lad he often used tombstones as slates to work on his maths.

Three Brides and a Boat Puzzle

It's from Tartaglia that we get this wonderful puzzle.

Three men and their new wives need to cross a river. Their small boat will hold only two people, and no man must ever leave his wife in the presence of another man, as in those days their morals were iffy to say the least. How can all six cross the river with their marriages intact?

Try it, before you read the solution in the Wow Factor Maths Index.

Fiore was so confident of winning the cubic equation contest with Tartaglia – which required solutions to 30 problems – that he suggested a prize of 30 dinners to the victor. The problems were all similar. Question 1 asked, 'Find me a number such that when its cube root is added to it, the result is 6.' This is equivalent to $x^3 + x = 6$, to which the solution is 1.63436. In those days, however, they only used fractions, and their workings became quite complex. Other questions looked for the solution to $x^3 + x = 5$ (answer 1.516), 12, 500 and 700. Some were more complex, such as $4x^3 + 3x = 40$.

Tartaglia, a university lecturer, had solved similar problems before, and easily won the contest, although he declined the prize at Fiore's expense. But Tartaglia was to have more trouble with our next character.

Jerome Cardan (1501–1576) – also known as Cardano – was a particularly colourful Renaissance character. He claimed his mother had tried to abort him, and that he had been a weak and sickly child, lacking physical energy. But he made up for his physical deficiencies with mental agility and crassness, leading a life of debauchery. He was vindictive and always claimed superiority over others. He graduated from Milan as a physician, and while in Scotland he cured Archbishop John Hamilton of asthma, correctly identifying his feathery pillows as the cause. When he wasn't being reasonably polite to those who could help him, he was an unscrupulous rogue, gambler and cheat – even a murderer. He was not a particularly good father either, though he did try to defend one son who had murdered his unfaithful wife, but who was eventually executed. In a row with his other son he cut off both his ears.

Yet he rose to be accepted at the Vatican, where he rather stupidly cast the horoscope of Jesus Christ, which was heresy. Nevertheless he survived that scandal and received a pension from the Pope, probably for casting him a glowing horoscope. Eventually he had wealth, fame and powerful friends, and only came truly unstuck at the very end of his life. He prophesied the date of his own death and, because he didn't want to disappoint and lose face, on that very day he committed suicide.

He was a perfect example of Renaissance Man. He went from ignorance, embracing mysticism, palmistry and magic for gain, to reading everything on mathematics he could find.

In science, Cardan was the first to describe the 'water cycle': where water evaporates from the oceans, forms clouds and falls as rain, in an ever-continuing cycle. In mathematics he recognised negative and imaginary numbers, and wrote a book on gambling and chance; he also knew of and used the mathematical triangle ahead of Pascal.

But he could not solve cubic equations, and he begged Tartaglia to disclose his method. But Tartaglia, who could also be a vindictive man, said no. Under the pretext of meeting a man of influence who didn't exist, Cardan lured Tartaglia to Milan. Over three days he coaxed the secret out of him, promising never to reveal it. But when he later discovered that Scipio had already solved it, he published it anyway in 1545, and it became the first solution for cubic equations in print. Tartaglia was livid, but he could do nothing about it. The method still used today is called Cardan's Rule, and the episode set a precedent. Because secrecy in science does no one any good, from this point on it became an established rule that credit for a finding or theory goes to the first person to publish it, irrespective of originality.

Cardan's pupil, Ferrari, solved quartic equations, which Cardan also stole and published first. But the older man then worked out the overall theory for both types of equation. It required negative numbers and *surds* (or *irrationals*). Eventually he could give a solution for $ax^4 + bx^3 + cx^2 + dx + e = 0$. Beyond that (such as factoring ax^5) was later proved to be impossible.

Other Renaissance Men of Note

The Italian influence in the mathematics of the period was very strong. Englishman **Cuthbert Tunstall** (1474–1559) wrote the first mathematical book in English in 1522, but it was taken from Pacioli's Summa.

Robert Recorde (1510–1558) was more original, and the earliest British Renaissance mathematician of note. Born in Tenby in Wales, he studied and lectured at both Oxford and Cambridge Universities, and became physician to the young Edward VI and Queen Mary. He wrote that the symbol + 'betokeneth too much' and – 'betokeneth too little', but these symbols seem to have been developed by German merchants, who – 200 years earlier – added + or – to sacks of goods that were overweight or underweight.

Recorde was, however, the first to represent 'equals' with the now familiar symbol of two parallel lines. At first he linked them with a third line to create a symbol rather like a 'z', but he then scrapped the extra line, saying of the remaining two that 'no two things can be more equal!' We have used it ever since.

He also wrote a book on algebra called *The Whetstone of Witte*, in which he suggested that mathematics sharpens the mind; and one on arithmetic, *The Declaration of the Profit of Arithmeticke*, although he admitted in the preface that there was little new material in it. Neither did it profit him a great deal: he died in his late forties in a debtors' prison.

Rafael Bombelli (1526–1572) was the last Italian mathematician of this era, and the man who, having discovered several ancient texts of Diophantus, would make the biggest advances in algebra. He used simple language to lay down the basic rules of algebraic operation – the language we still use today: 'plus times plus makes plus; minus times minus makes plus; plus times minus makes minus; minus times plus makes minus. So 8 x 8 = 64; –5 x –6 = 30; –4 x 5 = –20; and 5 x –6 = –30.

In addition to these basic ideas, Bombelli also made monumental contributions to complex number theory. But it became ever more complicated, and simplifying it had to wait for two later Renaissance figures, Stevinus and Vieta (who we shall meet in the next chapter).

This Way to the Revolution

Regiomontanus (1436–1476) was born Johann Muller, a miller's son. Clearly gifted, he was accepted at the University of Leipzig at the age of 11, and received a degree in Vienna at 16. Full of ambition, the lad looked around for a suitably grand name. His home town was Konigsberg or 'King's Mountain', so he chose the Latin version – Regiomontanus – as the name by which he would now be known.

His explanation of planetary movements was used by Columbus, although he also dabbled in astrology, and firmly believed in Ptolemy's notion that the Earth was perfectly still, while the Sun, the planets and the whole Universe revolved around it, even though this meant that the Sun must be travelling through space at a colossal speed. Surely, he said, echoing a familiar refrain, if the Earth spun round, birds would be blown away, clouds would get left behind and buildings would crumble.

He found fame when he introduced Arabic maths into Germany, including algebra and trigonometric tables printed on the new Gutenberg press. In 1472, after spotting what eventually came to be known as Halley's Comet, his fame spread. He was called to Rome in 1475 to sort out the calendar, which had been so messed about by the requirements of religious festivals that it was now wildly inaccurate. But before his work really got going, he died of the plague in 1476. It was another 20 years before someone else took on the job of correcting the calendar – but this time the results were quite revolutionary.

Nicolaus Copernicus (1473–1543) was 'Copper Knickers' when I was at school, and why not – he was the son of a copper merchant. His father died when Nicolaus was only 10, and the boy was brought up by his uncle, a high-ranking bishop in the Catholic Church. After gaining degrees in maths and art from Kraków University in Poland, and thanks to the influence of his uncle, he was summoned to Rome, aged 23, to tackle the job that had lain dormant for 20 years – to sort out inaccuracies in the calendar. The work was long overdue: Roger Bacon had pointed out the problems 200 years earlier.

To begin with, Copernicus studied the tables of planetary motion laid down by Hipparchus, but known as the Ptolemaic system. They were incredibly complicated, requiring circles within circles that acted on other circles in the most complex way.

Despite this, in around 1350 the astronomer Giovanni de Dondi had made a clock in Padua that traced the movements and timings of the seven moving bodies that apparently circled the Earth. It was an incredibly complex machine, which not only gave the time and date, but featured seven faces for the Sun, the Moon, Mercury, Venus, Mars, Jupiter and Saturn, each of which copied and displayed their relative positions in the heavens (see plate section).

The Sun's path around the Earth was fairly easy to recreate, as it was a circle, albeit one that rose and fell as the

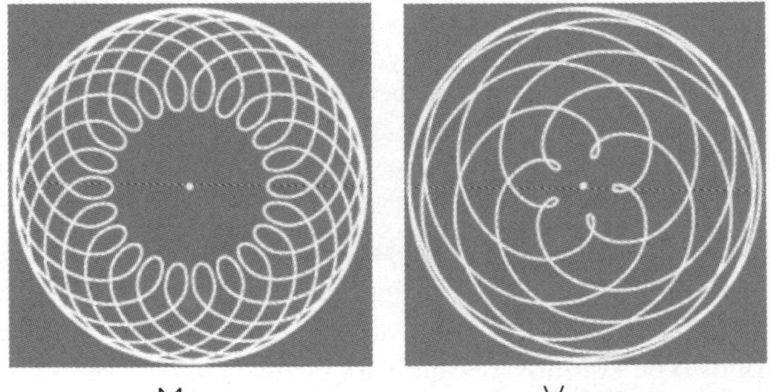

Figure 10.12

year progressed. But the movements of Mars and Venus were far more complex. Figure 10.12 shows those movements as viewed from Earth. Dondi's clock used a wheel inside another wheel to recreate the path of Mars including its sometimes retrograde path, while Venus sat on the edge of one wheel that turned around another hypothetical wheel. It was quite ingenious.

The calendar's accuracy had been so neglected for years that to get it back on track once more Copernicus would have to rework the maths for the orbits of all seven moving heavenly bodies. The prospect scared Copernicus to death, until he had an idea... He had read that the Greek mathematician Aristarchus had suggested that the Ptolemaic system was all wrong, and that while the Moon went round the Earth, the other five planets, and the Earth itself, actually revolved around the Sun. 'And why not?' thought Copernicus. The Sun is clearly different from all the other bodies, giving the Earth its heat and light, its night and day, its seasons and varying climates. Even the light from the Moon is actually reflected light from the Sun. So the Sun clearly has more influence on the Earth than the Earth does on the Sun.

In around 1512, and back in his homeland, Poland, Copernicus – without telling anyone else – tried to make a mathematical model that had the hot flaming Sun at its centre. Immediately he saw that the movements of Saturn, Jupiter and Mars were now much easier to explain and model. He also noted that the angle between the Sun and Mercury was never greater than 24 degrees, and the angle between the Sun and Venus never greater than 45 degrees, because they both clearly had smaller orbits than the Earth and were therefore always nearer to the Sun (see Figure 10.13).

It was now far easier to calculate the Earth year, and although he wasn't a particularly strong mathematician, Copernicus worked out its annual orbit around the Sun to within 28 seconds. The equinoxes could now be explained if the Earth's axis was set at an angle of 23.5 degrees to the Sun, as Eratosthenes and Aristarchus had explained. Hipparchus (see Chapter 5) had

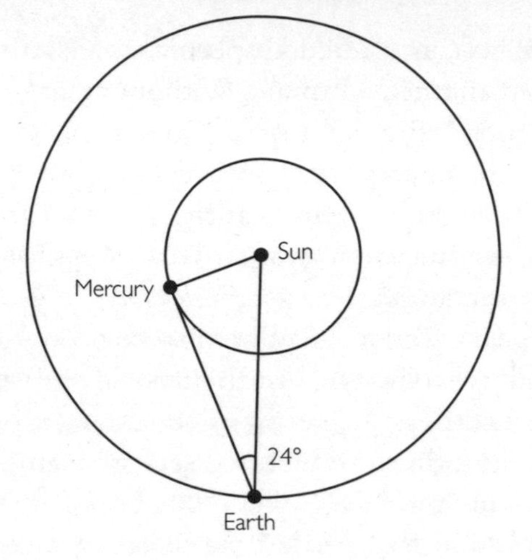

Figure 10.13

already discovered the equinoxes, but with his Earth-centred system could never explain them.

Copernicus's ideas were far from totally clear. He still believed that planets revolved in perfectly circular orbits, even though that didn't completely tie in with his observations. But he could see clearly that he had made a monumental discovery, revealing a far simpler and more easily explained model of how the Earth and the other seven heavenly bodies behave.

He must have been totally jubilant at his discovery – yet he told no one. The Church was all-powerful; to dispute a system believed to have been created by God himself was to court disaster. So Copernicus simply kept quiet about his discovery for more than 30 long years. Then at last he wrote a summary of his ideas and circulated it among European scholars, creating both interest and enthusiasm. Among them was **Georg Joachim von Lauchen** (1514–1574), whose dad had been a physician beheaded for sorcery when Georg was 14. Following this nasty incident Georg thought it wise to change his name, and called himself **Rheticus** after the province of his birth. He was the first to compile trigonometric tables and relate trigonometric functions to angles rather than the arcs of circles.

In 1539 Rheticus studied Copernicus's ideas for 10 weeks and published another summary, without mentioning him by name, for safety. But he urged Copernicus to write and publish his revolutionary discovery himself, and even produced his biography, which sadly no longer exists.

In 1543 Copernicus, now aged 70, grasped the nettle and wrote *De Revolutionibus Orbium Coelestium – The Revolution of the Heavenly Orbs*. Today we remember the American, French and Industrial Revolutions, but the term 'revolution' actually came into use through Copernicus's book and model – which were to revolutionise the way we see and comprehend our constantly revolving Solar System.

Copernicus concentrated on the Solar System only, and at first cited Aristarchus as the man to have first had the idea of a central Sun. But he then crossed out the reference, taking credit for the discovery himself.

Rheticus agreed to oversee the publication of the work, but when he was offered a lucrative post in Leipzig he left, placing Andreas Osiander – a Lutheran minister – in charge.

It was now 22 years since Martin Luther's excommunication, and the Protestant faith had grown substantially since then. But Luther and his followers still believed in the traditional religious doctrines.

In fact, Luther – just as strongly as Rome – still supported the Ptolemaic system with the Earth at the centre, and Osiander was a devout Lutheran. In an effort to protect Copernicus, but without telling him, Osiander added a preface to the book, saying that it was not Copernicus's intention to refute the Ptolemaic system. This was, he suggested, just a simpler way of computing mathematical tables of planetary movements.

In 1543, on his deathbed, Copernicus was shown the published book. But he did not examine it closely and died shortly afterwards. Only a few hundred copies were printed – most of which still exist. A chap called Rheingold saw the manuscript and produced excellent planetary movement tables based on it, though he too was still not convinced that the Sun was at the centre of the system – or perhaps he was too frightened to agree with it…

For another 65 years scholars continued to believe that Copernicus had shrunk from declaring his system the only true one. That was until 1609, when Johannes Kepler discovered that Osiander and not Copernicus had written the preface.

As we see in the next chapter, Kepler had proved for himself that a Sun-centred system was a true explanation of how the planets and the Earth behave. But it was yet another quite brilliant genius who would be determined to give his very life, if necessary, to prove that Copernicus was right. But that didn't happen until 1632, and the final fateful confrontation between the Church and Galileo Galilei.

CHAPTER 11

The New Age of Mathematical Discovery

The Renaissance got up steam in Italy, but in its second period – from around 1580 – the major action shifted mostly north, as the European movers and shakers were gripped by the idea of expanding their understanding of mathematics, science and the world around them.

The Flying Dutchman?

There is a mythical tale of a ghostly ship called the *Flying Dutchman* which is doomed to sail forever and never make port. A real-life claimant to the name 'The Flying Dutchman' could have been **Simon Stevin** (1548–1620), who was born in Bruges. He had a gift for mathematics, which got him his first job – as a tax inspector. But his ambitions grew with age, and in 1583, aged 35, he joined Leiden University, after which his career took flight. He became quartermaster general of the Dutch navy under the command of Maurice of Nassau. Much of Holland had been reclaimed from the sea, so Simon devised a practical plan to control the dikes and sluices that kept the water back, so that, should an enemy attack, whole areas of the country could be flooded quickly.

He also designed a sail-driven carriage to carry 28 people, which could be steered along a beach; in a high wind it almost flew across the sands and could easily outpace galloping horses. Possibly learning from this experience, Stevin then devised a new system to help ships' captains achieve a desired course by allowing for wind. Although he wasn't completely clear on the maths, others refined it, and today the system is essential for airline pilots, helping them reach their destination and calculate their arrival time by producing a parallelogram of forces.

A Parallelogram of Forces

For every flight, a pilot needs to understand precisely where he is going. Knowing his average speed, he can draw a line from his present position to his destination and set the angle of that course. If he knows the current wind speed and direction, he can add those to his course diagram. By drawing the lines in proportion to their speed, he can use them to form a parallelogram.

To make his destination, he needs to aim his aircraft along the diagonal line of the parallelogram, and this will account for the wind. And if all the line lengths are in proportion to speed, he can calculate how much longer the flight will take in a headwind – or, with a tail wind, how much shorter the journey time will be (see Figure 11.1).

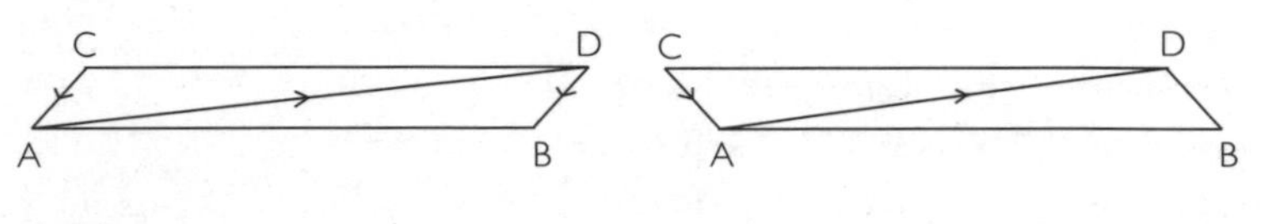

Figure 11.1

Stevin also extended Heron's ideas (see Chapter 6) by calculating the forces required to push a ball up a slope (see Figure 6.8), using an extension of his parallelogram of forces idea. On a slope, he drew two lines to represent the force of the ball resting on the slope and the pull of the downwards force (remarkably, this was before gravity was understood). Drawing a dotted line through the centre of the ball parallel to the slope, he competed a parallelogram. This diagonal dotted line is shorter on the shallow slope; on the steeper slope it is much longer, showing how proportionally more force is needed to push the ball up the steeper slope (see Figure 11.2).

Stevin went on to explore all kinds of engineering problems, and developed an idea eventually known as Stevin's Epitaph. He formed a chain of weights around a right-angled triangle, and by calculating the forces on each of the weights he could prove that the chain must remain at rest (see Figure 11.3). As far as he was concerned, this confirmed that

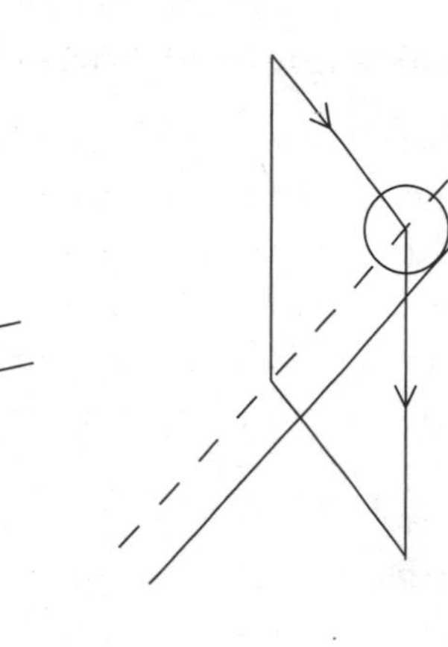

Figure 11.2

perpetual motion is not possible. As he put it, 'For those who cannot see this, may God have pity on their eyes, for the fault is not the thing, but the way they see it.'

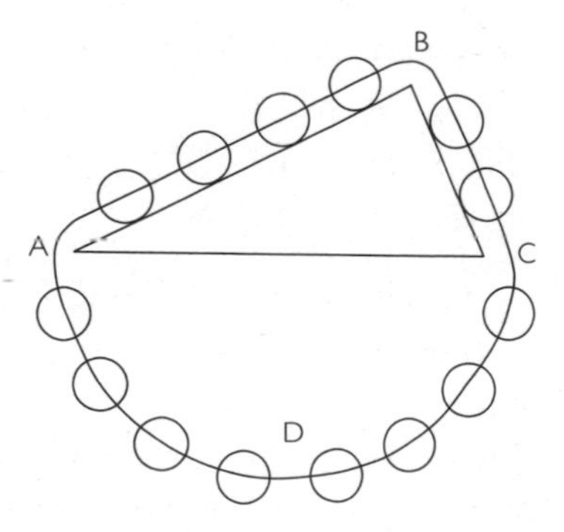

Figure 11.3

One thing Stevin did a lot of was to drop light and heavy objects from great heights to prove that things made of the same material always land at the same time, irrespective of size. Of course, if you drop a feather, it will float on the air and fall at a slower speed, which is why the experiments concern 'objects of the same material'. Stevin also realised that throwing one object sideways at the time of release does not alter its speed of descent. Being a navy man he would have known this, perhaps because of an old trick seamen used to play on new recruits. They would say to a newcomer, 'They'll be sending you up the rigging soon. When it's your turn, you hang back till the ship is moving really fast. Then, if you falls off, the ship will have sailed out of the way by the time you reach the deck.' Of course it was a fallacy, as everyone knew: irrespective of the speed of the ship, anyone falling from the rigging always lands directly below the point from which they fell. This was all before the phenomenon was tested by Galileo, who we shall meet soon, and finally proved by Pierre Gassendi, in 1642.

Stevin was the first scholar to use decimal fractions; he even suggested that a decimal system of weights and measures would be a good thing, though we had to wait another 200 years, and the French Revolution, before those ideas were taken up.

Being a navy man, Stevin was interested in why and how things float, a subject he explored in earnest. In doing so he extended the science of hydrostatics, first explored by Archimedes, and in 1586 he devised a new law: the pressure of a liquid upon a given surface depends on a) the height of the liquid above the surface; and b) the area of the surface. The shape of the container the liquid happens to be in plays no part in varying that pressure.

If a plastic bottle is filled with water and the top removed, then pinholes in the side of the bottle will produce jets of water. The jet from a pinhole high up the bottle will have a steeper and less forceful curve than that from a pinhole lower down. This is determined purely by the volume and resultant pressure of water above each pinhole.

The Vital French Mathematician

François Viète (1540–1603) was a French lawyer by trade, for whom mathematics was only a hobby, though he tackled it pretty intensely. Following Archimedes' lead, for example, he calculated how to create a polygon with 393,216 (6×2^{16}) sides, and accurately calculated π to 10 decimal places. This was the best estimate yet, and four computations further than Tsu Keng-Chih had managed – although to be fair he'd made his calculation a thousand years earlier.

Viete also discovered the first infinite series and its product – an important step in the history of mathematics. He placed a square of unit size inside a circle and asked how their areas compared. The area of the unit square was $1 \times 1 = 1$, the area of the circle was πr^2, and from Pythagoras'

Theorem r was $\frac{1}{2}\sqrt{2}$. Viete then showed that π can be expressed as:

$$2/\pi = \sqrt{1/2} \times \sqrt{(1/2 + \sqrt{1/2})} \times \sqrt{\{1/2 + \sqrt{(1/2 + \sqrt{1/2})}\}} \ldots$$

ad infinitum – a discovery now known as Viete's formula.

A generation later the English mathematician James Gregory found a simpler infinite series to produce π:

$$\pi/4 = 1 - 1/3 + 1/5 - 1/7 + 1/9$$

Leonard Euler later produced

$$\pi^2/6 = 1 + 1/2^2 + 1/3^2 + 1/4^2 \ldots$$

Viete gained international recognition in a rather unusual way. Dutchman Adriaan van Roomen had a maths problem, and wrote to all the top European mathematicians seeking a solution. The French King, Henry IV, was showing the Dutch Ambassador around Fontainebleau and became very upset when the Dutchman said, 'You have much to be proud of. It is a pity you do not have any great mathematicians.' Viete was sent for and told of the problem: 'Solve a polynomial equation of 45^0'. Viete looked out of thc window, deep in thought, then gave the answer. 'It is the chord of a circular arc of 8 degrees.' In effect he had simply divided 360 by 45. However, the next day he gave the answers to 22 other problems that van Roomen had set. In return, he asked van Roomen for a solution to Apollonius' 'Four Touching Circles' problem (which features in the Wow Factor Mathematical Index). Van Roomen, unfortunately, was out of his depth, and Viète's fame as France's greatest mathematician was established. But more fame was to come.

It was his 1591 book, *Introduction to Analytic Art*, that earned him the title 'father of modern algebra'. For him, algebra became analysis: he used the word 'analysis' to mean solving algebraic problems, and the word 'algebra' for manipulating equations. He also used letters (consonants and vowels) to represent unknown entities, and explained how algebra could be used to solve geometric problems.

The final sentence of Viète's book reads 'there is now no problem that cannot be solved'. While this commendable statement demonstrates just how much progress he'd made in algebra, it was probably slightly premature!

Viète also became an expert cryptanalyst, cracking the code used by the Spanish in their war with France in 1589. For two years the French seemed to know the movements of the Spanish forces in advance, and King Phillip II complained to the Pope that they were committing heresy by using sorcery. But it wasn't magic, of course – it was mathematics.

The Birth of Logarithms

John Napier (1550–1617), Eighth Lord Merchiston, was a Scottish aristocrat with both time and money on his hands. As a youth he travelled throughout Europe, and learning about the early Renaissance Men clearly enriched his thinking. A staunch Protestant, he worried that the successes of Phillip II of Spain might one day threaten Scotland. So, taking inspiration from Archimedes, he designed many defensive weapons: burning mirrors, a war chariot, a prototype submarine and an artillery system to wipe out all life more than 1ft (30cm) high within a radius of more than a mile. As in the case of da Vinci, none of these weapons was ever produced. But many locals considered Napier, given his dabblings in astrology and divination, to be unbalanced, and maybe even into black magic.

But this didn't discourage him in the slightest. In the field of mathematics he produced several new ideas to make calculation much simpler for ordinary folk, including the first use of the decimal point. He famously produced Napier's Bones (see plate section) – a calculating device that used rods with multiplication tables etched on their various sides, which made the grid multiplication system both practical and simple.

The system is an extension of Al-Khwarizmi's grid multiplication (see Chapter 8). Each rod is a times table as the 7 rod shows here – double figures are split with a diagonal line. With the 1, 2, 3 rods placed side-by-side on the board, it would be clear that 123 x 2 = 246, x 3 = 369 and 4 = 492, where the diagonal 8 and 1 would be added.

While on his travels, Napier had learnt of the principle of exponents, possibly from Chuquet's work, or that of the

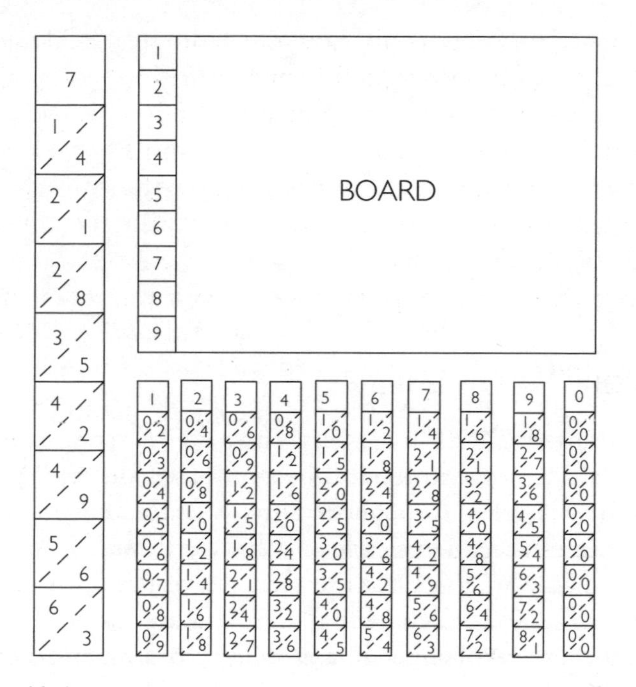

Figure 11.4

German mathematician Michael Stifel, who we met in the previous chapter.

Stifel had used fractional exponents for negative numbers. Napier sought to give values for all numbers by using decimal fractions. So 3 coming between 2 with an exponent of ^{1} and 4 with an exponent of ^{2}, would have an exponent of $^{1.5}$. So 3 x 16 = $^{1.5+4=5.5}$, which is the exponent for 48. The complexity as you get deeper into fractional decimal exponents didn't seem to bother Napier one little bit.

In Napier's own book, published in 1614, he suggested logarithmic valuations for every number to 7 decimal places, between 0 and 1.0000000. He spent endless hours working out fractional decimal exponents for all numbers, and collated them into what he called his table of *logarithms*, taking the name from *logos* and *arithmos*, or 'ratio numbers'. In all it took him 25 exhausting years to complete.

His two greatest admirers were Johannes Kepler (who we'll meet shortly) and **Henry Briggs** (1561–1630), Professor of Geometry at Gresham College in London, the future home of the Royal Society. Briggs travelled to Scotland to see Napier, and on entering the room neither man spoke for almost 15 minutes as Biggs simply stood there, admiring him. At last he congratulated Napier, saying how eager he was to learn how he came upon the discovery of logarithms, which were so very simple and had been there all the time, just waiting to be discovered.

Briggs suggested a more practical system, which Napier agreed to; sadly, however, he died before it could be implemented. So Briggs adjusted the system himself, setting log 1 = 0 and log 10 = 1 and completing tables for all numbers from 1 to 1,000. By 1624, Briggs – now at Oxford University – had extended the tables to cover numbers up to 100,000, providing logs that were accurate to 14 decimal places.

Logarithms made complex calculations much faster, and were quickly adopted in Germany (via Kepler) and in Italy (via Cavalieri). Logs became the computer system of the day for everyone working in astronomy and navigation. To multiply any two large or complex numbers, you simply looked up and added their logarithms, then converted the new log back to the answer.

In the early 1620s, however, one of Briggs's students, **Edmund Gunter** (1581–1626), took the concept a huge step further. He devised a logarithmic ruler 2ft long, and used a pair of dividers to make calculations along it. It also appears that, shortly afterwards, **William Oughtred** (1574–1660) simply slid one of Gunter's rulers against another to produce the very first slide rule, which over the next 350 years was to become as common a tool for engineers

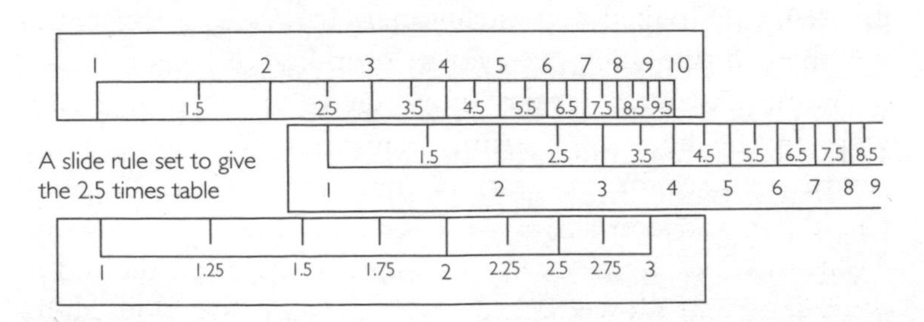

Figure 11.5

as the stethoscope (invented in 1816) would become for doctors (see Figure 11.5).

While studying at Cambridge, Oughtred stayed up most nights studying mathematics, and in his 1631 textbook *Clavis Mathematicae* (*The Key to Mathematics*) he introduced *x* as his multiplication sign and :: to represent proportion, as well as the trigonometric abbreviations *sin, cos* and *tan*, although these were firmly established much later by Euler.

Richard Delamain (1600–1644), a student of both Gunter and Oughtred, later claimed to have produced a circular version of the slide rule. This triggered a long-lasting row with Oughtred, who claimed to have come up with the idea first – he had simply neglected to announce his invention for 10 years (this was probably true). Oughtred was something of a luminary, teaching many fine mathematicians, including Christopher Wren. According to one story, Oughtred (then aged 85), on hearing that Charles II had returned to England to re-establish the monarchy, died of pure joy. (For an explanation of the early slide rules and how they worked, see the Wow Factor Mathematical Index.)

The Finest Naked-eye Astronomer

Tycho Brahe (1546–1601) was an aristocratic Dane who was kidnapped by his rich uncle when he was just a year old. As the uncle was childless, Tycho's father accepted the situation, and that was that. The boy entered the University of Copenhagen aged 13, and the following year saw an eclipse of the Sun, immediately dropping his first career of entering politics to follow astronomy. He eventually became history's greatest ever naked-eye astronomer. It was perhaps just as well he had two good eyes, for he soon had only half a nose.

As an aristocrat, Tycho (now aged 19) and his friends were in the habit of dressing magnificently and sporting swords, more for decoration than protection. On one particular occasion, however, after a good deal of drinking, Tycho got into a row over some mathematical point, and in the duel that followed he lost part of his nose. Now badly scarred, he studied metallurgy, and fashioned a new nose of gold and silver, similar in colour to the tone of his skin; he also had a copper one for everyday use. To hold them in place he concocted a sticky cream. The portrait of Brahe in his book of 1598 clearly shows his disfigurement.

Still aged 19, Brahe observed that the expected close encounter of Saturn and Jupiter was taking place a month away from the time it had been predicted. To correct the erroneous 300-year-old tables set out by Alfonso of Castile in 1252, he began to produce better astronomical instruments. He also started casting horoscopes as a way of raising funds when no one, as yet, would sponsor pure astronomy.

In 1572 Brahe spotted what for him was a new star, but which we now know was a very old star finally exploding. This nova became brighter in the sky than Venus, until it finally faded 18 months later. Brahe estimated that the new star was 3 billion miles from Earth, and that the Universe was 6,100,000,000 miles in diameter. As we now know, this estimate was less than the size of our own planetary system, but at the time it was the most expansive estimate to date.

Now aged 26, Brahe was something of a star himself. Fearing that he would be lured away to Germany, then the centre of astronomical activity, Emperor Frederick II gave him an island 3 miles square off the east coast of Denmark, and the money to build the world's first true observatory. Brahe named the island Uraniborg and placed the magnificent south-facing observatory on a diamond-shaped site (see plate section). It contained his huge astronomical instruments, and it is said that Brahe – ever the showman – wore full court dress while making his night-time explorations.

In 1577, after tracking a comet, Brahe realised that its orbit was not circular, but followed a much narrower path. Using parallax to measure it against fixed stars, he saw that it was much further away than the Moon. This was a new discovery: it had been assumed till then that comets passed through the Earth's own atmosphere, like shooting stars.

Magnificent though Uraniborg was, its instruments weren't steady enough for Brahe. So, next door, he built Stjerneborg, in which all the instruments were set into or under the ground to ensure absolute stability. These instruments included his 'Great Equatorial Armillary Sphere', set in a small stepped amphitheater so that observers always had somewhere to stand while they located any heavenly body over a huge range (see plate section). His new devices worked well, and soon his positioning accuracy for heavenly bodies was within 2 minutes of arc, or 1/30 of one degree. He also measured the length of the year to within one second. These were calculations that would never be bettered with the naked eye.

Other observers soon confirmed Brahe's measurements and, finally, in October 1582, Pope Gregory XIII announced the new Gregorian calendar. Since Roman times an extra 10 days had been accumulated because of the inaccuracy of the calendar, so on its introduction the new calendar leapt forward ten days. This caused riots across Europe, because people believed they'd had 10 days of their lives stolen from them, and wanted them back.

Despite learning so much about the heavens, however, Brahe couldn't accept Copernicus's theory that the Sun is at the centre of our Solar System, and that the Earth is continuously revolving. This would mean that his observatory was constantly moving sideways at around 600 miles per hour, which seemed impossible. And if the Earth was moving around the Sun, surely he would notice some slight movement in the stars throughout the year, and that didn't seem to be happening.

But he did go halfway, and drew up his own Tychonian system (see Figure 11.6) with the Earth still at the centre and the

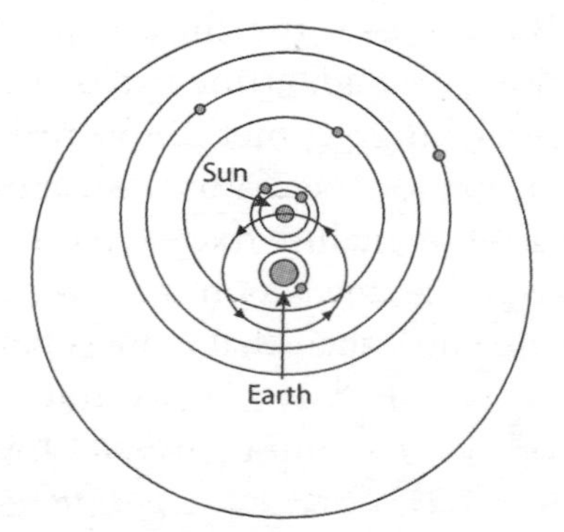

Figure 11.6

Moon revolving around it. However, the other planets all revolved around the Sun, which in turn orbited the still and unmoving Earth, every day.

Because he could not detect any relative movement of the stars, using parallax, he placed all the stars in one gigantic sphere equidistant from the Earth, a system Brahe firmly believed in right until his death.

Despite the huge cost of the observatory and its staff – about 1 per cent of Denmark's annual income – King Frederick II maintained his support. But Brahe's astronomical successes didn't improve his cantankerous and bombastic aristocratic behaviour. He argued with everyone and was a despicable landlord, making life a misery for the peasants on Sven.

Frederick had the patience of a saint, but when he died in 1588 his teenage son Christian IV took over, and he wasn't as supportive. One of Brahe's duties was to manage the chapel in which Frederick and his father before him were entombed. Totally wrapped up in his astronomy, however, he gave the place barely a second thought and it fell into disrepair. When Christian discovered this he snapped, and Brahe's days at Uraniborg were numbered.

In 1597, having lost face, as well as court battles with non-aristocrats, and the financial support of the King, Brahe and his family trundled south with wagons full of his 3,000 books and the astronomical instruments that were light enough to be transported. The Czechoslovakian Emperor Rudolph II welcomed him to Prague, where he made what some call his greatest discovery, when he met and befriended **Johannes Kepler** (1571–1630).

Kepler had the mathematical ability that Brahe lacked, and the Dane enlisted his help in compiling new and accurate tables of planetary movements (what would eventually become known as the Rudolphine Tables). This was only now possible because of Brahe's incredibly accurate figures, especially for the orbit of Mars.

But Brahe was now nearing the end of his life. At a royal palace dinner in 1601 he was desperate for the loo, but didn't want to rise before the host – something that just wasn't done in those days. When he did finally leave, he'd developed a condition that meant he couldn't urinate at all, and he died a few days later in agony, probably of a ruptured bladder.

Brahe's instruments at Uraniborg and Prague were never used again; within a decade astronomers had telescopes at their disposal, and could easily prove or deny his findings. In fact, they merely confirmed the incredible accuracy of his observations. Brahe's last words were reported to have been 'Oh, that it may not appear that I have lived in vain.' He needn't have worried: the man who inherited his astronomical papers used them to explain, once and for all, exactly how the Solar System works.

Johannes Kepler, in contrast to Brahe, was certainly no aristocrat, and was poor for most of his life. At the age of three an attack of smallpox crippled his hands and severely weakened his eyes, stacking the odds against his ever becoming a major astronomer. But his natural mathematical ability showed itself early on. At the University of Tübingen his slight stature made him a target for bullies, whom he appeased by casting their horoscopes, a skill that became a financial lifeline on several occasions during his career. In fact, mysticism was his lifelong travelling companion, and actually helped him make many of his scientific discoveries – both the correct ones, and those that were quite erroneous.

Aged 23, teaching science at the University of Graz, Kepler had already decided that Ptolemy's view of the Solar System was wrong, and that Copernicus was right. In fact, it was Kepler who discovered that Osiander had written the preface to Copernicus's book without the author's knowledge, watering down his entire argument. Kepler was convinced that the Sun was the powerhouse driving our Solar System; at the same time, however, he tried to use astrology to prove various Bible stories, coming to the dubious conclusion that the Earth was created on Sunday 27th April, 4977 BC.

Kepler loved a good mathematical discussion. He devised the Kepler Conjecture for packing fruit, having discussed the stacking of cannon balls with English mathematician Thomas Harriot in 1606. They found that cubic packing (Figure 11.7) and hexagonal packing, both take up the same density of around 74 per cent of the space provided.

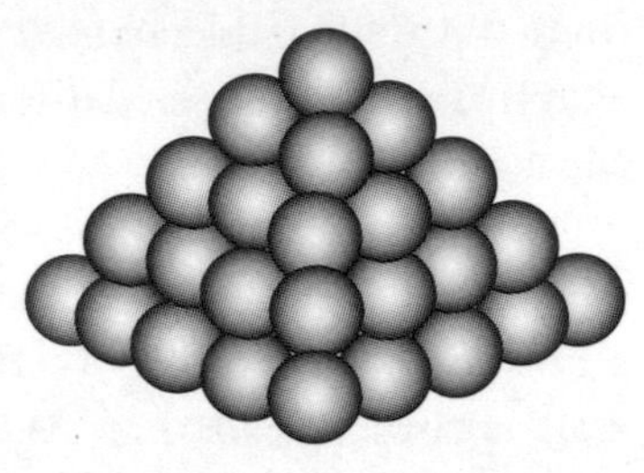

Figure 11.7

Kepler also explored the mathematics of barrels. Their curved shapes make them easy to move around, even when full, by rolling them, or 'walking' them balanced on one rim. Over the years coopers (barrel-makers) had made their barrels in several sizes by tradition. But now traders needed to be able to guarantee their capacity. Kepler imagined the barrels cut into very thin horizontal slices. He then measured each circular cross-section and added all the slices together to accurately calculate a barrel's capacity. This was an early attempt at differentiation, something that was to influence Newton and Leibniz as they formulated their ideas on calculus.

Kepler saw mathematics in everything, and was the first to declare that snowflakes, though they come in many forms, always have six-fold symmetry – kick-starting the science of crystallography. But his overriding concern was just how our planetary system stacked up. One day, while teaching geometry to some young students, he noticed something…

He drew an equilateral triangle with a circle inscribed in it, and another circle subscribed around it; he then noticed that one circle was twice as wide as the other. 'That's odd,' he thought, 'because Saturn's orbit is about twice Jupiter's orbit. I wonder if the spaces between consecutive planets always double as they move further from the Sun?' On checking, he realised that the six known planets were not that neatly spaced. But again his geometric mind took a great leap forwards. There are five platonic solids, and from the Sun, five spaces between the then known planets. Could there be a link between the two? (To see the details of exactly what Kepler discovered, see thc Wow Factor Mathematical Index.)

This beautiful idea appeared in Kepler's *Mysterium Cosmographicum*, a book full of strange fancies and theories, published in 1596. He made a paper model of his nesting planets, but hoped that someone might make a true metal model set in a huge semicircular punchbowl (see Figure 11.8).

Kepler's name among academics throughout Europe was now established, and pictures of the nesting planets cropped up everywhere. No one seemed to notice that the system didn't exactly work (it was about 5 per cent out). In fact, Kepler himself rejected it some years later, when he discovered that the planets didn't actually move in circular orbits after all.

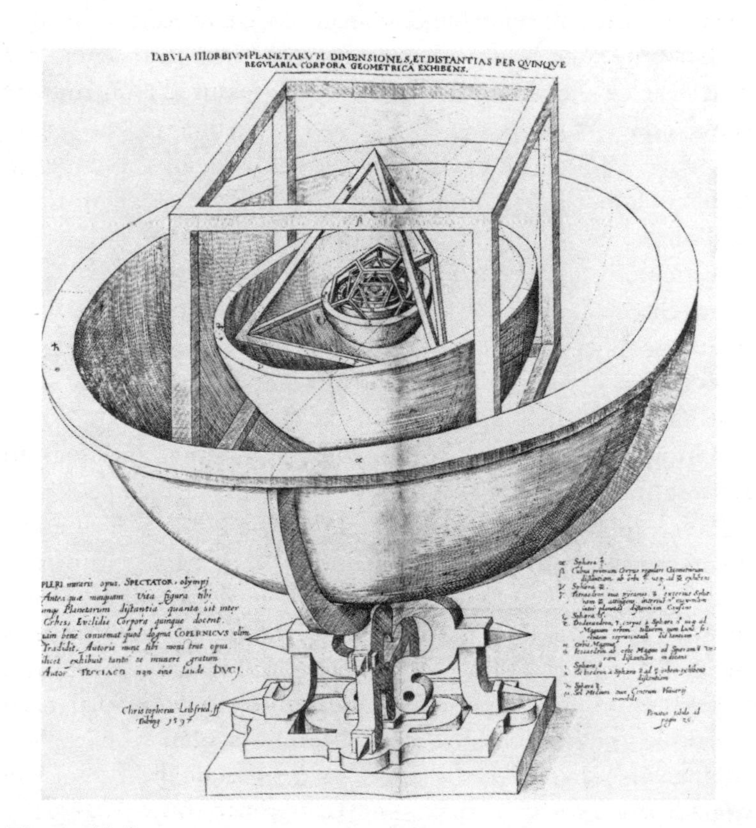

Figure 11.8

Having read about Pythagoras' idea of the Music of the Spheres, in 1599 Kepler had another stab at trying to divine a mystical reason for the planets' orbits. He gave each planet a musical note and produced a chord. Using the tonic *sol-fah* system – in which the notes in a scale are represented by syllables: *doh*, *ray*, *mi* and so on – he gave the Earth the note *mi*, which he claimed stood for 'misery'. He then changed the key and the Earth became *fa*, which he reckoned stood for 'famine' (he had a rather dour view of the world). Fortunately, very few people seem to have gone along with this pessimistic idea.

When Brahe died in 1601, Emperor Rudolph made Kepler 'imperial mathematician', and gave him a huge rise in salary. Kepler also bought Brahe's planetary measurements from the Brahe family for 20,000 florins, so at last they were in his hands. Sadly for everyone concerned, however, Rudolph was very slow at honouring his financial commitments, and both Kepler and the Brahe family had to wait a long time for their money.

Kepler's 1604 book, *Astronomiae Pars Optica* (*The Optical Part of Astronomy*) included the inverse square law for the fall-off in intensity of a point source of light. A ray of light has a particular intensity at a certain distance. As it spreads out, at twice that distance its intensity will only be a quarter of what it was originally; at three times the distance, it will only be one-ninth as intense. This finding would prove invaluable to Isaac Newton later (see Figure 11.9).

Using Brahe's figures, Kepler at last saw that his ideas about nesting planets and platonic solids didn't work after all, because the planetary orbits proved to be eccentric. By closely analysing the figures for Mars he deduced that its orbit was an ellipse, with the Sun at one of the two foci. And although Mars's orbit was only a slight ellipse, it was certainly not a circle. Very soon Kepler had confirmed that this eccentricity applied to all six planets to varying degrees, and that each ellipse had the Sun at one focus. Indeed, Kepler was actually the first person to use the terms *focus* and *foci*, and to explain their importance.

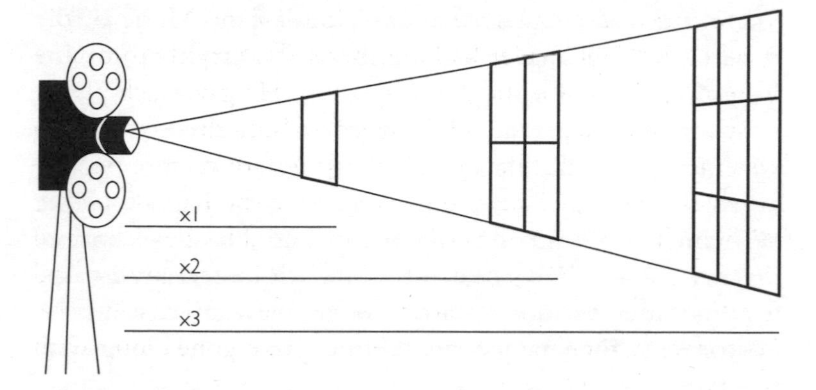

Figure 11.9

All this featured in Kepler's 1609 book *Astronomia Nova* (*The New Astronomy*), which put an end to Greek ideas of planets revolving in perfect circles or the Earth being at the centre of the Universe. Kepler also included a completely new idea in his second law of planetary motion: 'A planet, as it revolves in an ellipse, will always sweep out an equal area in any given period of time.' (see Figure 11.10). Kepler's simple proof of this was the first ever example of integral calculus, and yet another idea on which Leibniz and Newton would build, and which would ultimately help Newton explain gravity.

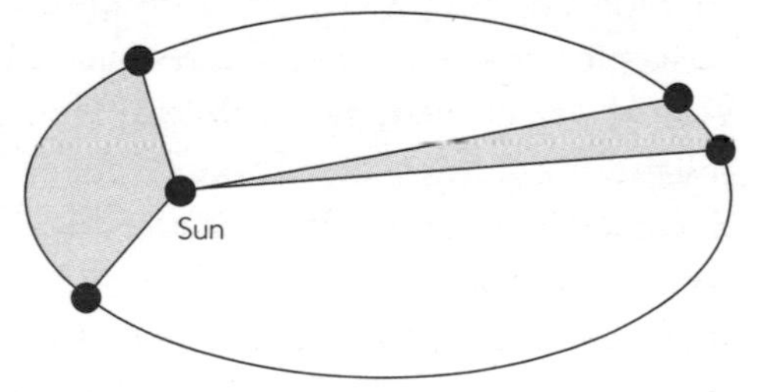

Figure 11.10

Kepler even devised a toy to describe this action – it's quite easy to make: try it for yourself. Hold a narrow cardboard tube in one hand and attach a tennis ball to a string threaded through the tube. With a little practice, by tugging the string as the ball approaches the bottom of its swing, then releasing it slightly as it rises again, you can get the ball to revolve in a continuous ellipse. You will also soon see that the closer the ball is to the end of the tube, the faster it flies and the harder it pulls, while it moves slower and pulls less when further from the focus – echoing Kepler's second law.

From this Kepler realised that the Sun had some controlling effect on each of the planets, holding them in their respective orbits (there is no string on a planet), and he sided with William Gilbert in thinking that it must be some sort of magnetic force. He received some help in studying the heavens after Galileo (whom he never met) sent him a telescope; he then made a small black tent that he could erect anywhere, and he would stick the telescope through a small slit. Even with his weak eyes he was able to see the four moons of Jupiter, which until then he had thought didn't exist. He called them 'satellites', the Latin term for the 'hangers on' of princes and the wealthy. Not long afterwards he confirmed that they too move around Jupiter in elliptical orbits. He could never have imagined that today satellites orbit the Earth in their thousands.

Although Kepler and Galileo wrote to each other, their correspondence stopped around 1610, possibly because Galileo was unhappy with Kepler's continued reliance on mysticism. However, it could have been because Kepler designed the first refracting telescope, a strong influence on future telescope makers, including Newton.

Kepler's life as a Protestant had always caused him problems, and even more so from 1618, following the outbreak, in Prague, of the Thirty Years' War between Catholics and Protestants. Worse for Kepler, however, was the arrest that year of his aged mother, on a charge of witchcraft. She had given a potion to a prostitute to abort a child, but the woman now blamed her for a problem that had arisen a couple of years later. After holding a sword to the old lady's throat, the woman's brother had her arrested. Although she was never tortured, she was held for 14 months until Kepler secured her release and the charges were dropped. Sadly, however, she died soon afterwards, aged about 68, probably due to the shock.

But nothing could stop the ever-prolific Kepler. In 1619 he produced his most mystical book yet. Emperor Rudolph had died in 1612, and Kepler now knew no one of influence who might fund his efforts in Central Europe. So he dedicated the book to King James I and had it published in England. In this book, *Harmonices Mundi* (*The Harmony of the World*) he said: 'Why waste words? Geometry existed before the creation and is co-eternal with the mind of God.'

In this book Kepler took a section of Archimedes' work a giant step forwards. He saw that platonic and semi-platonic solids could be *tessellated* by placing a pyramid on each face, with the sides of each new face equal in length to the original sides. He also observed that the pyramids could sometimes be indented, placed inside the original shape rather than outside. From his explorations into geometric shapes he even discovered two new stellated solids, which he suggested might have magical properties and could be worn as amulets (see Figure 11.11).

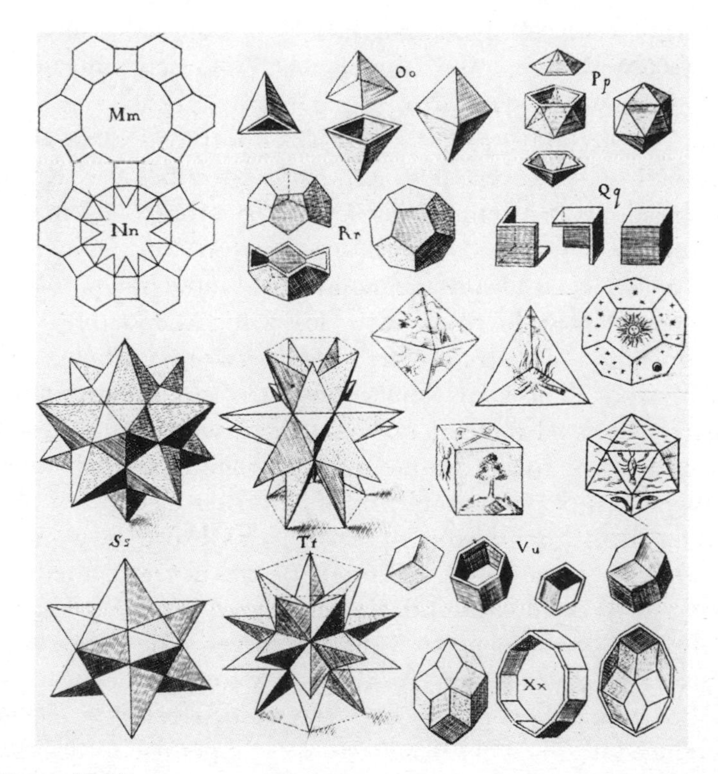

Figure 11.11

Our picture shows a collection of the models made by Kepler for his mathematical explorations. The images on the lower left show a stellated dodecahedron and icosahedron, all extensions of Archimedes' work.

Although Kepler's book contained much mystical pseudo-science, it also featured a mathematical gem that became the basis of his third law of planetary motion, and which he called his Harmonic Law. Always attempting to understand the mathematics involved in every observation, and still trying to tie in harmonic concepts, Kepler compared the orbital time of each planet with its average distance from the Sun, and discovered a rather complex but beautiful pattern, which you can see in the Wow Factor Mathematical Index.

In 1627, after 30 years of work, Kepler completed the Rudolphine Tables of planetary motion, which he dedicated to Tycho Brahe. Making it all possible had been Napier and Brigg's logarithms, which had cut out much of the long-winded mathematics. Kepler had even explored a different system using a tertiary base sequence (1, 3, 9, 27, 81 and so on), but it came to nothing.

It has often been suggested that this tertiary system could be used in weighing: with just these five weights you can weigh any item from 1 to 121 units, by placing appropriate weights on either side of the scales.

To weigh 5 units place weights 1 and 3 on one side and 9 on the other, so the difference is 5.

To weigh 32 units requires 27 and 9 against 1 and 3.

To weigh 97 units requires 1, 27 and 81 on one side and 3 and 9 on the other.

Another of Kepler's works was *Somnium*, a story in which a man dreams that he visits the Moon. It was published after his death, but given his wide scientific knowledge, this work could claim to be the very first work of true science fiction rather than just fancy.

Like Brahe, Kepler always believed that the stars were all roughly the same distance from the Earth, in a belt perhaps just a few miles thick. It was Italian monk **Giordano Bruno** (1548–1600) who first suggested that stars were really other suns and that there are millions of them, because space is infinitely large. He believed that any one of those stars might have planets like those in our Solar System that might support life. Bruno also developed a system of mnemonics to aid memory.

Sadly, although most of Bruno's ideas were correct, they were unacceptable to the Roman Catholic Church. So to escape persecution he left Italy to tour Germany and England for several years. Unfortunately he fell out with religious leaders of all faiths everywhere he went. He was eventually captured in Venice and had to endure a seven-year trial in Rome, during which he argued his position brilliantly and refused point blank to withdraw any of his beliefs – much as Socrates had done. Bruno was tortured terribly until, still unrepentant, he was finally burnt at the stake.

The Man Who Saw Further and Clearer

Galileo Galilei (1564–1642) was born in Pisa just three days before the death of Michelangelo. From now on the core of Italian learning would pass from fine arts to science. This was also the year of Shakespeare's birth; in retrospect, one might wonder if Galileo did as much for science as Will did for drama and poetry. He is certainly rated as one of the true greats of scientific history.

His natural ability was clear while he was just a youth at the monastery of Vallombrosa, so his father Vincenzo rejected a monastic future for him and sent him to Pisa University. A celebrated music theorist, Vincenzo played the lute and other instruments and, like Pythagoras, showed by experiment that to raise the pitch of a stretched string by an octave, the tension or load on the string has to be increased four times. This was familiar to instrument makers, but Galileo would find it recurring in his earliest scientific experiments.

In 1581, aged just 17, Galileo saw something that everyone had seen, but no one else had thought about. He noticed that a suspended brass lamp holder in Pisa Cathedral swung from side to side, moved by currents of air. When the door was open, breezes swept in and increased the swing, but Galileo saw that the time it took for the lamp to swing was always the same, irrespective of whether the swing was narrow or wider.

To be absolutely sure, at each visit he began timing the swing with his pulse. According to one story – probably a myth – an earthquake made the lamp swing, but the idea of using your pulse in those circumstances is surely a bit far-fetched. What he definitely did do was to take the problem home. There he made two identical pendulums and carried out an exhaustive series of experiments by impeding the swing of one of them. This features in the Wow Factor Mathematical Index.

Galileo tried changing the weight on one pendulum by adding or removing musket balls tied to the very end. Immediately he was using good scientific practice. As the balls were all identical, he could judge the result according to their unit weights – except that, in reality, he couldn't. He actually found that changing the weight made no real difference to the swing.

If it wasn't the weight, then, what was controlling the timing of the swing? Galileo discovered that it was all to do with the length of the string. By trial and error he deduced that it actually requires just a quarter of the original string

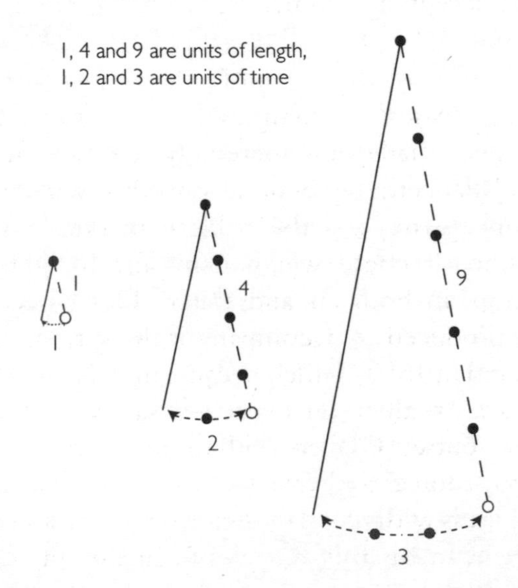

Figure 11.12

length for the swing speed to double exactly (see Figure 11.12). Try it.

This discovery was to have huge repercussions in explaining the force of gravity. Galileo was just 17. Surprisingly, though, it wasn't until near the very end of his life, when he was in his mid-seventies, that he suggested a pendulum might be used to regulate a clock. By then he was completely blind, and dictated the concept to his son (see plate section). But it was actually Huygens who gained the credit for producing the very first pendulum clock in 1656, 12 years after Galileo's death.

Although Galileo was inspired by Roger Bacon and William Gilbert, his greatest influence was Archimedes. Like them, he took nobody's word for anything, but instead always tried to confirm things through careful scientific experiment. However, Galileo was to surpass them all, and earn the moniker 'The First Great Scientist'.

By using his pulse as a timing device, he saw that people have different pulse speeds, and that illness often affects the pulse rate. So he designed a *pulselogia* – a pendulum used to measure a patient's pulse by shortening or lengthening the string. Using coloured marks on the string, he could suggest the patient's condition. He asked some of his tutors whether it would be accepted commercially; one of them promptly stole the idea.

Aged 22, after reading Archimedes' ideas on centres of gravity, and measuring volumes by immersing things in water, Galileo made an incredibly delicate hydrostatic balance, La Bilancetta. He wound very thin wire around the counterbalance arm, and the balance marker moved with high precision. It could weigh items much lighter than a postage stamp, in both air and water. This device and the booklet he produced to accompany it drew the attention of academics, and in 1589 Galileo became mathematics professor at Pisa University, albeit on a very low salary.

From the outset Galileo had to teach the writings of Aristotle, but soon found that his common sense seemed to be at loggerheads with the man. Aristotle had never conducted an experiment in his life, but merely applied his mind and recorded a conclusion that seemed to fit. He had written, for example, that a weight of 100 pounds falling 100 cubits would fall much faster than a weight of 1 pound, which over the same time would fall just 1 cubit. This is of course utter nonsense.

Stevin had performed experiments with falling objects just a few years earlier in the Netherlands, and although Galileo may have heard of his results, he rigorously tried them all for himself. One story has it that he dropped objects from the top of the Leaning Tower of Pisa, and some have described large crowds gathering and shouting and cheering as the objects hit the ground. The 57m tower was already 200 years old at that time and it had started leaning while it was being built, settling at an angle of 17 degrees. Galileo himself, however, never mentioned the tower, nor did he record his findings formally (as far as we know) until the last few years of his life.

It was clear that objects speed up as they fall, but how could Galileo time them? In those days many churches had clocks, but they were not accurate enough even to have minute hands, let alone measure seconds. 'Perhaps,' thought Galileo, 'I could slow it all down…'

He began to roll balls down a straight sloping plank to see if he could determine exactly how they speeded up. He tried several methods to time them, including using a water jar that dripped at regular short intervals. Florence's Galileo Museum contains a model in which a series of tiny bells is placed over a narrow trough so that a rolling ball just touches the clappers and rings each one. Galileo tried a number of other methods, including sticking single hairs across the plank, so that he heard a slight thump as a rolling cannonball bounced over each hair.

Galileo began to say 'Tick, tick, tick' at regular intervals. He would say 'tick' as he released the ball and 'tick' as it hit the first hair, then 'tick' once more, trying to record with chalk where the ball was at that time and placing another hair at that distance. In total he conducted this experiment for more than four months, but gradually came to understand exactly what was happening. However far the ball rolled in the first unit of time, it covered three times that distance in the second equal time unit, and five times the original distance in the third equal time unit. In fact, he had encountered this result before…

Going back to his pendulum – whatever the string length, to reduce the swing speed by a half, he needed to lengthen the string by an additional three times its original length. With the string now four times as long, the swing speed was halved. The rolling ball was doing the same thing, but in reverse.

Slowly Galileo homed in on the solution. In equal time units the ball covered a unit distance, then three times that distance, then five, then seven and then nine – all the odd numbers in order. On a steeper slope, the ball rolled faster, but the increasing distances were exactly the same (see Figure 11.13).

Time Units	1	2	3	4	5	6	7
Distances per unit of time	1	3	5	7	9	11	13
Distances added	1	4	9	16	25	36	49

So the total distance covered is always the square of the time taken. As we now know, Leonardo Da Vinci had recorded this fact some 100 years earlier; unfortunately he had told no one about it.

Galileo dropped iron, ebony and even gold objects from great heights. More importantly, however, he also tried throwing one object sideways while releasing another so that it fell straight down. Like Stevin he found that sideways movement made no difference to the speed at which the

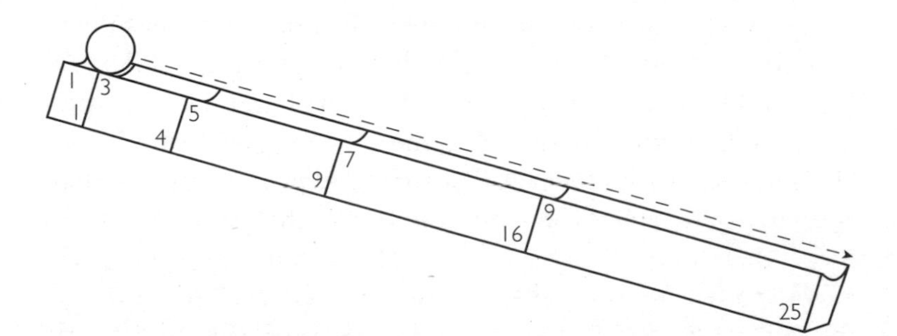

Figure 11.13

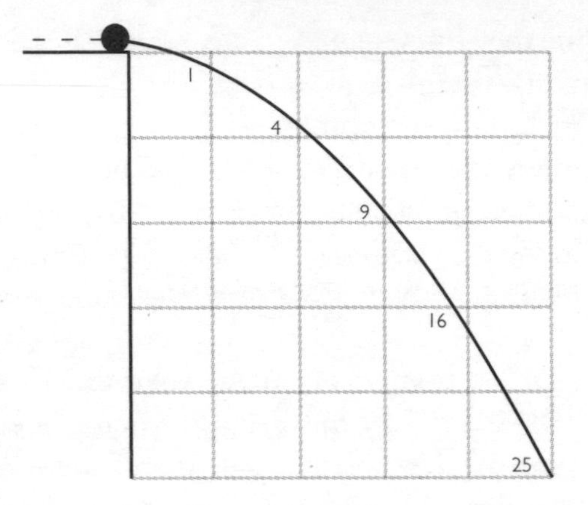

Figure 11.14

object fell. So now he imagined bowling a ball off a cliff and suggested how it would fall once it had gone over the edge (see Figure 11.14). But this time he had the maths of his rolling ball to help him.

Galileo explained that, in still air, the ball would continue to move sideways in equal unit distances, but that it would accelerate downwards at exactly the same rate as a ball falling straight down. Assuming that it dropped 1 unit in the first unit of time, by the end of the second unit of time it would have dropped 4 units, 9 units by the end of the third, 16 by the end of the fourth and so on. Most importantly, Galileo saw that the path of the ball would be a parabolic curve.

At last he had an explanation for why, as the Earth revolves, we do not all fly off into space. We are all unaware of our sideways movement, because everything – including the atmosphere – is pulled downwards by this as yet unknown force. To my mind, this was his greatest ever scientific achievement.

I like to think that, to celebrate, he went out and fired a cannon, because soon after his successes with falling balls and sloping planks, these were the experiments he began. Galileo had formed a friendship with General Francesco del Monte, who helped him land his second, more lucrative, university position. This was in Padua – in the area governed by Venice in 1592 – where he stayed for 18 years.

Soon, with del Monte, he was designing fortresses and methods to defend them against attack. They then designed methods of attack to test and overcome similar defences. Echoing Vitruvius, Galileo suggested that the corners of a fortress should be extended outwards, so that cannon and arrows could be brought to bear on attackers using ladders to scale the walls.

Galileo was also asked if he could give a better idea of how a cannonball flies, to give gunners a better chance of hitting their targets. Thanks to his ball-dropping and rolling experiments this proved easier than even he had imagined – he already had the mathematics.

Tartaglia had wrongly come to the conclusion that a cannonball flies in a straight line until it 'runs out of push', after which it falls to earth. Anyone experiencing cannon fire knew that a cannonball arrives at a very similar angle to the one at which it was fired. Galileo was to finally reveal all.

A cannonball explodes from a cannon's mouth in a straight line – but it doesn't hold that course because there is a force (as yet unknown) pulling it down towards the Earth. It flies in a curve – mathematically the same curve as the falling objects Galileo had rolled off his platform.

Let's say that after a short period of time and lateral distance covered, a cannonball has fallen one unit from the line along which it was fired. In the second equal period of time or distance, it will have fallen not two times as far, but four. After three periods it will have fallen nine times as far, then 16, then 25 times from the original line of flight. Galileo

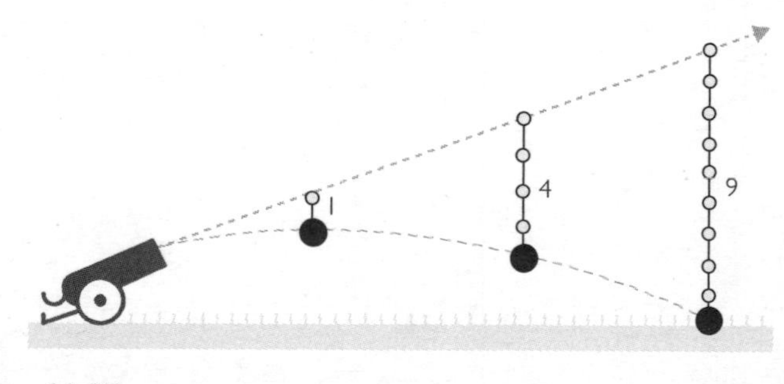

Figure 11.15

demonstrated that due to this downwards force, a cannonball also always flies in a parabolic curve (see Figure 11.15).

He wasn't finished, however. He went on to explain that, on level ground, the greatest distance will be achieved if the cannon is set at a 45-degree angle. Surprisingly, if the cannon is set at 30 degrees, or a reciprocal 60-degree elevation, then the 30-degree parabolic trajectory is a lower, shallower curve and the 60-degree trajectory is a higher, bowed path – but on level ground both balls will hit the ground the same distance away (see Figure 11.16).

Galileo saw that a cannonball fired straight upwards achieves a height determined by the force of the explosion behind it. As the ball rises, its speed reduces until it stops rising, at which time it starts to descend again. As with an arched flight, the slowing and speeding up are the same both ways, and the ball will hit the ground at more or less the same speed it left the cannon. This is something that should be remembered by people who celebrate by firing guns into the air…

Very soon, Galileo saw the value of these collective discoveries, and in 1597 he incorporated them all into a much improved military compass. Previous cannon-aiming devices were simply a wood or metal square with a plumb bob hanging from the 90-degree angle, which was then inserted into the cannon's mouth.

Galileo's instrument was much more elaborate, incorporating Greek methods for calculating proportions, areas, volumes, squares and cube roots – even for calculating

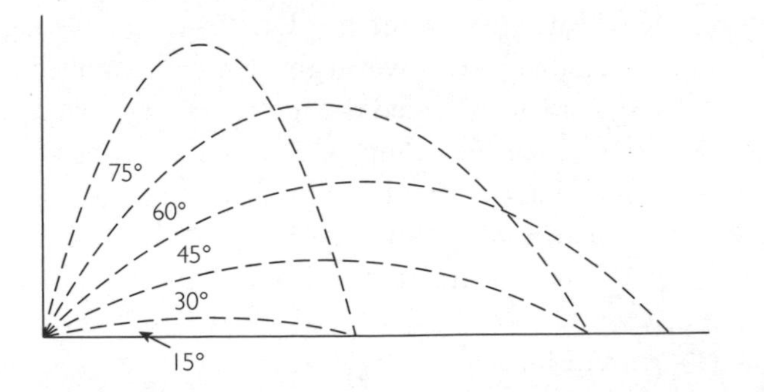

Figure 11.16

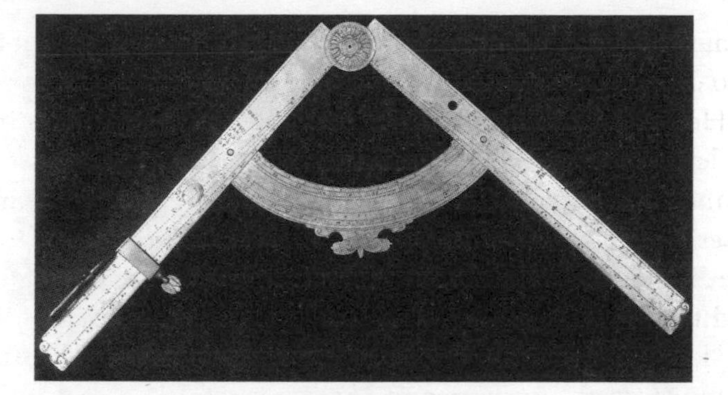

Figure 11.17

interest. Lines in proportion were traced on the moveable legs of the compass, with four scales marked on the quadrant (see Figure 11.17). Military students could use this versatile tool to draw polygons, calculate areas and even survey territory. These military compasses proved so successful that Galileo set up his own factory to produce them, and by 1603 he was at last a relatively wealthy man.

Another of Galileo's early successes at Padua was a long thin tube in which a gas expanded and contracted according to changes in a person's temperature. It was wildly inaccurate, but this first ever thermometer would later be improved by one of Galileo's disciples, Torricelli.

From a Totally Different Viewpoint?

Galileo was already certain that, along with Aristotle's teachings, Ptolemy's model of the Universe was wrong, and that the Copernican model was right. He had received a copy of Kepler's *Mysterium* and liked the idea of nesting planets with the Sun at the centre. Although he wrote to Kepler praising his bravery, he declared that he himself was just too scared to take on the Italian establishment. This attitude wouldn't last.

Without doubt the great turning point in Galileo's life occurred in 1609 (when he was 45). The assistant to Dutch spectacle maker Hans Lippershey lined up two lenses and saw that far-off things seemed to be much nearer. News of this nascent spyglass spread to Italy, but didn't reach Galileo for some months until his lifelong friend Gianfrancesco Sagredo mentioned it. Galileo was livid that he hadn't been told at once, but in a few days he had produced his own telescope, and succeeded in achieving a magnification of 32 times. He was now on the threshold of becoming the most famous man in the entire world.

After dragging the great and the good of Venice to the top of the bell tower in St Mark's Square, they could clearly spot ships that were still a full day's sail from port. The advantage for traders was immense, and soon everyone wanted one of Galileo's telescopes. But their greatest scientific impact occurred when he turned them on the heavens.

He saw at once that the Moon was not the prefect sphere the Bible would have us believe. By recording changing shadows caused by the angle of the Sun, he estimated the height of the Moon's main mountains and troughs. He saw that the same 'face' of the Moon was always in view, although it did swing from side to side slightly, so that a

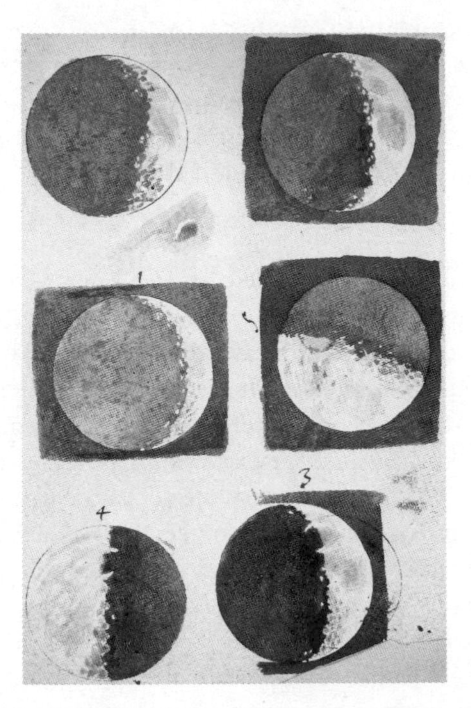

Figure 11.18

little more than half of the surface could be seen over time (see Figure 11.18).

He could clearly make out the definition on the dark portions of the Moon, and declared that this was due to 'Earthshine', or sunlight reflecting from the Earth, just as we experience moonlight. He also saw clearly now that Venus, previously called 'The Horned One', was often seen as a crescent, because its apparent shape varied like the Moon's over time – so it had to be revolving around the Sun.

Of even greater impact was his discovery that Jupiter had four moons all revolving around the planet at different speeds. Later, in 1616, Galileo drew up charts of their timings, and suggested to King Phillip of Spain that these might help calculate longitude. But the clocks of those times weren't accurate enough, and after 16 years of trying Galileo gave up. It wasn't until John Harrison, more than a hundred years later, produced clocks of quite amazing precision that sailors could accurately find longitude.

Clearly Jupiter with its moons was similar to the Earth, thought Galileo, so the Earth could surely not be the centre of the Universe? Mercury, Venus, Earth, Mars, Jupiter and Saturn, he deduced, were all planets revolving around the Sun. What's more, all the stars in the heavens – of which Galileo could now see more than 10 times as many as with the naked eye – were other suns, many of which could have planets like ours. Copernicus had clearly been right.

Galileo included all his heavenly discoveries in his book, the *Starry Messenger*, in 1610. It clearly blew Ptolemy's model of the Universe out of the water. But while scientists rejoiced, the Church began to get very worried indeed.

Galileo's fame and fortune were secured, however, so he left Padua for his beloved Florence. But there he was no longer under the protection of Venice. It was to prove a terrible mistake. A year later, in 1611, Galileo chose to visit Rome to show off his telescopes and other scientific findings. Now a huge celebrity, everyone including the Pope was thrilled with him, as he lectured with confidence and great humour. But the men of the Church, though entertained, were not swayed from what they already believed.

As well as huge confidence, Galileo also had a short temper. He railed against anyone who tried to claim scientific principles without first proving them, as he demonstrated not long after returning from Rome, in a public row in Florence over 'Why Ice Floats'.

His antagonist, Lodovico delle Colombe, produced a bowl of water with a film of ice on the top and declared, 'Ice is condensed water and it floats because it is flat.' The wild, red-haired Galileo could not control his anger as he pushed the ice below the water and released it, only for it to rise to the top again. He then smashed it into fragments, all of which still floated. Then, in words to this effect, he thundered: 'Look, you bungling oaf – ice does not float because it is flat. It forms as a flat film, yes, this being the frontier between the water and the cold air. But ice of any shape floats, not because it is condensed but because – on the contrary – it is rarefied, or less dense than water.'

Galileo's next scientific escapade was to leap from cold to hot. He explored the Sun when it is at a low position, near the horizon, and not so bright as to damage his eyes – although this dangerous practice would ultimately make him blind. He then let his telescope cast the Sun's image onto a piece of paper, which is by far the safer method. He discovered *sunspots*, which were sometimes larger or smaller, but which all slowly moved across the Sun's face from south-west to north-east, showing that the Sun was also revolving and, like the Earth, was tilted on its axis.

As usual, Galileo wrote letters about his findings, and convinced many academics of his views. It was clear, he pointed out, that nothing in the heavens is stationary, remarking that had Aristotle had the benefit of a telescope, he too would have come to the same conclusion. Still the Church held to the belief that everything in the Universe was as God had made it, and could not change. In fact, from 1611 onwards thoughts of the Church acting against Galileo were stirring, instigated by Cardinal Bellarmine, the man who had overseen the execution of Giordano Bruno.

Galileo's explorations continued and he felt compelled to announce every new discovery, irrespective of who it upset. Then in 1615 he was summoned to Rome. After cordial meetings and discussions, the Pope officially announced that certain propositions were forbidden. Among them were any suggestion that the Sun and not the Earth is at the centre of the Solar System, or that the Earth moves in a double motion, around the Sun and on its own axis. The declaration condemned the Copernican system, without actually mentioning Galileo. The Pope even met Galileo a few days later, gently but firmly suggesting that he drop his scientific pronouncements in this field entirely– which he did, for a while.

But Galileo could not see how his ever-growing list of proven scientific facts could continue to be ignored. In 1623 he published *The Assayer*, a book in which he pleaded that men of science might be allowed to assay or 'weigh in the balance' each new scientific discovery based on its merits, rather than being tied to ancient beliefs that did not fit the facts. It was in this book that he remarked, albeit briefly, that 'Everything in the Universe is written in the language of mathematics and its characters are triangles, circles and other geometric figures. Without an understanding of that language, one is wandering in a dark labyrinth.'

Surprisingly, Galileo made one scientific error in *The Assayer.* He wrote that the Sun and the spinning Earth caused the tides, rather than the Moon, as Kepler had suggested. From this time onwards Galileo never again mentioned Kepler or how much he had learnt from him. The scientific truth about tides would have to wait for Newton.

But Galileo made every effort in the book to help the Church accept his science, linking all his discoveries to passages in the scriptures. He sent *The Assayer* to one Father Riccardi in Rome, who gave it the Vatican seal of approval. Pope Gregory XV, however, never saw the book as he was already dying. The new Pope was to be Cardinal Maffeo Barberini, who was a patron of the Academy of Lynxes, a society of academics to which Galileo belonged. Barberini had even once written a poem in support of Galileo. Surely this was great news at last?

Sadly no, for as soon as Barberini changed his name to Pope Urban VIII he also shed his liberal attitude, declaring, 'I am greater than any other Pope as I am the only one "living", and it is on me that the Church depends for protection.' In 1624, Galileo showed the new Pope his microscope, a refined telescope in reverse, and first conceived in 1610. Soon the whole of Rome was talking about bugs, having seen them enlarged to monstrous sizes. Nevertheless, *The Assayer* was promptly condemned and banned for supporting Copernicus.

But Galileo refused to give in. Over the next eight years he produced his masterpiece, a *Dialogue Concerning the Two Chief World Systems*. In it he made every attempt to offer the alternative views on each topic, so that readers could then decide which they preferred. Rather as Plato had done, Galileo chose to make the book a series of discussions between three men. Sagredo, a wealthy wit, would listen to the arguments and ask questions to create a discussion. Salviati was a scientist, who represented Galileo. Both these men had actually been Galileo's closest friends, and Salviati had lived and worked with him for years (although sadly both were now dead). A third character was required, however, one who took the viewpoint of Aristotle, Ptolemy and the Church. As this person found it difficult to take on new and sometimes complex scientific ideas, even when proven, Galileo chose to call him Simplicio. Being the Italian word for simpleton it was an outrageous idea. Worse still, many people in the Church immediately drew the conclusion that Simplicio, in defending the faith, was meant to be the Pope himself.

Galileo gave the Copernican Salviati the best of the fight, in vigorous Italian rather than Latin, though the book was soon being translated into other languages, including Chinese. By May 1632 the first reactions in Rome seemed favourable, with Simplicio appearing as a good sport. But certain Jesuits named in the book flew into a rage, and Father Riccardi, the book's papal censor, acted to prevent further sales and confiscate all printed copies. He was too late.

The book was dedicated to the Grand Duke of Tuscany, and his envoys lobbied that it receive papal approval after it had been edited using the customary method: waxing over offending words. With a smile the Pope replied: to achieve that, he said, the whole book would have to be thrown into a tub of molten wax. Galileo was summoned to appear before the Inquisition.

His trial began on 12 April, by which time Galileo, almost 70, crippled with arthritis and nearly blind, was lodged in the Vatican, albeit in rooms with servants to attend him. He conducted his own defence stoically, but after being shown the Inquisition's instruments of torture, on 22 June 1633 he confessed exactly as they demanded. The story goes that he then whispered under his breath, 'And yet it moves!,' although no one is sure of this.

Galileo was the world's most renowned scientist, and the eyes of the world were upon the court – and, indeed, the Church. Galileo was condemned to lifelong house arrest, although – as soon became apparent – he could choose where that house might be. This sentence showed quite extraordinary leniency for the times. He stayed in Rome for a little while, then Siena, before at last requesting a move to his own home in nearby Arcetri. While there he was denied leave to visit a doctor in nearby Florence and he went completely blind in 1637.

But he never stopped working; indeed, for future generations this house arrest was a godsend: Galileo at last had time to write or dictate all the science he had discovered but never formally written down. One example is his work on the *cycloid* – the path made by a point on the edge of a rolling disc (see Figure 11.19).

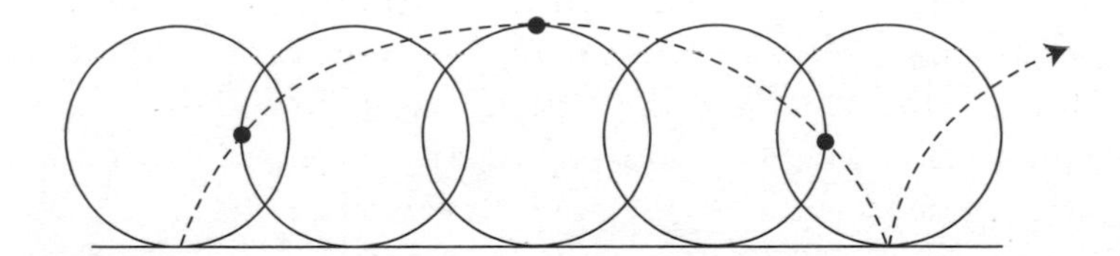

Figure 11.19

Galileo wondered how the area of the cycloid related to the area of the circle. With his assistant, Torricelli, he drew the cycloid and its rolling circle on paper, cut out the pieces out and weighed them. The weights were all the same – so the area of a circle is exactly one-third of the area of the cycloid it generates. This was exactly the same method Archimedes had used to find the area under a parabola (see Chapter 4).

The cycloid is formed by a point on the edge of a rolling circle, but what shape would be produced by a point *inside* the circle? When the circle is rolled along, this point produces a wavy line called a *curtate* line, which is described by the piston rod of a steam engine attached to the drive wheel of a locomotive (see Figure 11.20).

But train wheels have flanges that keep the train safely on the track. A point on the edge of this flange produces a *prolate* line, which has a quite amazing property. As the train progresses, this point moves below the line of the track, forming a backwards loop for a brief moment. So even when the fastest-wheeled train in the world, currently the French Train á Grande Vitesse, is going forwards at its maximum test speed of 574.8kph or 359mph, at the same time parts of every single wheel are actually going backwards.

Galileo continued to work till the end. He contemplated just what infinity means and how strange an animal it can

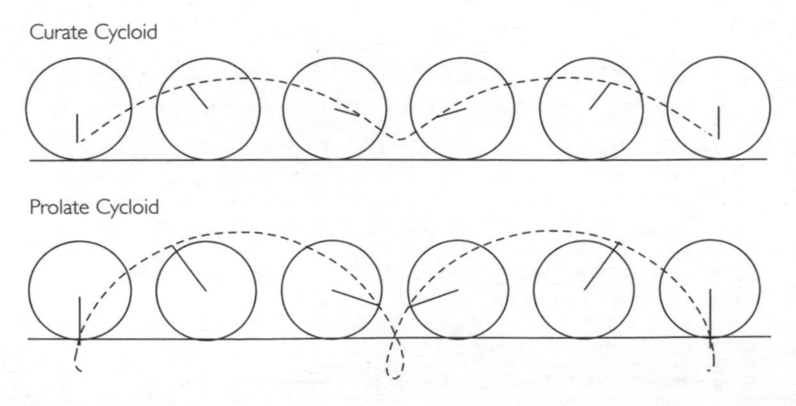

Figure 11.20

be. As every number can be squared, so that $1^2 = 1$, $2^2 = 4$, $3^2 = 9$, $4^2 = 16$ ... , this must mean that in an infinite world there are just as many square numbers as there are numbers. This strange anomaly was to dog mathematics right up to modern times.

Though the now aged Galileo could no longer travel, scientists and scholars from all over Europe flocked to see him, until he died in January 1642. In the same year, in England, a man was born who would be greatly influenced by the life and work of Galileo. His name was Isaac Newton.

CHAPTER 12

How to Calculate Anything and Everything

While Galileo in Italy struggled to establish the true science governing our Solar System, exploratory mathematics further north was blossoming, with many new practitioners expanding new ideas in many directions. Soon their combined aim was to so improve the power of mathematics that it might become possible to calculate or measure absolutely anything.

René Descartes (1596–1650) was an unlikely person to change the world of both mathematics and philosophy forever, because he began life as the most idle chap imaginable. His mother died when he was just a year old, and René looked likely to follow her soon afterwards. He had a chronic cough, and was so weak that his guardians allowed him to stay in bed until lunchtime – a habit he maintained for almost all his life.

At his Jesuit college, La Fleche (or 'The Arrow'), the monks rose before 5 a.m. (perhaps believing that to stay in bed longer was a sin of the flesh?). When they dragged young René out of bed, however, he simply crawled back in again, persisting until they let the sleeping dog lie. But Descartes's mind was far from idle: during those somnolent mornings he thought through the problems of the day, proving to be the college's most able scholar.

His early career options were limited to the Church or the army. Aged 23 he chose the latter, and joined the Bavarian Army (perhaps it only went into battle in the afternoon…?). It was at this point that he began his deep philosophical thinking, devising the concept '*Cogito ergo sum*' – 'I think, therefore I am.' Philosophy would prove to be a major driver in Descartes's life for the next 30 years.

But it was also at this time that he 'dreamt up' a hugely influential mathematical concept. It appears that his grey cells didn't operate well when he was cold, so one winter day he climbed into a wall cavity above a stove and fell asleep. He dreamt of a fly droning round a room, and amazingly saw how he could describe its position at any one time by using mathematics.

He reasoned that all he needed was three lines: one running left to right, another to show where the fly was on the vertical, and the third to indicate whether it was towards the back of the room or the front. These became known as *Cartesian coordinates*, an invention that would ally geometry and algebra, making maths far more powerful. Descartes himself, however, never expanded the idea fully, confining himself for the most part to just two dimensions using x and y axes.

In fact, he wasn't the first to think in this way. Menaechmus, who tutored Alexander the Great, seems to have first used coordinates, while Frenchman **Nicole Oresme** (around 1325–1382) had used the terms *latitudo* and *longitudo* to form a graph that compared time (going upwards) with distance (going left to right) to illustrate a journey and its average speed (see Chapter 8). Oresme also used the system to measure relative values for hot and cold, dark and light, and sweet and sour.

In 1637 Descartes wrote *Discourse on the Method*, at the last minute adding an end section called *The Geometry*, which was to have a huge impact on future mathematics. As an introduction, he gave just two geometric examples. First was the mesolabe compass used by Hippocrates of Chios (see Chapter 3), with which it was possible to multiply or divide any two numbers, or compare their ratios, using parallel lines across two lines set at an angle.

Secondly, he showed how, using a simple straight line and a semi-circle, and starting with unity, it was possible to find the square root of any number (as we also saw in Chapter 3) (see Figure 12.1).

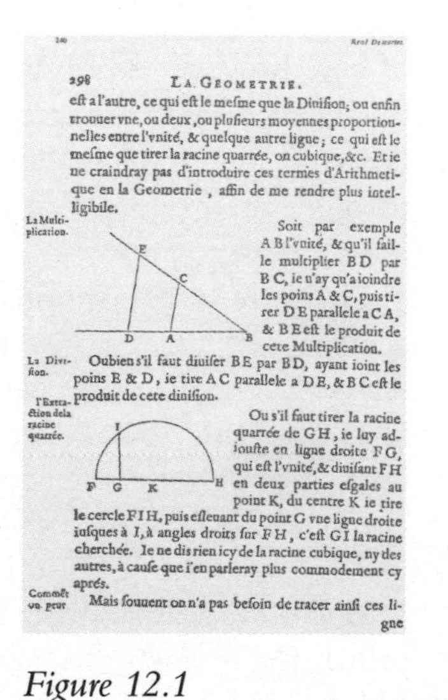

298 LA GEOMETRIE.

eſt a l'autre, ce qui eſt le meſme que la Diuiſion; ou enfin trouuer vne, ou deux, ou pluſieurs moyennes proportionnelles entre l'vnité, & quelque autre ligne; ce qui eſt le meſme que tirer la racine quarrée, ou cubique, &c. Et ie ne craindray pas d'introduire ces termes d'Arithmetique en la Geometrie, affin de me rendre plus intelligibile.

La Multiplication.

Soit par exemple A B l'vnité, & qu'il faille multiplier B D par B C, ie n'ay qu'a ioindre les poins A & C, puis tirer D E parallele a C A, & B E eſt le produit de cete Multiplication.

La Diuiſion.

Oubien s'il faut diuiſer B E par B D, ayant ioint les poins E & D, ie tire A C parallele a D E, & B C eſt le produit de cete diuiſion.

l'Extraction de la racine quarrée.

Ou s'il faut tirer la racine quarrée de G H, ie luy adiouſte en ligne droite F G, qui eſt l'vnité, & diuiſant F H en deux parties eſgales au point K, du centre K ie tire le cercle F I H, puis eſleuant du point G vne ligne droite iuſques à I, à angles droits ſur F H, c'eſt G I la racine cherchée. Ie ne dis rien icy de la racine cubique, ny des autres, à cauſe que i'en parleray plus commodement cy aprés.

Comment on peut

Mais ſouuent on n'a pas beſoin de tracer ainſi ces lignes

Figure 12.1

Despite this beautifully simple opening, *The Geometry* became far more complex as it progressed. Descartes remarked that, 'To solve any problem we first suppose the solution to be affected and give names to all lines that seem needful for its construction – those unknown as well as those known. Then we must try to unravel the difficulty in any way we can. If we cannot, we must add arbitrary lines of known length ... and build equations which we are able to solve.'

It all sounded complex, but it showed the way for other gifted mathematicians to build on the idea, and it spawned an age of mathematical invention that culminated in the calculus of Newton and Leibniz, and arguably the Industrial Revolution itself. For at last engineers had the mathematics to calculate and compare not just lengths and angles, but forces, pressures and torques to determine exactly why and to what extent a new machine might be better than those that had gone before.

Descartes also revolutionised the way algebra and algebraic formulae were written. Here we show the progression from Regiomontanus in 1464 to Descartes in 1637, when at last we see a version of an algebraic formula that any maths student of today would easily recognise.

Regiomontanus 1464: 3: Census et 6 demptis 5 rebus acquatur zero

Pacioli 1494:	3 Census p 6 de 5 rebus ae 0
Stevin 1585:	3 (2 in a circle) – 5 (1 in a circle) + 6 (dot in a circle) = 0
Viete 1591:	3 in A quad – 5 in A plano + 6 acquatur 0
Descartes 1637:	$3x^2 - 5x + 6 = 0$

To show the simplicity of his geometric ideas, Descartes demonstrated that complex curves could be drawn using a mechanical version of the mesolabe compass (see Figure 12.2).

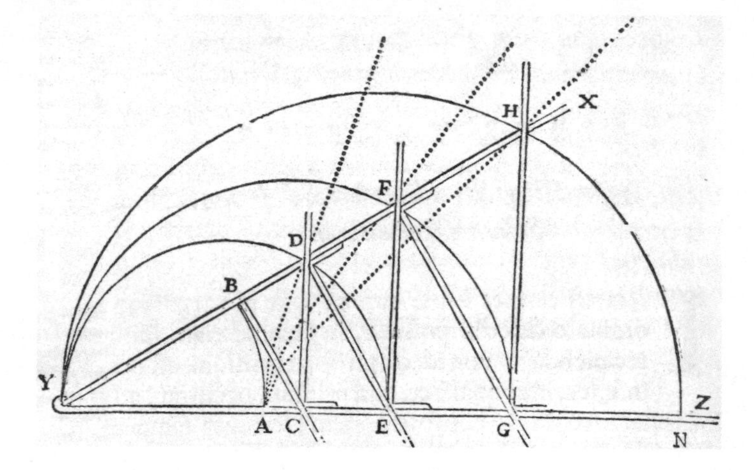

Figure 12.2

Turning Maths into Pictures

Most importantly, Descartes's marriage of algebra and geometry turned complex mathematics into pictures. He explained that the properties of any curve could be expressed using an algebraic equation, or in a geometric picture.

The principle is all built on a graph, in which the base is a horizontal x line running left to right from negative to positive, and the vertical y line is positive above the x line and negative below it. This produces four quadrants, in which both x and y are positive at top right and negative at bottom left.

As an example, using coordinate geometry, the formula $x + y = 0$ produces a straight-line graph running top left to bottom right through the point where both x and y are zero (see Figure 12.3a).

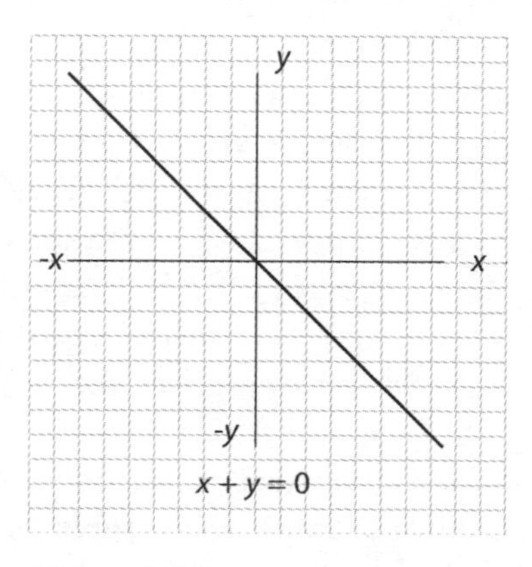

Figure 12.3a

However, $x^2 + y^2 = 25$ produces a perfect circle with a radius of 5 units. As you can see, there are four points where either x equals 0 and $y = +$ or -5; or vice versa. But there are eight points where x and y are a combination of 3 and 4. These are the only points on the circle where x and y are whole numbers. At the point exactly halfway between the 3 and 4 points, the value for both x and y would be 3.535 units, or the square root of 12.5 (see Figure 12.3b).

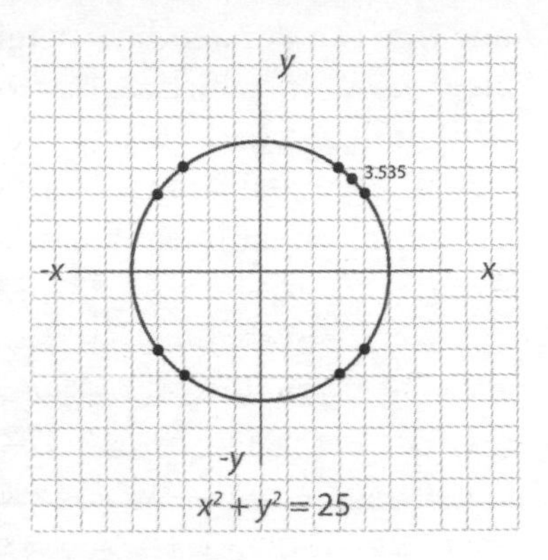

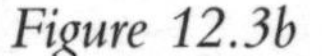
Figure 12.3b

Descartes began to find simple formulae for many other geometric shapes – the formula $x^2 + 4y^2 = 36$ will produce an ellipse (see Figure 12.3c). In this graph the y axis is always 10 times the x axis. The formula $y = x^2$ produces a parabola with the most amazing property (see Figure 12.3d). Choose any two digits, say 3 and 4, and find them each side along the x axis. Join their two dots on the curve with a straight line and it will cross the y axis at point 12, which is -3 x 4. Magic.

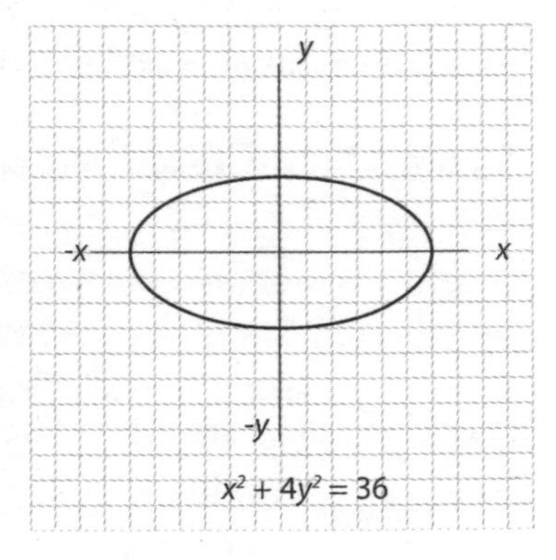

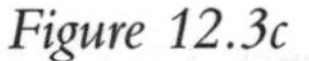
Figure 12.3c

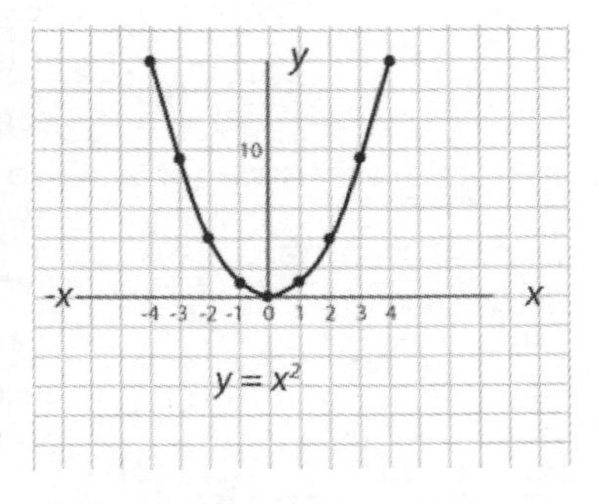

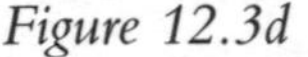
Figure 12.3d

Most importantly, Descartes's saw just how each type of formula behaved when plotted as a graph. A first-degree curve, in which x and y are simple and not squared, produces a straight line that can cut another straight line only once. A second-degree curve, in which at least one factor is a square, will cut a straight line twice. Cubic curves, produced by the sort of formulae the Italian mathematicians tussled over in Chapter 10, can cut a straight line three times, and are often 'S' shaped. And so on…

Very soon other mathematicians were finding more and more fascinating shapes produced by ever more complex formulae. One of Descartes's personal favourites he called the *folium*. It has the formula $x^3 + y^3 = 3axy$ (see Figure 12.3e). Some favourite curves devised by other mathematicians are in the Wow Factor Mathematical Index.

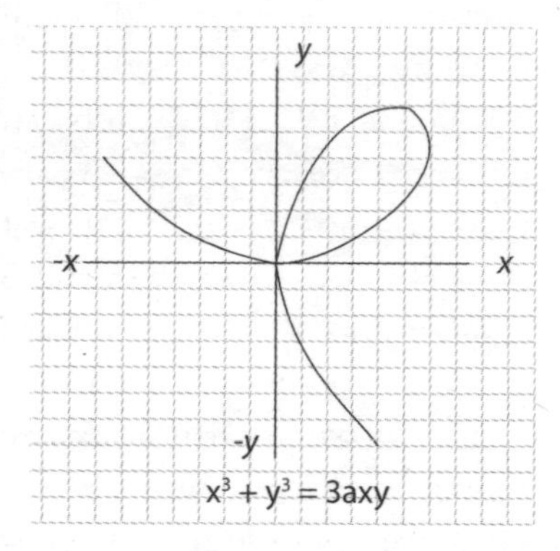

Figure 12.3e

Descartes's basic concept was to completely change every aspect of mathematics by making it possible to show complex maths as illustrations. Today the images we see on our computer game screens all have their original foundation in the coordinate system given to the world by René Descartes.

Reflecting on Rainbows

Descartes observed and explained mathematically how rainbows are formed. When you see a rainbow, he reasoned, the Sun, your head and the centre of the rainbow's arch are all in a perfectly straight line, so only you can see the rainbow that you see (see Figure 12.4). One step sideways and you are looking at a different rainbow. Parallel sun rays pass over you and hit water droplets in the air, then reflect off the back wall of the droplet and leave it, heading straight for your eye. But this only happens at particular angles, which is why your rainbow is in a circular shape – you only see a rainbow arc because the Earth gets in the way. From an aircraft you can sometimes see a completely circular rainbow on the clouds below, with a shadow of your aircraft right in the middle.

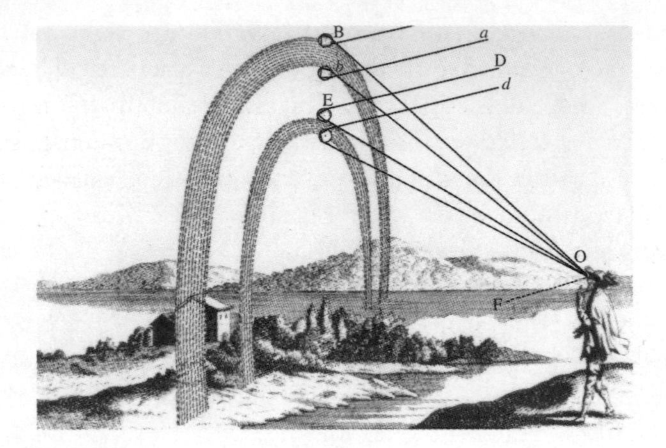

Figure 12.4

The angle at which the rays bounce in the water droplets determines their colour. Blue rays bounce at an angle of about 40 degrees, while red rays appear slightly higher in the sky as they bounce back at an angle of 42 degrees – called the *caustic* angle because it is the brightest and hottest.

Descartes seems to have concentrated only on these most intense red rays; it was left to Isaac Newton to explain the rainbow in full. Descartes did point out that sometimes a second, dimmer rainbow can be seen outside the first one, and upside down. Here, he reasoned (correctly), the sunlight, on its way to your eye, reflects twice within each water droplet. Each bounce causes a slight loss of light intensity, so the outer rainbow always appears fainter. You may also notice that between the red bands of the two rainbows, the sky is always darker. In 1800, Sir William Herschel (after reading Newton's *Opticks*), explained that here you are seeing the unseeable: infrared (which also shows black in burning coals).

Descartes also explored how light refracts as it passes from air into another medium such as water or glass. But here he got rather confused, likening light to tennis balls, and asserting – mistakenly – that it has an infinite speed. Before Descartes even got involved in these considerations, however, Dutchman **Willebrord Snell** (1580–1626) had discovered exactly what happens – even today the explanation is still called Snell's law. When light travelling through air hits a medium like water, the light rays are refracted or bent to some degree towards the vertical because water is more dense than air. The angle of approach is called the *angle of incidence* (i) and the corresponding angle in the water is called the *angle of refraction* (r). The *refractive index* between air and water (or i/r) is about 1.33.

Our eyes cannot comprehend that the light is being bent to a more vertical angle in the water, which is why the water in swimming pools seems to us much shallower than it really is. Snell showed how, from observing how a ray of light (A-A1 in Figure 12.5) bends as it hits water at an angle, you can work out exactly how light will bend for any other angle (B-B1 in the diagram). For shallower angles (such as C-C1), however, the light will tend to bounce off the surface instead of penetrating the water.

Following Galileo's trial, Descartes dropped his plans to publish a book in support of the Copernican system like a red-hot brick. Later, while travelling in Italy, he didn't take the opportunity to visit Galileo, possibly to avoid upsetting the great and good of the Catholic faith (which, after all, was his faith). He then lived for 20 years in the Netherlands which, although Protestant, was tolerant towards Catholics; elsewhere in Europe 30 years of religious wars were still rumbling on. Many have since found it odd that Descartes chose to live so

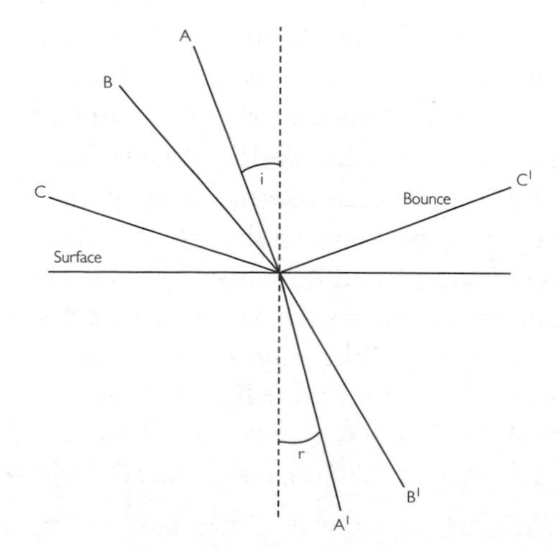

Figure 12.5

far from the major centres of scientific advancement, unless it was as some kind of observer – or even a spy – on behalf of the Church. But he was always in touch with other scholars and mathematicians, usually via an old friend from his schooldays, **Martin Mersenne** (1588–1648).

In contrast to Descartes, Mersenne chose the other career option on leaving La Fleche, becoming a monk with the Minim friars. Ultimately, however, he was to become a man of immense importance. In this age before scientific journals, Mersenne became a conduit for information, writing thousands of letters to scientists throughout Europe, from Scotland to Istanbul and Scandinavia to Spain, keeping each up to date with the others' achievements. From now on, progress in all aspects of mathematics moved faster, aided by the better communication of new discoveries from many different sources.

Mersenne openly backed both Galileo and Descartes, while denouncing mystical ideas such as astrology and alchemy. He asserted that the true way to knowledge was discovery through rigorous experimentation. He also pointed out, with surprise, that Galileo had failed to use a pendulum to time his rolling balls, although he did suggest to Huygens (more on him later) that it might make a suitable regulator in a clock.

Mersenne also devised a formula for producing prime numbers (so-called *Mersenne primes*), although his system fails with very large numbers. Nevertheless, his work encouraged others to explore the theory of numbers further – especially perfect numbers. The number 6, for example, is perfect because its factors (1, 2, 3) add up to 6; as is 28, because its factors (1, 2, 4, 7 and 14) add up to 28.

Mersenne's formula is 2^{p-1}(Note it is 2 to the power of 2 - 1). This always produces perfect numbers when p is prime.

Value for p =	2	3	5	7	13	17	19
$2^p - 1$ =	3	7	31	127	8,191	131,071	524,287
$2^{p-1}(2^p - 1)$ =	6	28	496	8,128	33,550,336	8,589,869,056	137,438,691,328

These are the first seven perfect numbers, whose factors add up to give the number itself.

Descartes's Two Major Errors

Descartes was some 30 years younger than Galileo, whose telescopes had opened up our view of the cosmos. But so far no one had discovered mathematical laws that explained exactly how planets revolve around stars for immense periods of time. Descartes decided that space must be filled with areas of force he called *vortices* (see Figure 12.6).

Descartes believed that each star pushes outwards to hold the planets in their continuous orbits – without such a force, he reasoned, sooner or later each planet must plunge or spiral down to collide with its own sun. But he was way off course: his forces were working in the wrong direction. It took a whole series of scientists to explain the truth of it all, culminating – as we shall see – with Isaac Newton.

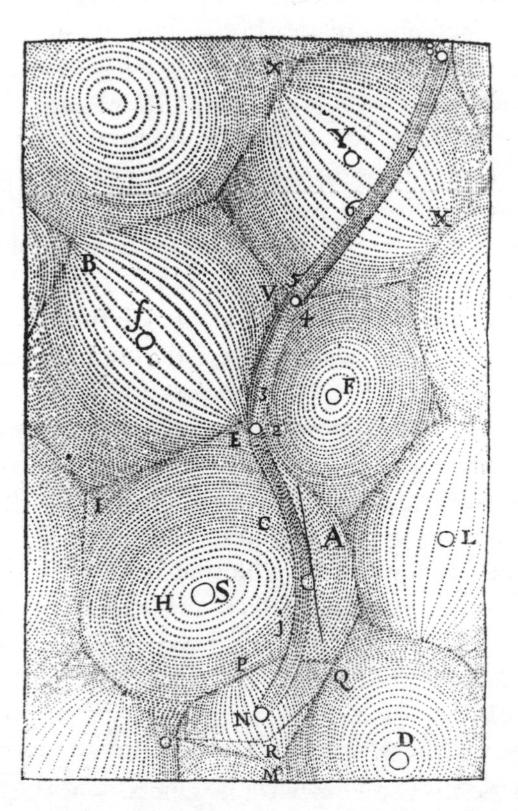

Figure 12.6: Descartes's Vortices

Sadly, the last part of Descartes' life was something of a tragedy caused by a terrible error of judgement. Queen Christina of Sweden – then 20 years old – begged him to join her court, making him an offer he couldn't refuse. He accepted, but it proved to be a fatal mistake: she was spoilt and insensitive to the point of madness. Descartes arrived in Stockholm on 1 October 1649, but the Queen ignored him until January – possibly because of pressure from her own court, as Sweden was strongly averse to Catholicism.

At last, and possibly just to be difficult, Christina summoned Descartes (who was still slumbering in the mornings) and demanded he teach her maths and philosophy three days a week from mid-January, starting each day at five o'clock in the morning. The winter of 1650 was the harshest in Sweden for 60 years, but Descartes had to be in a coach by 4.30 a.m. On 1 February he fell feverishly ill; on the 11th he was dead.

Though the cause of death appeared to be pneumonia, some suggest he was poisoned with arsenic by those fearing that his Catholicism would rub off on the Queen – a concern that may have been justified: four years later Christina not only converted, but abdicated her throne and left for Rome. Outrageous as ever, on her arrival she went straight to the new Pope, telling him she hoped he'd turn out better than the last four Popes, who had all been idiots.

As for Descartes, when his body arrived back in France it was missing its right index finger, the one he traditionally raised when saying, 'I think – therefore I am.' Oh, and his head was missing too, but that turned up again in 1809.

Today, with the universal tool that is the computer, we can design and create on screen any image our minds can contemplate, using software that makes hugely complex mathematics so simple that children, and even grown-ups, can use these amazing machines. This is all thanks to the coordinate geometry, first suggested by René Descartes, which underpins it all.

Bonaventura Cavalieri (1598–1647), a disciple of Galileo, was perhaps the last important Italian mathematician of this period. He introduced Napier's logarithms to Italy, and while Descartes was working out his geometry, Cavalieri was also breaking new ground. Both Archimedes (with the parabola) and Kepler (with wine barrels) had sought to measure areas or volumes by dividing things into incredibly narrow pieces. Inspired by Pappus, Cavalieri remarked, 'A line is an infinite number of points, a surface an infinite number of lines and a volume an infinite number of slices. By knowing the number and the size of any of these, you can calculate the whole thing.'

Evangelista Torricelli (1608–1647) impressed Galileo with his own book on mechanics, and became the old man's secretary and companion for the last three months of his life. Galileo had believed that a water pump raises water because, in Aristotle's words, 'nature abhors a vacuum'. At the time, however, owners of mines complained that no pump could lift water more than 30ft above its natural level. Galileo had shown that air had weight (something else that Aristotle got wrong) by weighing an empty pig's bladder, then filling it with air and weighing it again. Torricelli wondered if the weight of the atmosphere had something to do with the pump's height limit of 30ft…

In 1643, shortly after Galileo's death, Torricelli filled a long glass tube with mercury (which is 13.5 times more dense than water), blocked the end of the tube with his thumb and inverted it in a bath of mercury. No matter how long the tube, the mercury always settled in a column about 760mm (30in) above the level in the bath, leaving a vacuum (now called the *Torricelli vacuum*) in the top, closed end of the tube. Even when he leant the tube to one side, the level still settled at about 760mm (30in), leaving a smaller vacuum (see Figure 12.7).

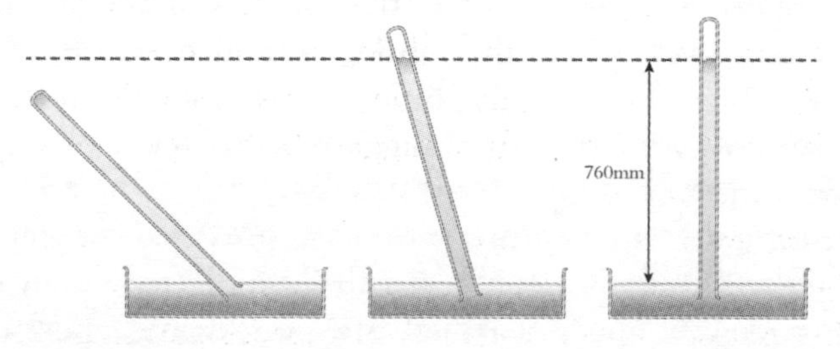

Figure 12.7

So what was stopping the mercury in the tube from descending down the tube? Torricelli thought it must be the weight of the atmosphere pushing down on the surface of the mercury in the bath. Over several days, he observed that the mercury level in the tube rose or fell a little, indicating that the weight – or pressure – of the atmosphere varied from day to day. He had invented the first ever barometer.

Otto von Guericke (1602–1686) was the flamboyant Mayor of Magdeburg, and an engineer who had helped to rebuild the city after it was ravaged during the Thirty Years' War. Following Torricelli's lead he also made barometers, and was the first to use them to predict changes in the weather. In 1657 he designed an air pump: he had two metal half spheres made, each about a metre wide, with very flat rims so they could fit together to create a ball. Each also had a handle. To seal them properly he used thin leather between the two rims, moistened to make it pliable and sticky. For a good hour, with the hemispheres held tightly together, he pumped the air out through a valve, which was then closed. Eight horses were then harnessed to each hemisphere, and were forced to pull the hemispheres apart. But they couldn't.

When the valve was released and air was allowed in, the two heavy hemispheres simply fell to the ground with a clang. Von Guericke had demonstrated that the pressure of the atmosphere was all that held them together, even against the combined power of 16 horses (see plate section).

A Blaise of Glory?

Blaise Pascal (1623–1662), despite being a sickly child, was clearly a genius from an early age: aged nine he had discovered Euclid's first 32 theorems for himself. Still only 16, he wrote a book on conic sections that included what has since been called Pascal's theorem. It states that if you draw an irregular hexagon in a circle, each pair of opposite sides, provided they are not parallel, will meet at a point, and all three of these points will fall on a straight line. Amazingly, the same thing applies if you inscribe the hexagon in an ellipse.

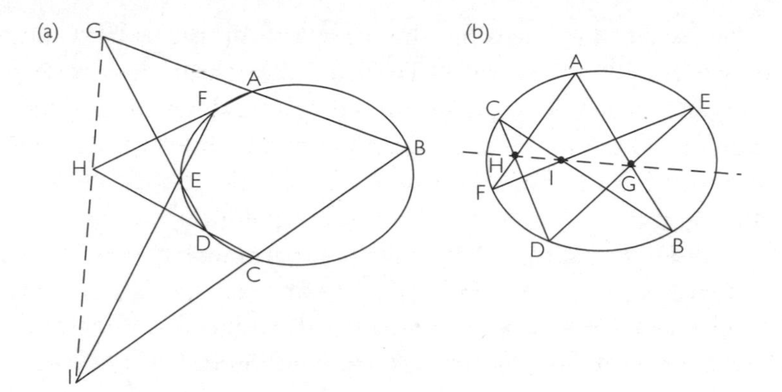

Figure 12.8

Figure 12.8a shows points A, B, C, D, E and F on an ellipse. If you extend sides AB and DE they will meet at point G; while CD and AF will meet at point H, and sides BC and FE will meet at I. In all cases, the meeting points G, H and I will always fall on a straight line.

Pascal also found that with any hexagon contained within an ellipse, if alternate points are connected, so that each point has two lines from it meeting two other points, then once again the three crossing points of those six lines will always all on a straight line, this time within the ellipse (see Figure 12.8b). He was only 16 when he met Gerard Desargues (1593–1662) who advised him to explore and extend these discoveries in projective geometry. This he did and we have an example of that work in the Wow Factor Mathematical Index.

Aged 19, the precocious Pascal invented a cog-wheeled calculating device based on the mechanism inside weaving machines (see plate section). It worked perfectly, but it was too expensive to sell widely. At the age of 24 Pascal was now famous, and René Descartes came to see him. It was then that Pascal, still a delicate man, heeded Descartes's advice and took to staying in bed in the mornings.

To explore the nature of the atmosphere, from his bed, Pascal asked his brother-in-law to trek up the Puy de Dôme (a local mountain) with a mercury barometer strapped to a mule. At a height of a mile above their starting point, the mercury level in the top of the tube had fallen by about 3in, so the pressure of the air at that height was much lower. Pascal concluded that the atmosphere must have a top to it, above which there must be nothing. Aristotle had said that nature abhors a vacuum. In fact, discounting stars and their planets, it now seemed that the entire volume of space was filled with nothing but nothing. But the laws explaining how it all worked were to remain elusive for a few years yet...

Pascal also expanded on Archimedes' ideas about levers to demonstrate the mathematical principle behind two different pistons (see Figure 12.9). Here the large piston is about eight times as big as the small one. Push the large piston down one unit and the small piston will rise eight units – the water pistol principle. But push the small piston down eight units and the large piston and its load will rise one unit. Today this mechanical advantage enables modern heavy building machinery to exert immense forces using varying oil pressure.

Pascal's name is often linked with **Pierre de Fermat** (1601–1665) – about 22 years his senior – who worked independently to devise a theory of probability. Both men were asked to solve a problem for the Chevalier de Méré, a wealthy gentleman whose gambling habit was making him less wealthy by the day.

One problem they were asked to solve went like this: a gambler was involved in a three-dice game with simple

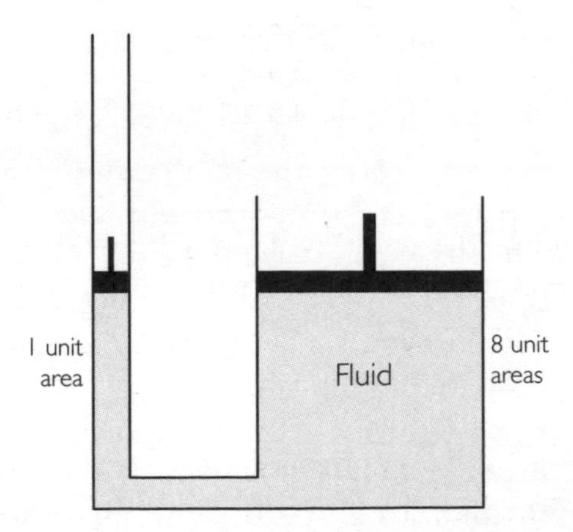

Figure 12.9

rules. After rolling all three dice, if the total was 12, the gambler won, but if the total was 11, he lost. With three dice there are six possible ways to make either number, so why, asked the gambler, was he consistently losing over time? Pascal and Fermat explained: he had not taken into account the number of possible ways that each of the six variations can be thrown.

Why, With Three Dice, a Throw of 11 Occurs More Often Than a Throw of 12

If all three dice show different numbers, there are six ways to throw a number.

For example, 641 can be thrown as 641, 614, 416, 461, 146, 164: six ways.

If two of the three numbers are the same, there are only three ways to throw a number.

For example, 551 can be thrown as 551, 515, 155: three ways.

But if all three dice show the same number, there is only 1 way to throw that number: 444.

Possible ways to throw	11	Possible ways to throw	12
6 4 1 variations	6	6 5 1 variations	6
6 3 2	6	6 4 2	6
5 5 1	3	6 3 3	3
5 4 2	6	5 5 2	3
5 3 3	3	5 4 3	6
4 3 3	3	4 4 4	Only 1
Total	27		25 x 4 = 108/100

If we multiply the result by 4, we can see that the player throwing for 11 always has an 8 per cent advantage over the person throwing for 12.

Pascal wrote *A Treatise on the Arithmetic Triangle*, although it wasn't published until after his death. In it he famously explained what we now call 'Pascal's triangle'.

The triangle had been discovered by the Chinese, who laid it on its side, working from right to left – as we can see from the lettering on this version, from AD 1303 (see Figure 12.10a).

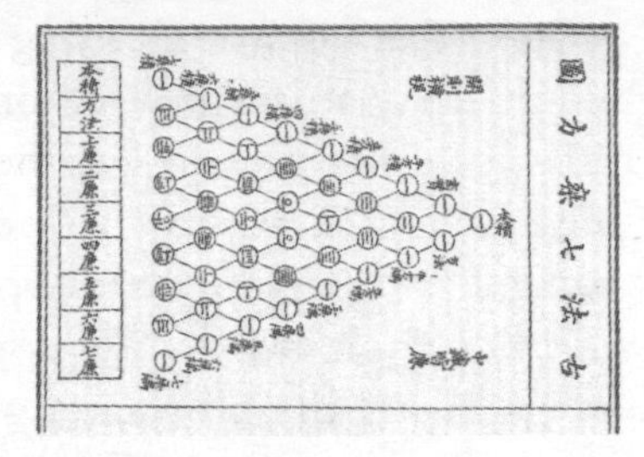

Figure 12.10a

Pascal first produced it as a square array (see Figure 12.10b), in which the first row contained 1s, the second row contained the natural numbers: 1, 2, 3 etc., and the third row contained triangular numbers: 1, 3, 6, 10, 15 etc. Today we normally see it as a triangle, as shown in the right-hand diagram, starting with a single 1 at the top and then progressing downwards. After the first 1, each successive row is an expression of the binomial (a + b) to the power of that row.

So the fourth row down is $(a + b)^4$, which gives $1 + 4 + 6 + 4 + 1 = 2^4 = 16$.

With this table, calculating the possible number of chances that something will happen is child's play. As an example, how many ways can you take two fingers from five? There happen to be 10 ways – try counting them.

Pascal's triangle automatically gives you the answer. You have 5 fingers, so from the top 1, count down 5 rows. To check how many times you can take 2 from 5, count 2 in and you arrive at 10. Sure enough there are 10 ways of taking 2 from 5.

But every time you take 2 from 5, you leave 3, so this time count down 5 rows and in 3 and you arrive at 10 once more.

1	1	1	1	1	1
1	2	3	4	5	6
1	3	6	10	15	21
1	4	10	20	35	56
1	5	15	35	79	126
1	6	21	56	126	252

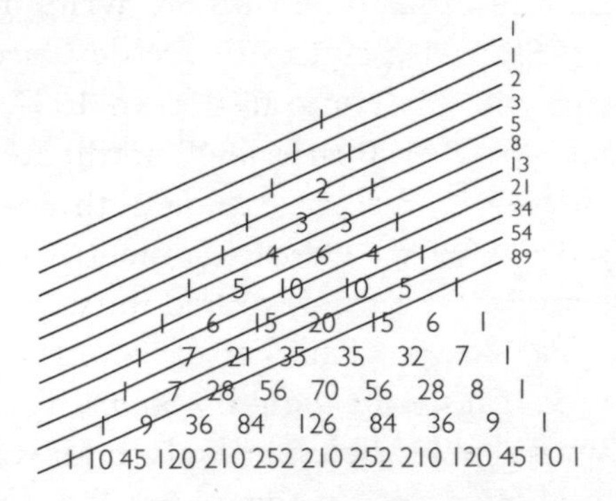

Figure 12.10b

One other fascinating thing about Pascal's triangle is that it features the natural progression of Fibonacci numbers. In the right-hand figure, notice the parallel diagonally slanting lines. The first passes through the top 1 and the second through the 1 on the left end of the next row. The third line passes through two 1s; the next through a 2 and a 1, which equals 3, and the next through a 1, a 3 and a 1, making 5. Sure enough, successive lines give totals of 1, 1, 2, 3, 5, 8, 13, 21, 34 etc – the Fibonacci series.

In 1654, after surviving a severe coach accident in which the horses were killed, a much-shaken Pascal became consumed by religion, dropping mathematics like a hot brick. Due to his mathematical reputation he was asked to work out the odds against there being a God. He decided, 'Should I arrive at the Pearly Gates not believing, and should He be there to meet me, it would prove to be a very unfortunate error.' After that, he returned to mathematics on only one occasion, when, consumed by a raging toothache, he tackled and solved much of the maths around cycloids (which we featured in Chapter 11).

Fermat's Elegant Proof That Never Was

Pierre Fermat was quite an amazing mathematician, especially when you consider that maths was only ever a hobby to him, born of reading Archimedes, Pappus and Diophantus. Because he didn't take it too seriously, he neglected to write down many of his discoveries, and he published nothing in his lifetime. Instead he chose to write many letters to Mersenne and Huygens, and to Pierre de Curcavi, to whom he sent a paper on analytical algebra in 1637, a year before Descartes published his Cartesian coordinates. In fact, immediately afterwards Fermat explored three-dimensional coordinate lines, whereas Descartes mentioned three dimensions but seems to have only worked in two.

To impress others, Fermat could be devious, and occasionally made claims he could not justify. Once, while exploring Pythagorean triples, he had a brainwave while reading a book by Diophantus. In the margin he wrote, 'Just found a simple and elegant proof that while $a^2 + b^2 = c^2$ is common, there can be no possible solution for $a^3 + b^3 = c^3$, or for any higher equal powers.'

When English mathematician Andrew Wiles finally proved this to be correct in 1995, the proof ran to 130 pages. It's pretty certain that there never was a 'simple and elegant proof', as Fermat had suggested, and as far as we know he never wrote anything further on the subject.

Independently of Pascal, Fermat worked on his own theory of probabilities, and one example clearly explains the complexity of his methods. Two players are playing pitch and toss, and each has four throws left. One requires two heads to win, the other three tails. What are the odds for each player?

First, Fermat tabulated every outcome for four tosses of a coin: there are just 5 possible outcomes, but 16 versions of them, according to the order in which they occur.

Outcomes	hhhh,	hhht,				hhtt,						httt				tttt.
Versions	hhhh	hhht	hhth	hthh	thhh	hhtt	htht	htth	thht	thth	tthh	httt	thtt	ttht	ttth	tttt
2 throws	h	h	h			h										
3 throws				h	h		h		h						t	t
4 throws								h		h	h	t	t	t		t

Note that of the 16 possible outcomes, there can only be one winner for each. So each group of four has a letter underneath it, which signifies when the bet is won by 'h' or 't'.

Heads wins 11 times while only five solutions favour tails. So the odds are 11/5, and heads is the best bet by more than 2 to 1. But heads has four ways to win in just two throws. After three throws the odds are 8/2, or 4/1 in heads' favour. Only if the match is still undecided by the fourth throw are the chances even.

The Point about Fermat Points

Fermat once challenged Torricelli to find the solution to a problem regarding triangles that was equivalent to asking, 'If you have three towns, where would you place straight roads to connect them so the total length of the roads was the minimum possible?'

Fermat made a triangle linking the three towns and then drew an equilateral triangle on each side. He then connected each far point to the opposite angle of the triangle. These lines crossed at what became known as Fermat Point. Now roads from each of the three towns to this point would give the shortest total connected road length (see Figure 12.11).

A Belgian physicist called **Joseph Plateau** (1801–1883) solved this problem using a solution of soapy water. If you place three thumb tacks between two pieces of Perspex and dip the whole lot in soapy water, bubbles will form as walls linking the three tacks, neatly solving the problem. The three bubble walls always meet at an angle of 120 degrees. If you used four or more tacks you will get more bubble walls, but they will always meet in threes with 120 degree angles – Fermat Points (see Figure 12.12).

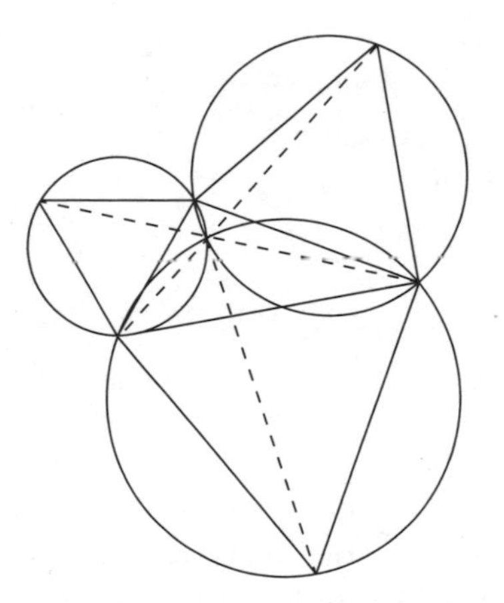

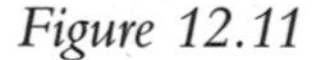

Figure 12.11

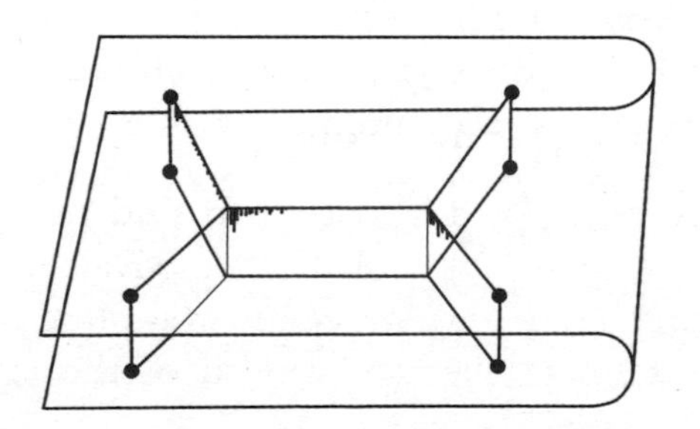

Figure 12.12

Fermat's failure to publish his work often meant that – his celebrated theorem aside – he discovered simple and elegant solutions to problems that we only found out about after his death. From gathering the evidence found later in his correspondence, it seems that he independently discovered the principles of integral calculus ahead of both Newton and Leibnitz, who both got the credit. In doing this he discovered a very neat fact about tangents to a parabola. Take a parabolic curve with an axis running through it. Can you find the tangent's exact point of contact with that curve? In Chapter 4, where we met Archimedes, I showed you how to fold a parabola with a piece of A4 paper, and how to understand the relationship between the curve and its directrix.

Just as the peak of a parabola is equidistant from its focus and its directrix, Fermat demonstrated simply and elegantly another remarkable property of these curves. If you take the point where any tangent to a parabola crosses its axis and repeat that distance along the axis inside the parabola, a line drawn at that point at right angles to the axis will meet the curve exactly at the point of contact of the tangent (see Figure 12.13). So neat!

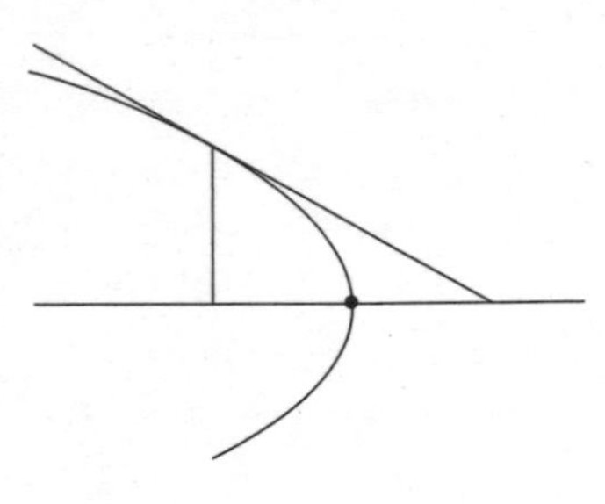

Figure 12.13

Terribly British

John Wallis (1616–1703) was an English mathematician taught by William Oughtred, who turned out to be a good gambler with a knack for making the right choices at the right time throughout his life. When the English Civil War broke out, he sensed that the Parliamentarians would win and, like François Vieté, he gained favour with the victors by using his skills to decipher coded Royalist messages – which turned out to be a wise move. In fact he was a Royalist at heart, and – taking a big risk – tried to use his influence to plead for clemency for King Charles I, without success. The King lost his head, but Wallis kept his: the Parliamentarians forgave him, and even offered him a

Left: *Saint Jerome in his Study* by Albrecht Dürer. The vanishing point is well to the right of the picture (p. 278).

Left: *Melancholia I* by Albrecht Dürer, painted in 1514 as the lower line of the magic square shows (p. 279).

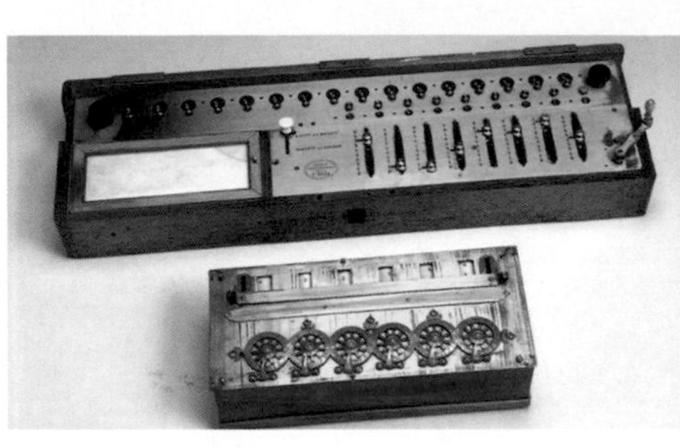

Above: Pascal's first calculating machine (below) with later version above (p. 345).

Above: A similar hydrostatic balance to the one produced by Galileo (p. 317).

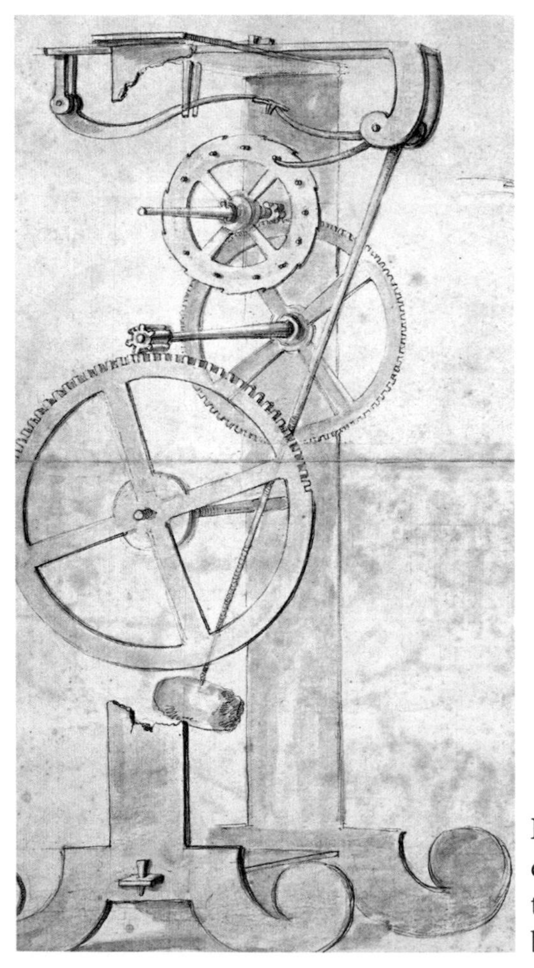

Left: Galileo's clock escapement, dictated to his son when he was blind (p. 317).

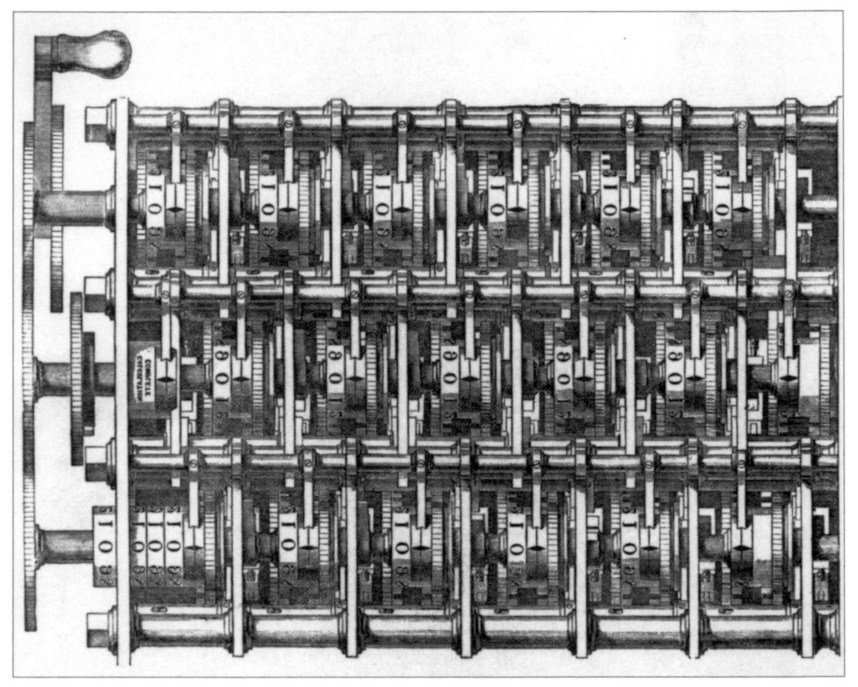

Above: Charles Babbage's analytical machine. It worked in base 10 (p. 409).

Below: Geissler tubes and various glowing gases (p. 391).

Below: Newton's prism and rainbow experiment (p. 369).

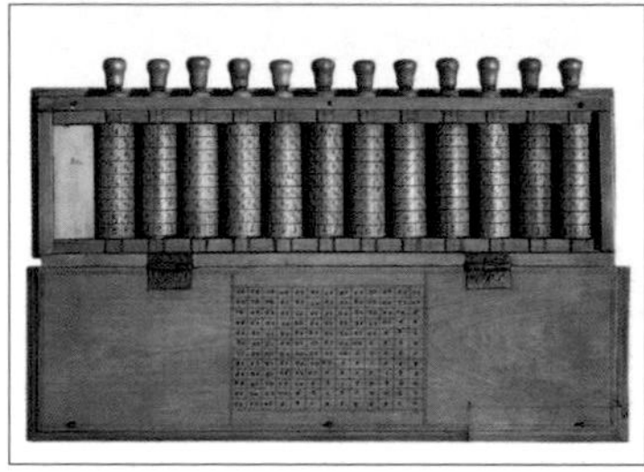

Above: Plan of The Winton Beauty of Mathematics Garden, Chelsea Flower Show 2016 (p. 403).

Left: Napier's Bones – an 'upmarket' version (p. 300).

Below: Hooke's revolutionary micrographia drawing of a flea (p. 362).

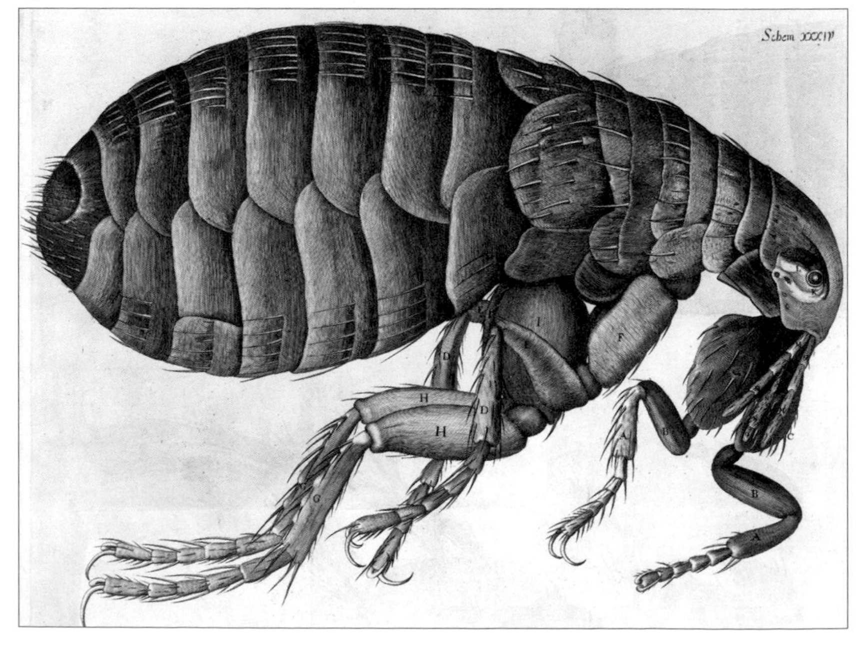

Left: Gaudi's hanging cathedral design, as seen through a mirror (p. 429).

Left: The Menger sponge – a Sierpinski gasket in three dimensions (p. 428).

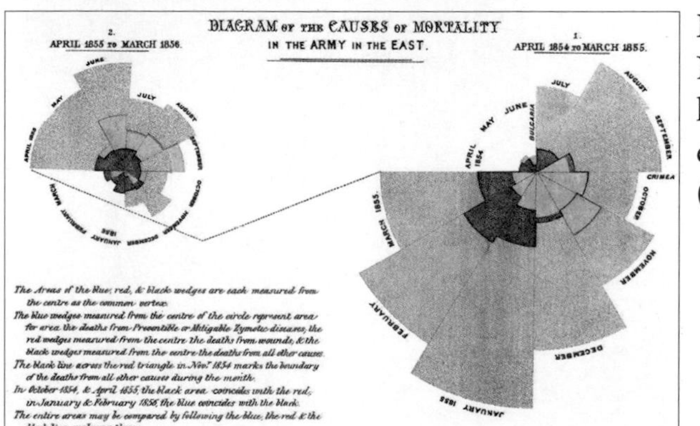

Left: Florence Nightingale's highly detailed early pie charts (p. 410).

Right: Two examples of Julia sets. In both cases, both patterns and colours are created by ever-changing mathematical equations (p. 433).

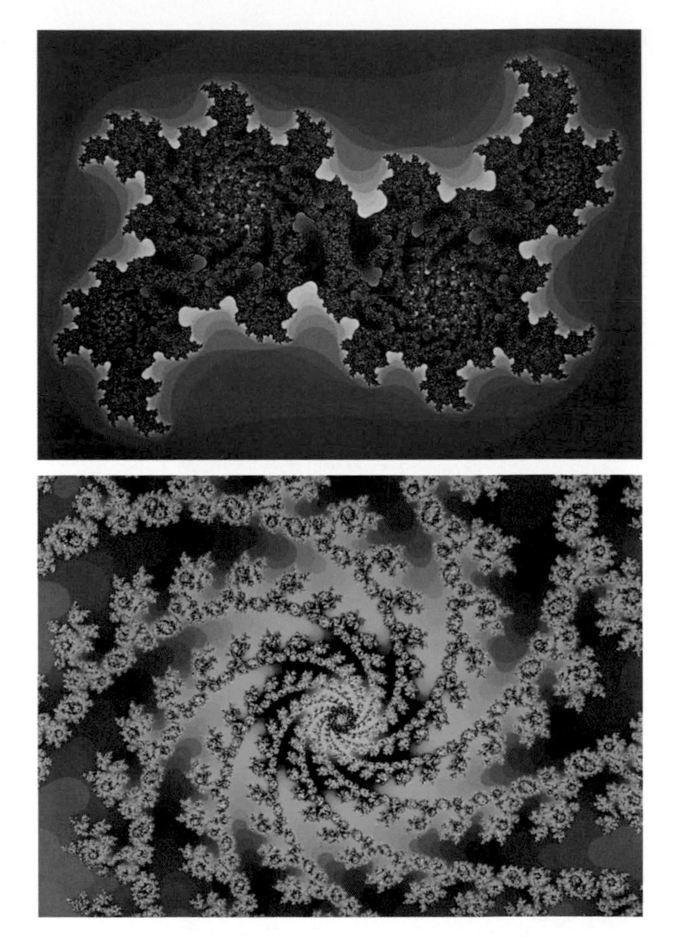

Below: Two examples of Mendelbrot sets (p. 433).

Left: Three examples of tiling geometry. Gaudi's hexagonal tiles, Barcelona. They form three hexagonal patterns (p. 457).

Left: Coloured tile mosaic floor from the baths of Caracella in Rome, 3rd century AD.

Below: This modern Lisbon wave pattern seems to rise and fall like the sea.

Left: Wall tiling from the Alhambra, Spain, with changing coloured tiles on a repeating lattice (p. 457).

Left: Multi-coloured design, with an explosion of pattern ideas.

Below: Anything achieved in tile can also be replicated in wood carving – this is also from the Alhambra.

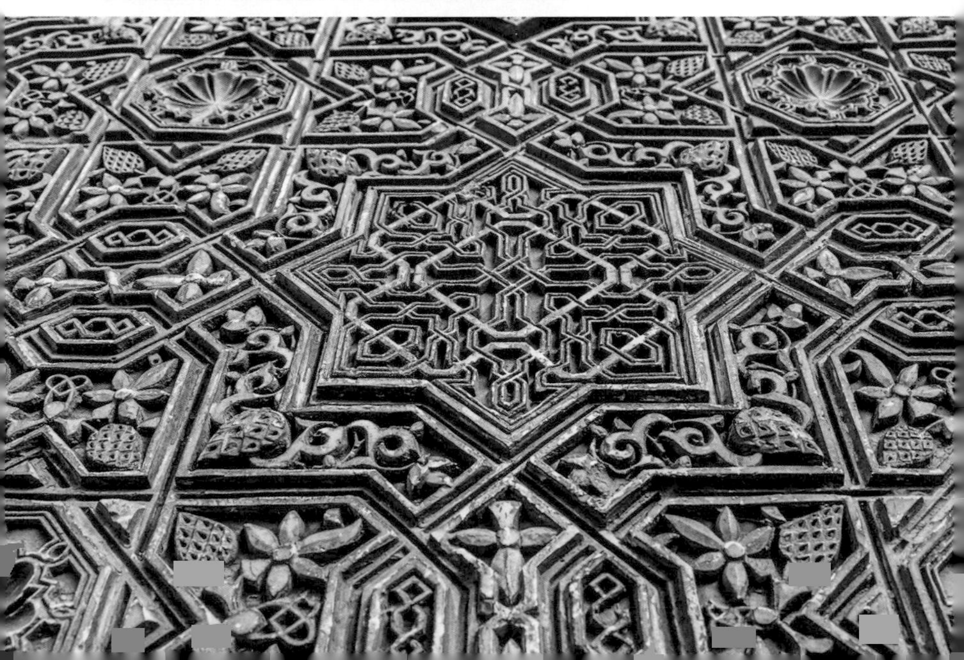

professorship at Oxford. When Charles II regained the throne in 1660, Wallis's Royalist leanings were remembered and he was made King's chaplain.

Wallis was brilliant at mental arithmetic: he once worked out the square root of a 53-digit number, correct to 17 decimal places, in his head. He was the first to use exponents to represent negative numbers and fractions, writing $x^{1/2}$ as equal to $\sqrt{x}$, for example. He was also first to use ∞ as the symbol for infinity. And in seeking to solve the ancient 'squaring the circle' problem, he remarkably devised an infinite series that gradually homed in on an accurate value for π:

$$\frac{\pi}{2} = \frac{2 \times 2 \times 4 \times 4 \times 6 \times 6 \times 8 \times 8 \quad .....}{1 \times 3 \times 3 \times 5 \times 5 \times 7 \times 7 \times 9 \quad .....}$$

Most importantly, Wallis made Descartes's *The Geometry* much more understandable, which was a great help to his friend Isaac Newton. He was a founder member of the Royal Society and was 'terribly British', showing contempt for the Gregorian calendar and delaying its adoption in Britain for about 50 years, just because it originated in Rome. He hated both Descartes and Leibnitz simply because they were foreigners, and had a blue fit (aged 70), when he learnt, after Leibnitz had published the calculus, that Newton had come to the same conclusions with his fluxions much earlier, but had sat on the idea for 20 years.

The Monumental British Architect

But for the Great Fire of London of 1666, **Christopher Wren** (1632–1723) might have been remembered not as an architect but for his mathematics, because he was a student of Oughtred and certainly no slouch at the subject. He increased our understanding of cycloids by discovering that the length of a cycloidal curve is exactly four times the diameter of the circle that generates it, and would neatly form a perfect square around it.

Of course, Wren is most famous for designing today's St Paul's Cathedral, and for having the task, in partnership with Robert Hooke (more on him later), of rebuilding the whole of central London, which had been ravaged by fire. Dreaming of achieving what Agrippa had managed in rebuilding Rome, both men went to work with great energy and confidence. They planned grand, wide, tree-lined avenues to replace the ramshackle narrow streets clogged with outwards-leaning Elizabethan half-timbered houses that, in some cases, almost met over the middle of the street. They designed parks and grand squares too, but almost all their efforts were in vain. The original property owners demanded that no one reduce the plots they owned before the fire by so much as an inch.

In rebuilding St Paul's (1675–1710), Wren grasped this enormous opportunity with both hands, revealing his genius. His tombstone, inside the cathedral itself, states simply (in Latin), 'If you want to see his monument, look about you.' It was not 'all his own work', however – Hooke was on hand to help, and it was he who was inspirational in creating the building's celebrated dome. As Brunelleschi had done, Wren and Hooke found designing a dome a highly complex undertaking. Hooke, however, in his work at the Royal Society, was intimately aware of every new idea in science, and he had become a great friend of Huygens.

It was Huygens who saw that a chain hanging down from two points at the same height did not 'quite' form a parabolic curve: the curve is slightly narrower and sharper at the bottom, a shape that Huygens dubbed a *catenary* (see Figure 12.14). As the catenary is shaped by gravity, Hooke reasoned that when upside down it would form a very strong self-supporting shape. He and Wren placed a mirror on the floor and hung a chain from higher and lower points until, in the reflection, they saw the shape they agreed would be most suitable as their central core support.

Figure 12.14

St Paul's inner dome is a conventional half-sphere shape, as is the outer dome, though set much higher. But a rigid

Figure 12.15

rounded top cone of stone, very close to a catenary, set between them was used to bind the two domes together, and it worked (Figure 12.15).

This method inspired the amazingly complex La Sagrada Familia Cathedral in Barcelona, designed by Antoni Gaudi in the twentieth century: its many arches were all first planned by observing hanging strings in a mirror on the floor and are almost all upturned catenaries.

You can see an example of how Gaudi designed his buildings upside down in the plate section.

Out of This World Discoveries

Christiaan Huygens (1629–1695) was a Dutch physicist and astronomer. His father was an important government official, and Descartes was a friend who encouraged the lad's love of mathematics. In 1657 Huygens published a book on probability that focused on the odds of people's life expectancy, which had considerable influence on the emerging life-insurance business.

Huygens loved astronomy and physics more, however. He befriended **Baruch Spinoza** (1632–1677), who became a noted philosopher, often opposing the philosophy of Descartes. Spinoza was an excellent lens grinder, but he died aged only 44, probably from inhaling glass dust. With Spinoza's help Huygens made a telescope 23ft (just over 7m) long, and used it in 1656 to discover a huge cloud of gas and dust, the Orion Nebula.

He also discovered that Saturn wasn't a triple planet as Galileo had assumed from its apparent shape, but was in fact a unique planet surrounded by a huge but very thin ring that did not touch the planet at any spot. The ring was set at an angle, and seemed to disappear every 14 years, when its position was end-on to the Earth. He additionally discovered that Saturn had a large moon, which he called Titan.

This discovery meant that the Solar System now contained six planets and six satellites (including the four around Jupiter and our own Moon). For Huygens this was so neat that he declared that there were no more moons to be discovered. Soon, however, thanks to **Giovanni Cassini** (1625–1712), four more moons were found orbiting Saturn alone. Cassini also saw that Saturn's ring had a gap, splitting it into an inner and an outer ring, both of which orbited the planet, and both of which were made up of literally billions of rocks, all in effect miniature moons. From that moment Huygens sensibly stopped making astronomical predictions.

In 1656 Huygens saw that a pendulum's swing was not quite regular if it swung widely from side to side. Eventually he solved this problem by using two upturned cycloidal curves to control the pendulum's swing, making it far more accurate (see Figure 12.16). Using gradually falling weights pulling on chains to drive the mechanism, Huygens then produced the first ever longcase or grandfather clock. This

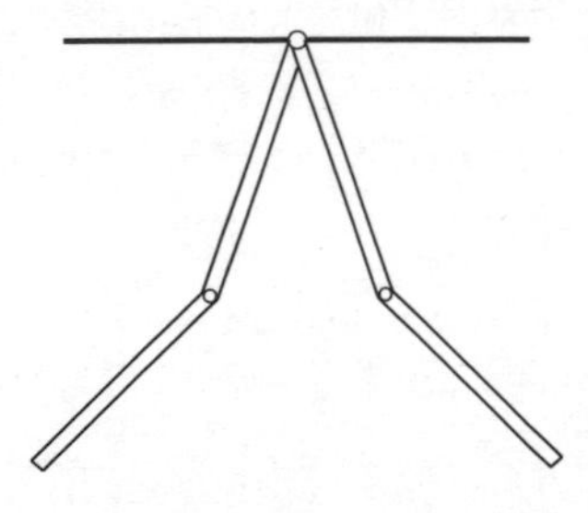

Figure 12.16

invention opened the floodgates to a rush of new clockmakers, all endeavouring to produce ever more accurate clocks and watches.

Almost immediately, the invention of more accurate clocks led to more accurate astronomy, and at last someone improved upon Eratosthenes' measurement of the Earth. He was **Jean Picard** (1620–1682), one of the charter members of the French Academy of Sciences, founded in 1666 by Louis XIV, and financed by the crown (unlike the Royal Society in England). Using Huygens's pendulum clock in 1671 he declared the Earth's circumference to be 24,876 miles. Today we know it to be 24,859.75 miles over the Poles and 24,901.47 around the Equator. You'd have to be pretty picky to pick holes in Picard.

To achieve his accurate measurements, Picard went to Denmark to check the figures Brahe had gathered from the site of the old observatory, which by then had been destroyed. They were spot on. Picard was assisted by a young Dane, **Ole Roemer** (1644–1762), who showed him the site. Picard was so impressed with Roemer that he brought the gifted young astronomer back to Paris with him.

Roemer soon took the art of measurement to new heights by determining the speed of light, which is quite enormous. René Descartes had made the error of declaring light to be infinitely fast. Galileo had believed that light 'must have a speed', and first tried to measure it by deploying men with lanterns on distant hills, and asking them to cover and uncover their lights. Unfortunately his results were worthless.

Only now, with accurate timing devices, could the error be corrected. Roemer studied the timings of Jupiter's moons and their eclipses, observing that his timings varied according to the time of year. When the Earth was approaching Jupiter the eclipses came progressively earlier, and when the Earth was receding from Jupiter they came slightly later. This meant that light took a definite time to cross the Earth's orbit.

Unlike Galileo's hills on Earth, Roemer's 'hills' were Earth and Jupiter, some half a billion miles apart. He calculated the speed of light at 298,000km per second, an incredibly close first attempt – we now know that the correct value is 299,792km per second. Surprisingly, other than Picard and Huygens, no one in France seemed really excited by this. In England, however, Hooke, Halley, Flamsteed and Newton were all very impressed.

The Royal Society

The first scientific society was Italy's Academy of Experiments (Accademia del Cimento), established in 1657 by Giovanni Borelli and Vincenzo Viviani, both of whom had been pupils of Galileo. Sadly, it lasted just 10 years, perhaps best illustrating how the centres of European learning had by then moved north.

Even before then, however, from 1645 onwards like-minded British scientists had started to gather regularly in colleges and London coffee houses. In 1662, under the charter of Charles II, they formed the Royal Society. As England's first 'society' it has never required an addition to its name, which is often shortened simply to 'The Royal'. **Robert Boyle** (1627–1691) and Christopher Wren were two founder members, while Robert Hooke became active secretary, responsible for helping speakers and authors present their findings to a wider audience. This was of great benefit to Hooke, whose knowledge of many scientific subjects – and the mathematics involved – grew and grew.

Boyle's father, the Earl of Cork, was the richest Englishman of the times, but was constantly on a war footing with the rebels in Ireland. So he sent young Robert off to Eton College where, by the age of eight, he was already speaking Greek and Latin; he later toured Europe with a private tutor. While in Florence, aged 15, Boyle heard of Galileo's death; seeing how influential and respected he was, young Robert was inspired to follow science.

By the time he was 18 his father had died, but Robert was now independent and wealthy. He managed to keep out of the English Civil War, studying at Oxford, where he developed an improved pump with his able assistant, Robert

Hooke – repeating Guericke's experiments. Establishing good scientific practice, Boyle also carefully described all his experiments, and showed them to others for confirmation.

Boyle was the first person to drop a feather and a lead weight in a vacuum to demonstrate that they both fell at the same speed – proving Galileo was right. He also showed that a ticking clock kept going in a vacuum, although it couldn't be heard. Boyle concluded that sound must be conducted via air, which must be composed of discrete particles – as Democritus had suggested (and which he had called 'atoms'). But if air can be compressed, thought Boyle, then those atoms must be surrounded by nothingness…

Boyle was also the first to collect gases, noticing that they compress by a simple inversion process: double the pressure and the gas takes up half the volume; halve the pressure and the volume doubles (this is still known as 'Boyle's law'). He referred to the 'springiness of air', because it matched perfectly the results of the experiments with springs that Hooke was performing at the same time.

When Boyle was still a young man alchemy was rife, and he himself believed that gold could be made from other substances, even advising the government not to ban the practice because it was sure to bring financial benefits. Slowly, however, he began to transform the discipline of chemistry, distancing it from alchemy and medicine, until it became a science in its own right. In 1661 Boyle published *The Skeptical Chymist*, in which he decried alchemical notions. In it he maintained that an element was a substance that could only be deduced by experiment; anything that could not be broken down into simpler substances must be an element; and two elements could be combined to form a compound.

In 1680 Boyle produced phosphorus, although the German chemist Hennig Brand had managed it about 10 years earlier. Both men arrived at the element by boiling huge amounts of urine – a messy business, although the reward was this magically glowing element.

Robert Hooke (1635–1703), the son of a parson, was a weak and sickly youth, but his natural cleverness got him to Oxford. But he had no money, and had to pay his way by

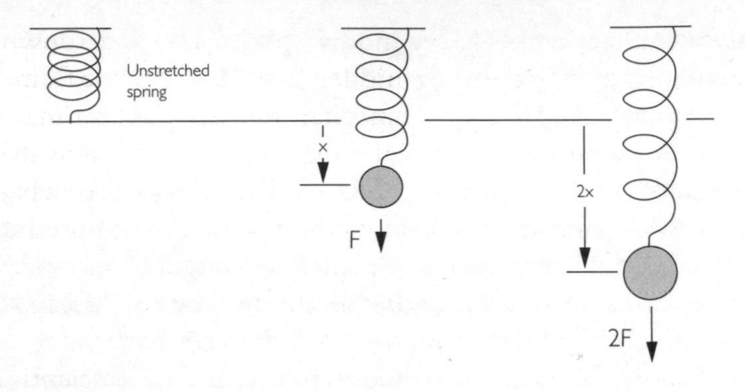

Figure 12.17

waiting on and cleaning for wealthier students – a humiliation that stayed with him all his life. Boyle, however, just eight years his senior, took him under his wing. Hooke's mechanical skills made him an ideal assistant, helping him to create Boyle's efficient vacuum pump.

He studied springs; 'Hooke's law' still states that 'Double the weight on a spring and the spring becomes twice as long; halve the weight and the length reduces by half' (see Figure 12.17). Hooke also produced the first hairspring, which moves backwards and forwards from its natural central state; immediately afterwards timepieces and even pocket watches began to emerge, with an accuracy at least as good as Huygens's cumbersome pendulum.

By testing theories that came to the Royal Society from blossoming scientists in many fields, Hooke gained great scientific knowledge. He saw that earthquakes are caused by heating and cooling within the Earth's crust, and that steam power would be a major force in the very near future. He even suggested an atomic theory, well ahead of John Dalton 150 years later.

Hooke also invented the universal joint, in which one shaft turns another, irrespective of the angle between them. You can still see them in action today on the undersides of trucks, where the drive shaft meets the differential that turns the rear wheels.

Hooke invented the sash window, the distinctive feature of English buildings from his time right through the

Georgian period. The window slides up and down, counterbalanced by weights hidden in each side of the frame. And because the rope in the mechanism has to last for many years, Hooke devised a special woven rope that today is still called *sash cord*. He used this cord – along with an ebony ball and tube – to make a version of the toy Kepler had used to help explain the movement of planets around the Sun. Today it can be seen in a glass case in the library of the Royal Society in London.

As soon as Galileo had discovered the telescope, scientists (including Galileo himself, as we have seen) tried using it in reverse. The science of *microscopy* reached new heights in Italy in 1660, when Marcello Malpighi became the first person to examine blood vessels in a frog's lung. He speculated that they were so thin that air might pass through their walls so it could be carried around the body – quite a revolutionary observation.

An even more prolific microscopist was Dutchman **Anton van Leeuwenhoek** (1632–1723), a draper who needed to study very fine cloth close up. He ground his tiny lenses – some as small as a pinhead – to incredible perfection and mounted them on a simple device. With these single-lens microscopes he studied living spermatozoa, which embarrassed him terribly; he also saw that fleas have parasites which in turn have smaller parasites. At the time, the writer Jonathan Swift quickly penned a poem – of which this is the modern version:

Little fleas have smaller fleas
upon their backs to bite 'em,
and these small fleas have lesser fleas,
and so, ad infinitum

Van Leeuwenhoek ground over 400 lenses in his lifetime. He sent 26 to the Royal Society, which made him a Fellow.

Hooke took Van Leeuwenhoek's lead, making a magnificent compound microscope, and in 1665 producing a truly wonderful book – surely the finest publication of the age – *Micrographia*, which features many superb drawings, including several by his close friend Christopher Wren. The most famous of all is Hooke's Flea (see plate section).

By observing thin slivers of cork under a microscope, Hooke saw that the rectangular spaces had been filled with fluid when the tree was alive. These he called *cells*, because they resembled the small, cell-like rooms in monasteries. It is still the term we use today.

Robert Hooke was clearly one of history's greatest scientists. Yet today he is more widely known, not for the sheer genius of his scientific contributions, but for the disputes he had with the only other scientist of those times who could be said to be greater than he was: Isaac Newton.

CHAPTER 13

A Mathematician With Gravitas

Without doubt, Isaac Newton is one of the greatest mathematicians of all time, and arguably Britain's greatest ever genius. His discoveries and originality of thought were extraordinary, and his life and works mark the threshold to our modern world of mathematics.

He was born prematurely on Christmas Day 1642 in Woolsthorpe, and was so small – it was said – that he could fit into a quart pot. His father had died before he was born, and when Isaac was three his mother married again and left Isaac with his aged grandparents. This may have been why Newton was always a solitary figure – a common trait among those of great genius. He experienced the company of women hardly at all until very late in life.

But he was an incredibly resourceful youngster, making kites, sundials, water clocks, a mill powered by a mouse in a wheel and possibly the first known cat flap. His strength of character, and his dislike of being second best, first showed itself when he outwitted the class bully as a young teenager.

When he was 17, his mother demanded that he quit school and earn his keep by working their farm. Isaac obeyed, but was a terrible farmer (probably because he chose to be). Luckily his uncle had spotted his academic potential and, as a member of Trinity College himself, arranged for Isaac to come up to Cambridge as a *sizar.* There he could pay for his education by serving more wealthy students, cleaning their rooms and emptying their chamber pots.

Newton's Solitary Two Years

Newton revelled in his studies and got his degree from Cambridge in 1665, but plague was now ravaging England: at its peak 20,000 people died in one week in London alone. Cambridge, like all places where people gathered together, closed its doors, and Newton returned home to Woolsthorpe. Crucially, however, before leaving he grabbed every book he felt might be of use to him. Over the next two years, and completely alone, he was to make some quite astounding discoveries in the most productive period of his immensely inventive life.

He read Galileo, repeating and confirming the Italian's many experiments with falling objects, and his assessment of the mathematical flight pattern of a cannonball. He also had a copy of Kepler's *Harmonice Mundi*, which Newton devoured, confirming its three laws of planetary motion and making the ball-and-string toy that helped explain it. It was at this time, free from outside influences, that he put the work of these two great men together, laying down his 'laws of universal gravitation'. This monumental discovery alone was to eventually earn him the title of the greatest ever British scientist: essentially for reading and combining the work of just two scientists, and perhaps reading just two books.

Much later, in his eighties, Newton mentioned to the writer Voltaire that the idea of gravity possibly came to him as he considered why an apple falls downwards from a tree. Voltaire published the story, and today 'the falling apple' is the most widely known Newton anecdote, which is a monumental shame given the richness of his many achievements. When the third-greatest mathematician of all time (after Archimedes and Newton), Carl Friedrich Gauss, heard this story, he said, 'Silly. He must have talked about apples because he saw he was talking to an idiot.' I am sure this is true: Voltaire was an opportunist, and certainly no mathematician, and I have never found a mention of apples in Newton's writing. Yet the story is fed to every young person as though they too are idiots…A falling apple would only have been of great interest to Newton if it had fallen from a plum tree. The idea is just too simple, and Newton's mind clearly explored and explained much more in those Woolsthorpe years, as shown by his own drawing of the time, which admittedly came to light only after his death (see Figure 13.1).

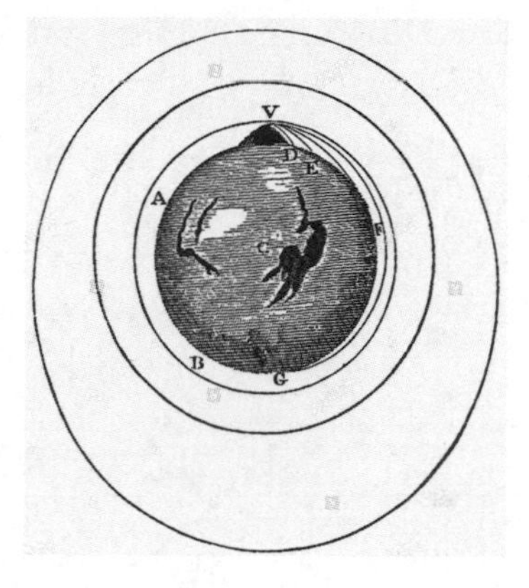

Figure 13.1

Newton asked himself, 'What would happen to a ball thrown sideways?' The answer: it would fall to Earth in a curve. Throw it harder sideways and it would fly further but still curve down to Earth. 'But', he thought, 'what if I could throw it sideways superhumanly fast – so fast that by the time it started to curve into its descent, the Earth itself was curving away? In that case the ball would travel right around the world at roughly the same height, come back and hit me on the back of the head!' That, he realised, is what the Moon is doing: not hitting him on the head. The Moon is falling and accelerating towards Earth all the time, yet travelling at just the right speed so it continues in a perpetual orbit. But in that case, he then asked, how is the flight of the Moon regulated?

If the Moon falls slightly closer to the Earth, gravity increases and it speeds up. But as it speeds up, it swings further out and in doing so, slows down. Now, as it slows down, it starts to fall towards the Earth again, so it speeds up and swings out once more. So the Moon self-regulates its orbit around the Earth and the Earth self-regulates its orbit around the Sun – constantly. Using the Kepler ball-and-string toy, you can 'feel' the difference in 'pull' when the ball is closer or further from its centre of spin.

Although Newton never went this far, today we know that the pull of gravity makes the Moon virtually weightless with respect to the Earth, and the Earth and other planets are virtually weightless with respect to the Sun. Their orbits are always slightly adjusting, with no great strain because of their comparative weightlessness. Today's astronauts in our space stations have to live with this weightlessness all the time they are in the Earth's orbit. But Newton didn't know this.

Tides in the Affairs of Men

Newton then explained how the tides on Earth are caused by the Moon – not by the Sun, as Galileo had suggested. In effect the Earth and the Moon are a twin planet. The mass of the Earth is about 80 times that of the Moon. But the centre of gravity of the Earth and Moon combined is on the axis line between both of them, but still well within the Earth. Gravity acts by pulling the fluid oceans in towards this axis line. As a result, the seas bulge out along the line, causing high tides directly under the Moon, but also on the exact opposite side of the Earth at the same time, though to a marginally lesser degree. The Sun is involved too, but only slightly, because it's such a long distance away, and also because of the Earth's comparative weightlessness as it orbits the Sun (see Figure 13.2).

The highest tides – the spring tides – occur when the Moon and the Sun are in alignment, round about the spring and autumn equinoxes. In midsummer and midwinter, the Sun is highest or lowest in the sky, and most out of line with the Moon, which travels around its own orbit, called the *ecliptic*. In mid-ocean, tides are hardly noticeable, but on shallow-water coastlines they can be 20ft high or more; in river estuaries under certain conditions, they can be as high as 40 or 50ft.

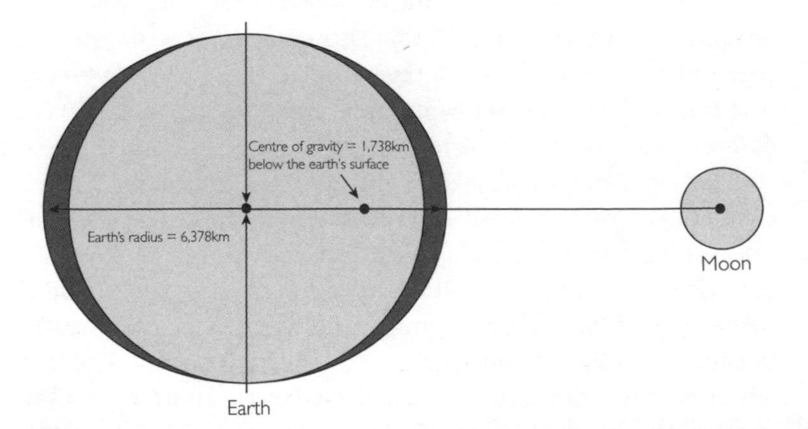

Figure 13.2

Around this time, in Woolsthorpe it seems, Newton made another remarkable breakthrough. He saw that gravity acts on stars and planets as though they were mere points, with all their masses concentrated in their centres. In this golden period of his genius, Newton developed a totally new understanding of everything in the Universe, or as he might have explained it poetically: 'Every particle in creation is attracted to every other's station, and that attraction is in relation to their dual-mass combination, and also in inverse relation to the squares of their separation.' (I've added my full poem, a piece of doggerel written in a style befitting Newton's age, at the end of the Wow Factor Mathematical Index.)

Measuring Things That are Changing

While still a young man, Newton made huge strides in mathematics. He had read Descartes's *The Geometry*, but did not particularly enjoy it, until his friend John Wallis showed him his clearer adaptation, of which Newton approved. He explored and proved the binomial theorem as an extension of Pascal's triangle, and calculated the negative-value version.

He also explored infinite series and, following his work on planetary motion – in which he had to calculate the positions of constantly changing objects – he invented a type of mathematics designed specifically to measure things that are continually fluctuating: his *fluxions*. Like Fermat, Newton saw that, on a curved graph that shows something constantly changing, we can calculate the ratio of change by exploring the angle of tangents to the curve at various points. If a curve is horizontal there is no change, but as the angle of slope of the tangent increases, so does the degree of change.

Cavalieri and Kepler had both referred to the idea of measuring things by slicing them into smaller pieces. Newton saw that if you divide the area under a curve into equal-width rectangles, there are small semi-triangular areas not accounted for at the top of each rectangle – and their total area will be that amount short of the area under the curve. The narrower the rectangles, the closer their combined area is to the area under the curve, but it is still short by the area of the semi-rectangles.

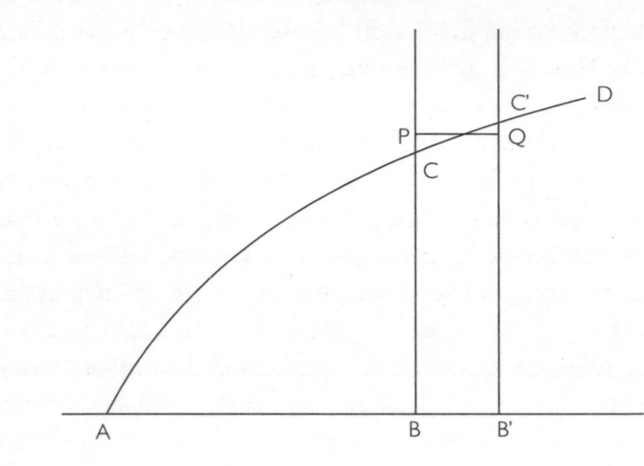

Figure 13.3

However, if you draw equal-width rectangles that straddle the curve so it coincides with the centre of the rectangles' top edge, then the two semi-triangles, one inside and one outside the curve, more or less cancel out. So now the total area of all the rectangles is very close to the area under the curve. In both cases, the narrower the rectangles, the closer their combined area is to the area under the curve (see Figure 13.3). (For more on fluxions and calculus, see the Wow Factor Mathematical Index.)

But Newton's greatest discovery in those plague years was yet to come – an explanation of gravity. For this great work, however, Newton surprisingly never used fluxions, instead relying on the tried and tested mathematics of the ancient Greeks.

Seeing Further Than the Rainbow

At Cambridge, before the plague years, Newton had experimented with light with his roommate John Wickins. He allowed the smallest chink of sunlight to enter through a hole in a thick curtain and strike a prism. On entering the prism the light was diffracted at slightly different angles, splitting the ray into different wavelengths. On exiting the prism the light was bent again, and fell on a wall some distance away. This produced an oblong rainbow of colours, to which Newton gave seven names: red, orange, yellow, green, blue, indigo and violet (which children were soon taught to remember using the mnemonic 'Richard of York gave battle in vain'). Newton may have been influenced here by Kepler, choosing seven because there are seven musical notes in a scale: he even suggested that five of the colours were 'naturals', while orange and indigo represented 'semitones'.

Next, using a second inverted prism, Newton merged the rainbow of colours back into white light again. He then isolated the red and blue ends of his rainbow image so they alone hit the second prism. Now the light passing through always remained the same colour. From this he surmised that white light requires the complete set of rainbow colours.

Newton gave a much fuller account of rainbows than Descartes: he included specific angles for each colour; described how light is refracted on entering a water droplet, then reflected and refracted again on leaving the droplet, and that there is a second reflection from the rear of the droplet in a secondary rainbow, causing a slight loss of intensity. Then, once he had a clear understanding of the rainbow's colours, he began the work of getting rid of them.

Getting Rid of the Rainbow

As Galileo had noticed, the images on early telescopes were often spoilt by the presence of rainbows. This was because the outer edge of a lens was essentially triangular, and acted like a prism, splitting light into its different wavelengths. Newton determined to get rid of these aberrations. In doing so he explored soap bubbles, different kinds of glass sandwiched together and clouds of water vapour, all of which produced curved rainbows. And although many other scientists had noticed these rings of colour, they became known as 'Newton's rings'.

Newton also wondered if the light we see is governed by properties of the human eye. He gazed at the Sun so often that he damaged his eyesight quite severely, despite knowing that the same activity had caused Galileo's blindness. Taking an equally dangerous approach, he also probed behind his eyeball with a bodkin (a thick, smooth-ended needle) to explore how the eye worked; that exercise got him nowhere, and thankfully he stopped it before doing irreversible damage.

Newton worked ceaselessly for two solitary years, until at last the plague had dissipated and he could return to Cambridge. His professor, Isaac Barrow, who had never actually taught mathematics before becoming the first Lucasian Professor at Cambridge, asked Newton for his help with a paper he was writing on optics. But after seeing Newton's knowledge of optics first-hand, Barrow realised he had much to learn, and he resigned the post in 1669, recommending that the 'genius' Newton take his place. Barrow reverted to theology, eventually becoming Master of Trinity College at Cambridge.

Now aged 27, Newton took to the new post like a duck to water, writing 97 lectures on algebra alone. But he was not a good teacher, never adjusting a lecture to suit the level of his audience or even enquiring after its ability. Few came to hear him, and he often ended up talking to the walls. Nevertheless he took his professorship very seriously indeed, and resolved to reveal his findings to the world.

Possibly influenced by Kepler's thoughts on Galileo's telescopes, as well as the possibility of using mirrors instead of lenses, Newton had fashioned a 6in tube into a reflecting telescope, grinding the mirror himself. His first effort magnified images by about 40 times. A wide tube provides an advantage because it gathers more light. A parabolic mirror at the far end, which is the full width of the tube, reflects the image back to the point of focus quite near the open end, where a second mirror set at 45 degrees directs the light towards the observer's eye. So instead of having to raise the telescope to his eye, as Galileo had done with his device, Newton could sit beside it and make notes as he viewed the heavens. Today all major optical telescopes

Figure 13.4

throughout the world are based on Newton's reflector design (see Figure 13.4).

Newton Visits London

Newton sent a copy of his telescope to the Royal Society, whose secretary Henry Oldenburg replied asking if he would demonstrate it for them. In 1671 Newton did just that, and in 1672 he followed up with a lecture on colour and his prism experiments. He was then shocked to hear Robert Hooke's claims that he had previously made a similar telescope, much smaller but more powerful, and that his own experiments had shown that the colours produced by a prism were a property of the prism itself, making Newton's theory of light incorrect. This was the opening shot in a continuous feud between the two men – as a result Newton resolved not to publish his explanatory book on optics, which he had by now already written.

We should remember that Robert Hooke was knowledgeable in just about every scientific area under investigation at that time, and took his position at the Royal Society very seriously indeed. In fact, it was his job to object to anything that was proposed, if he was not sure that the claims made were scientifically sound and justified.

We now know that Hooke, after seeing Newton's telescope, did then try to make one himself, but he couldn't get close to Newton's effort. It's possible too that Newton may have got wind of Hooke's attempt. Whatever the details, Newton was extremely miffed, and applied to leave the Royal Society – it took much cajoling from Henry Oldenburg to dissuade him from doing so.

In an exchange of letters that followed, Hooke made a strong attempt to smooth the troubled waters. Newton responded courteously, and even visited Hooke's lodgings while he was in London. In a reply to one of Hooke's kind and complimentary letters, Newton famously said, 'If I have seen further than others, it has been by standing on the shoulders of giants.' The giants he was referring to were clearly Galileo and Kepler, and possibly those who had gone before them, right back to Archimedes. Many have claimed that the statement was a backhanded insult directed at Hooke, who had a spinal disorder and was a small man. I prefer to think that that idea was probably false, because the two men were soon exchanging ideas. Newton offered to make an astronomic observation for Hooke, and told him where the straightest piece of water in England was (a drain in the Fens of East Anglia), possibly to help him measure the curvature of the Earth.

In fact, Newton faced another pressing problem at that time, which had nothing to do with the Royal Society. It is ironic that he was at Trinity College, named after the Holy Trinity of Father, Son and Holy Spirit. Newton was an Arian or Unitarian, believing in God alone, and did not hold with the New Testament at all. He had kept his views secret, but before Cambridge professors could progress to high office they were obliged to take Holy Orders, which he had no desire to do.

Luckily, when the Lucasian Chair had been established, taking Holy Orders had been declared unnecessary. But Newton still had to apply to King Charles II himself, citing this clause, before he finally received consent to continue without having to be ordained as a priest. However, this also meant that he could never progress to Master of Trinity College, or indeed Cambridge University itself.

Fruitless Isolation

This inability to progress at Cambridge, on top of academic criticism of his ideas on optics from several scientists other than Hooke, hurt and troubled Newton immensely. For most of the next 25 years he slowly shut himself off from London and the Royal Society, and indeed from any public activity outside his own college.

We now know that during that time Newton read and dissected every work he could find on ancient alchemy which, thanks to Boyle, was now being replaced by the science of chemistry. He also made extensive dissections of the Bible and other early religious works, as well as the works of Herodotus. He wrote some two million words on these two subjects alone – two-thirds of them on religion. These notes in their entirety were left in a box at Trinity College on his death, and were only rediscovered in the 1930s. Amazingly, this phenomenally painstaking effort proved to have absolutely no value at all.

It is my contention that Newton, having discovered so much by simply combining the work of Galileo and Kepler, was sure there must be other monumental revelations as yet undiscovered somewhere, which he only needed to identify and link together. Sadly, he found absolutely nothing of value, and the relentless search ruined his health. He increasingly neglected his diet and appearance, until in 1693, aged 50, he suffered a total mental breakdown.

As luck would have it, by this time his future fame had been secured, thanks to a diversion from alchemy and religion caused by a casual visit from Edmond Halley in 1684. In conversations in London coffee houses with Hooke and Wren, Halley had discussed what path an object might take if you assumed it was attracted to another body under the inverse square law. If you halved the distance between the two objects, then the force of attraction between them would be four times as much, while if the distance was twice as far, the attraction would fall to just a quarter of its original strength. Hooke claimed he already knew the solution, but when Wren offered him 40 shillings to explain it, he got no reply. All three men agreed that if anyone knew the answer, it would be Isaac Newton.

When Halley called on Newton in Cambridge and asked him the question, Newton replied instantly that the path such a body would take would be 'an ellipses'. 'How do you know?' asked Halley. Newton replied that he had calculated it to be so, and began searching for the proof, which he couldn't find among the piles of research notes that covered just about every surface in the room. He did recall, however, that he had completed this work some 18 years earlier in his solitary plague years.

Eventually Newton found his early calculations, but by now he had gleaned far more accurate figures by referring to Picard's size of the Earth and Roemer's speed of light calculation, enabling him to improve his previous measurements. In a few months he had produced a paper for Halley entitled 'On the motion of bodies in orbit'. When Halley saw not only the answer to his question but also the wealth of original mathematical ideas Newton had included, he grasped the immense importance of this work and implored Newton to write it down in full and publish it.

Over the next two years Newton produced the work that would prove to be the most celebrated and far-reaching book on science ever: *Philosophiæ Naturalis Principia Mathematica* or the *Principia*. He deliberately wrote it in Latin, possibly so that it would not be widely read and criticised by the 'little smatterers in mathematics' who had previously annoyed him. Only in the final year of his life did he agree to a version in English, but by then the *Principia* was universally acclaimed. But what did this monumental work contain? Simply, the greatest wealth of totally new mathematical ideas ever produced.

The *Principia* (1687)

The book was in three parts, with an introduction that featured Newton's three laws of motion:

1. An object at rest will remain at rest unless acted upon by a force.
2. A moving object will travel in a straight line unless acted upon by a force.
3. For every action, there is an equal and opposite reaction.

Galileo and Kepler had both stated the first two laws, but Newton brought everything together. The third law was the result of experiments with billiard balls (which Huygens, Wallis and Wren had also studied) – billiards being a mathematical game of angles, forces and the effects of collisions (for more on the mathematics of billiards, see the Wow Factor Mathematical Index).

Figure 13.5

Newton's Cradle, the famous toy that elegantly explains much of this science, emerged less than a hundred years later (see Figure 13.5). The mass of a ball raised to one side and released causes on impact the central three balls to remain stationary, while the furthest ball absorbs and carries the force, swinging out. On its return it passes its own force and momentum through the central three balls to the original ball, which now swings out, and so on until slowly energy is lost through friction and the swings diminish.

In its final form the *Principia* contained 550 pages, in three sections, and was written in such a way that readers couldn't really understand each section unless they had absorbed all the previous material. Book One concerned itself with bodies moving in straight lines and then in circles. Applying Kepler's second law, Newton went to great lengths to show that, obeying the inverse square law, an orbiting object actually accelerates towards its host continuously, which is why it remains in an elliptical orbit (see Figure 13.6). He explored and explained all this in studying the orbit of the Moon around the Earth, which Kepler had never tried. Newton found that his original work on the Moon's orbit agreed with and confirmed all of Kepler's claims about the planets orbiting the Sun, and specifically his second law.

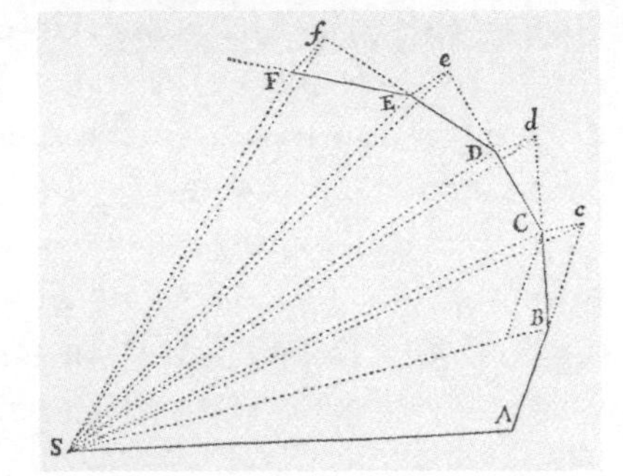

Figure 13.6

In essence, said Newton, if the Moon continued on course AB (as shown in the diagram), it would fly off into space, passing through point C. But it doesn't: instead it is pulled from B towards point C by an inwards or *centripetal* force. Now B and C are opposite corners of a parallelogram, whose diagonal BC is shorter than BC. So the Moon covers the distance BC slightly faster than it travels AB or BC, so it must be accelerating towards the Earth. If points A, B, C and so on were very close together, they would form a smooth elliptical curve.

Newton showed that the area swept out is proportional to the time taken, exactly in line with Kepler's second law. If the object had gone from A through B to c, the triangles SAB and SBc would be of equal area. But it moved from A to B to C, and triangle SBC is shorter than SBc. But as C is beyond the line Sc, so triangle SBC is greater than triangle SAB and is swept out in a slightly shorter time.

By knowing that the distance from the Earth to the Moon is about 60 times the Earth's radius, Newton then calculated in detail the Earth's gravitational pull on the Moon. He also demonstrated that the Earth's constant spinning made it an *oblate sphere*: slightly flattened at the Poles, compared to the bulge at the Equator, by about 17 miles.

Book Two of the *Principia* concerned fluid motion, looking at how bodies move through resistant mediums. Newton seems to have gone to enormous lengths to investigate this area: covering hydrostatics, dropping lead balls into wax, and – with Christopher Wren's cooperation – dropping objects from the maximum interior height of the almost completed St Paul's Cathedral.

Newton also measured the speed of sound, by timing the echo of his hand claps in Nevile's Court at Trinity College; he even calculated the wavelength of those sound waves. He observed that the speed of sound varies under different conditions, declaring that it did not behave exactly like light – in which case light was not a wave, as Huygens and Hooke had claimed, but was made up of tiny *corpuscles* travelling in straight lines. Newton ended Book Two with a full onslaught discrediting Descartes's vortices, arguing that planetary centrifugal force was total nonsense: everything is held together by centripetal force, or the inwards pull of gravitational attraction.

Book Three, by contrast to the other two, was some time coming, and Halley started to worry. To hurry Newton along he had to tell him that Hooke, having seen the first two books, was causing trouble by claiming he had discovered the inverse square law earlier, and was grumbling that he had not been mentioned often enough. The trouble subsided after both Halley and Wren rounded on Hooke, telling him he was making a fool of himself. Newton, meanwhile, ensured that every mention of Hooke's name was removed from the entire work.

At last Book Three, *The System of the World*, arrived. It truly was the icing on the cake. It explained that gravitation was universal, and that the same laws cover every object on Earth and in the entire Universe.

There were still pitfalls ahead, however. The Royal Society, unlike its Parisian equivalent, had no regular funding. It was in fact, in debt, and at that time was preoccupied with publishing a book about fish, so could not (or would not) help. But Halley was determined the book would not fail. He had inherited a fortune following the murder of his father in 1684, the same year he had visited Newton. He now took it upon himself to cover the cost of publication. At last the book appeared, in 1687.

The *Principia* was a sensational success, although it had its doubters. Many asked questions such as, 'Why does gravity act as it does?' or 'How does action at a distance happen?' to which Newton replied: 'I do not make hypotheses,' or 'What I have written are irrefutable laws. This is the way the Universe behaves!' Slowly his ideas became widely accepted, and remained unchanged until Albert Einstein extended and broadened them 300 years later with his theories of relativity. But Newton's fluxions received a strong challenge from the German mathematician **Gottfried Wilhelm Leibniz** (1646–1716), who claimed he'd thought of the ideas first.

Leibniz taught himself Latin aged eight and Greek at 14, then studied law and became a diplomat, advising Louis XIV and Peter the Great of Russia. As a diplomat he even attempted to reconcile the Catholic and Protestant Churches. He also loved mathematics.

Heavily influenced by Huygens, in 1671 Leibniz devised a calculating machine – as an extension of Pascal's device – that could also multiply and divide. He showed it in London, where Robert Hooke, as was his nature, claimed he could have done better, although the Royal Society made Leibniz a foreign member on the strength of it. Leibniz tried to work out a system of logic to aid computation as early as 1667. He suggested using the binary number system because it would make mechanical calculation much easier – all more than 250 years before computers.

He announced that he had published his work on the *calculus* in 1684, three years before Newton's *Principia*. Newton accused Leibniz (with his known diplomatic background) of spying when he had been in London in 1671. To be fair, although Leibniz may have heard of Newton then, he could not have known that he had devised these mathematical principles during the plague years five years earlier. The two men corresponded, each setting trick questions to test the other's true understanding. At last they agreed that they had both arrived at the calculus independently, and the row faded away.

However, Leibniz's notation, which built on the previous work of Cavalieri and Fermat, was far clearer than Newton's, whose concept was totally original. Slowly the German's system was universally accepted, and was eventually preferred to – and would eclipse – Newton's fluxions.

A Bevy of Bernoullis

The Bernoulli family originated in the Netherlands, but through the end of the 1600s and into the early 1700s it came to dominate European mathematics from its adopted home in Basel, Switzerland. Three members of the family in particular gained a considerable reputation: brothers **Jakob** (1655–1705) and **Johann** (1667–1748), and Johann's son **Daniel** (1700–1782).

Jakob travelled widely and, while in England, he met both Hooke and Boyle. Johann, 12 years younger, also took to mathematics rather against his father's wishes, and moved in with Jakob in 1683. Both men became devotees of Leibniz and his calculus, although it took them six years to get to grips with it.

This is a fine cautionary tale for all budding mathematicians and scientists. The Bernoulli brothers were clearly very keen and able, yet even they found calculus difficult. In study there is no gain without pain. Rather as professional sportsmen and women hone their skills, for mathematicians achievement is all about making the constant effort to gradually improve, and gain a better understanding, of exactly what they are capable of. As Einstein once said, 'The most amazing thing is not that we understand the world, but that we *can* understand it.' He perhaps should have added, '... if we try and actually *want* to understand.'

The Bernoulli brothers were in contact with Leibniz a great deal, and in turn taught one of Johann's students, **Guillaume François Antoine de l'Hopital** (1661–1704). It was he who first wrote a textbook on calculus that established Leibniz's version rather than Newton's as the accepted one – first in Europe, then in England, and eventually worldwide.

As often happens with most brothers, sooner or later, Jakob and Johann had an almighty row – their falling out was over how things fall down. Galileo (see Chapter 11) had explored falling objects, and had noted that a ball rolling down a slope starts slowly (because of friction) before speeding up. He wondered what the quickest path of descent might be between a higher and offset lower point, and ultimately suggested that it would be along the arc of a circle. But Galileo was wrong.

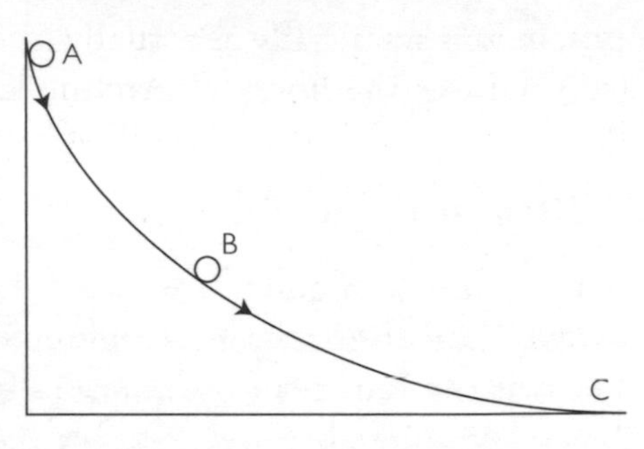

Figure 13.7

In 1673 Christian Huygens had solved the problem of how to regulate a pendulum by turning Galileo's cycloid upside down to form curved jaws that controlled the swing. While doing this, however, he also solved the *tautochrone problem*. This asks, 'Is there a curve such that a ball rolling down it will take the same time to reach the bottom irrespective of its starting point?' (see Figure 13.7)

So, what do you think? Our diagram shows a cycloid shaped slope with two balls, A and B, set at starting points of different heights. If released together, which one will reach point C at the bottom centre of the cycloid first? The answer is a surprise for most people first meeting this problem: it will be a dead heat. From whatever start point, any ball will take the same time to reach C.

The outcome of the row between the Bernoullis, 23 years later, had to do with a similar question, known as the *brachistochrone problem*. Johann decided that others might help him settle the argument, so he wrote to several academics seeking an answer. 'Find the curve along which a particle will slide down under the force of gravity in the shortest possible time.' He got the correct answer from Newton, Leibniz, l'Hopital and his brother, as well as working it out for himself. Once again the answer is 'a cycloid'.

Jakob Bernoulli is known for exploring and explaining the *logarithmic spiral*, which he called *spira mirabilis*, or 'miraculous spiral'. It was actually discovered by Descartes in 1638. Unlike the linear or Archimedean spiral (see Chapter 4), which is formed by a coil of rope or the grooves on a vinyl record, and in which the distance between successive coils remains the same, the logarithmic version grows at a steady geometric rate.

A nautilus shell is a good example of a logarithmic curve in action. As the creature inside the shell grows, so does the shell itself. The creature occupies the area close to the exit and the extra interior space is filled with gas. In this way, the buoyancy of the shell and the creature remains exactly the same as the animal grows, ensuring that it can live at the ideal depth for the warmth, survival conditions and abundance of food it needs.

Perhaps the most fascinating fact about the logarithmic spiral was discovered by Evangelista Torricelli in 1645. If you take a line from the centre origin of the curve to any point on the curve, the angle at which it meets the curve – and therefore the tangent angle at that point – will always be the same. If you construct a right angle at the centre origin and continue that line until it meets the tangent line, then the hypotenuse line of this triangle will be exactly the length of the spiral to that point (see Figure 13.8). This spiral was to become the inspiration for Benoit Mandelbrot as he began his explorations into fractals in 1951 (see Chapter 15).

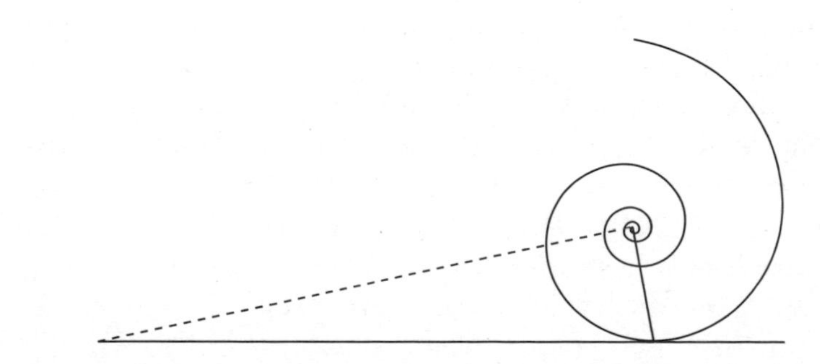

Figure 13.8

Jakob Bernoulli asked that when he died, the logarithmic spiral might feature on his tombstone. Although his wishes were carried out, the engraver was untutored in mathematics, and carved an Archimedian linear spiral instead. Enough, I would have thought, to have Jakob spiralling in his grave?

Two Rather Different Endings

It must have become clear to Newton quite early on that Leibniz's calculus was gaining the upper hand in Europe, but both men eventually wrote glowingly about the other's genius. It was always understood, however, that Leibniz had confined himself to one field of mathematics (calculus), while Newton's far greater achievements included a totally original and correct explanation of universal gravitation and the workings of the entire Universe.

Leibniz served under a number of Electors of Hanover for 40 years, including George, who eventually became King George I of England. But when he went to England, George declined to take Leibniz with him, possibly aware of the still strained atmosphere between English and German scientists. Sadly, and surprisingly, Leibniz – who had inspired so many budding mathematicians – was left very alone. On his death, his secretary was the only person to attend his funeral.

Isaac Newton's final years, on the other hand, were filled with new achievements and accolades. He became Cambridge University's Member of Parliament for eight years, although in all that time he was only known to have spoken once, when during a debate he rose to his feet and to ask if someone might close a window because it was causing a draught.

Newton's religious beliefs prevented him from rising to the very top at Cambridge, so his London contacts offered him the post of Warden of the Royal Mint, then housed in the Tower of London. He improved the running and security of the Mint, and in 1699 was made its Master. During his time he displayed considerable ruthlessness in dealing with forgers and 'clippers' (who clipped the edges of coins before melting the cuttings to make new counterfeit coins), and sent a good few to the hangman.

In 1703 Hooke died, and finally Newton published his *Opticks*, which he had suppressed following Hooke's claims of plagiarism back in 1672. Newton had waited 30 years and, despite the pleasant exchange of letters between the two men, had borne a long-standing grudge. Arguably, however, Hooke had a case: while he had acknowledged his possible sources among previous scientists, such as Galileo, Kepler and Descartes, Newton did not always give them full credit.

As soon as it was offered, Newton accepted the Presidency of the Royal Society, a position he had never sought while Hooke was alive. He held the post until his own death in 1727. Queen Anne knighted Newton at Trinity College in 1705. He was the first ever scientist to be so honoured, although it may have had more to do with the Prime Minister's posturing than a true recognition of what Newton had achieved scientifically.

In 1710 Newton supervised the Royal Society's move from Gresham College to premises in Crane Court, Fleet Street. Somehow, during the move, all pictures of Robert Hooke seem to have disappeared; today no one can be sure what this very inventive and multitalented genius even looked like.

The many images of Newton, meanwhile, and the value of his achievements, are known worldwide. The Italian-born mathematician **Joseph-Louis Lagrange** (1736–1813) summed up the value of Newton's work when he said, 'The *Principia* is the greatest production of the human mind – Newton was not only the greatest genius ever, but also the most fortunate, because as there is only one universe, it can only happen to one man that he first interprets its laws.'

CHAPTER 14

The Simple Mathematics That Underpins Science

As I mentioned at the very beginning, the story of mathematics is rather like an upturned pyramid: the subject becomes ever more diverse as we go along, perhaps like a young tree slowly growing towards maturity and spreading its branches exponentially. In all the sciences, mathematics is always at the root: first making it understandable, then proving it to be so. Of all the wonders of human endeavour, surely the greatest have been revealed in the advancement of science, and especially since the time of Newton.

A Matter of the Greatest Scientific Importance – Matter Itself

Henry Cavendish (1731–1810) was an aristocratic and eccentric Englishman who suffered from acute shyness all his life. Being a wealthy man he could dictate how he lived, and he cultivated a reputation as the archetypal mad professor. He was also frequently a secretive man, so most of his discoveries were only revealed long after his death, when James Clerk Maxwell inherited and studied his papers while at the Cavendish Laboratory in Cambridge.

The laboratory, named in honour of Henry, opened in 1874 and was financed by William Cavendish, who served as Chancellor of the University. Since then this laboratory alone has produced 29 Nobel laureates.

Henry spent his life exploring science, and followed every move in the development of electrical experiments. He made his own version of an electrostatic generator, in which a glass sphere or bottle was revolved while in contact with a piece of cloth, creating static electricity. In 1746 Dutch scientist Pieter van Musschenbroek found he could channel the static electricity via a metal rod into a flask of water. His assistant held the flask in one hand. When he touched the metal rod with the other hand, he received a violent electric shock. They had discovered the *Leyden Jar*, though to improve it lead shot soon replaced the water and the outside of the flask was covered in a metal foil.

In 1746, Abbé Nollet in France persuaded a whole monastery of monks to hold hands in a huge loop. The two monks at either end of the chain discharged a Leyden Jar, and the shock caused the entire chain to instantly jump into the air (they probably broke their vows of silence as well…). Soon Cavendish had made his own Leyden Jar, but because there was no way of knowing how strong a charge it might produce, he would give himself violent electric shocks and scored them accordingly.

It was this electrical exploration that began to explain 'matter' itself. So far, gold, silver, iron and other elements had given no real clues as to their structure. These secrets would only be revealed through the study of the very elements no one could actually see: gases. During one experiment in 1766, Cavendish discovered a gas that was highly flammable but incredibly light – just one-fourteenth of the weight of air. He had discovered the lightest of all gases, *hydrogen*.

Joseph Priestley (1733–1804) lived close to a brewery in Leeds, where a gas formed on the top of the open beer vats. A candle was snuffed out by the gas and a mouse held in it went to sleep, but it revived when it was taken out. So this 'heavier than air' gas wasn't flammable or poisonous, nor did it taint anything it came into contact with.

Priestley held two tumblers in the gas and poured water from one to the other several times. When he tasted the water, he now found it to be slightly fizzy. He had invented soda water, and become the very first fizzisist (ouch)! It was only later that he realised that this brewery gas was not a single substance at all, but a mixture of two: it was *carbon dioxide* (CO_2).

When he sealed the gas in a jar with freshly cut flowers, the flowers thrived: carbon dioxide, in essence, is plant food. The carbon was absorbed into the plant's main structure, while another gas slowly replaced it, eventually filling the jar. This gas was *oxygen*.

Scientists soon realised that all creatures do the reverse of plants: breathing in oxygen (or, in the case of insects, absorbing it through their skin) while taking in carbon from their food and burning it through muscular activity to release CO_2 back into the air. This maintains the balance of gases in the atmosphere, and releases water as a waste product. Quite simply, here was an explanation for the continuation of all life on Earth.

In 1774 Cavendish passed electricity through water, splitting it into hydrogen and oxygen – but he told no one about it. In 1800 German chemist **Johann Wilhelm Ritter** (1776–1810) repeated the experiment, and noticed that there is always twice as much hydrogen as oxygen in water, which could now be represented by 'H_2O'.

John Dalton (1766–1844), a quiet Manchester schoolteacher, set up a weather station 2 miles from his home, where he went every day for nigh on 60 years. He found that air was 80 per cent nitrogen and 20 per cent oxygen – but why, he wondered, in those proportions? Dalton also discovered that methane gas was made up of carbon and hydrogen, again in a 20 per cent/80 per cent split, just like oxygen and nitrogen in the air. But why? Eventually Michael Faraday observed that elements seem to have a *valence* (or 'basic number') when they mix with others. Hydrogen seemed to have a valence of 1, rather like having one hand to hold on to other elements, while oxygen had two and carbon had four, making carbon's ability to link up far greater than that of other elements. The simple mathematical form at the heart of chemistry was starting to be revealed.

Dalton also observed that hydrocarbons only burn in oxygen – to put out a fire, simply cut off its oxygen. He explained what happens when methane is burnt in air like this:

H H C H H	methane	when burned	HOH HOH	water and
O O O O	plus oxygen	produces	OCO	carbon dioxide

Before and after the reaction, the weights were always the same because nothing is lost and nothing is gained – but energy is released and for every molecule of CO^2 you get two

Figure 14.1

molecules of H^2O – water. This was John Dalton's Atomic Theory (which he devised in 1805). He also produced the first sketchy periodic table, placing the known elements in order of their weight (see Figure 14.1).

However, French scientist **Joseph Louis Gay-Lussac** (1778–1850) had already gone further, showing that, even though pure gases have different weights, they all take up the same volume and expand equally when heated. In 1811 Italian scientist **Amedeo Avogadro** (1776–1856) made the amazing suggestion that a set volume of *any* gas at a certain temperature must always have the same number of molecules. Carbon has a molecular weight of 12 and oxygen 16, so one molecule of CO_2 weighs 12 + 16 + 16 = 44.

Avogadro astounded the world by declaring that 44g of CO_2 must contain as many molecules as 1g of hydrogen: roughly 602,600 million million million of them. This is known as *Avogadro's number*, and it demonstrates just how quickly science and the mathematics governing it become more complex, while successive scientists learnt to cope with, explain and even simplify the complexity.

One such scientist was **Friedrich Kekulé** (1829–1896), who tended to solve problems by dreaming up the results – once while dozing on a double-decker bus in London. On one occasion, while dozing in front of the fire, he dreamt that carbon atoms were like monkeys holding onto each others' tails, and that six monkeys formed a ring. On waking he realised that carbon molecules with their four hands could join together in long chains, and still have hands free to form bonds with hydrogen or oxygen atoms – or indeed atoms of many other elements. Today we know that all living things, and those that once lived, like the constituents of fossil fuels, are built around chains or rings of carbon, in thousands of variations. Together these give us the central structure of all our food, plants, fuels, alcohols and acids, and the many thousands of man-made drugs we use today. This became known as 'organic science', the study of how carbon atoms form the basis of all the substances that make life on Earth possible.

I find it sad that the term 'organic' was hijacked in the 1980s to mean foods grown without synthetic fertilisers. Were they alive today, Kekulé and the many other early chemists would be bemused (to say the least) by many modern-day arguments that are based on alarmism and marketing ploys rather than proven science.

Numbering the Elements

Russian chemist **Dmitri Ivanovich Mendeleev** (1834–1907) was one of 17 children, but was clearly the most gifted. By the time he was 15 his parents' business had collapsed, and his now penniless mother grabbed his hand and walked him some 1,000 miles to Moscow, and then a further 400 miles to St Petersburg. Once they had arrived, the lad took an exam and was accepted into the company of academics; his mother then turned around and walked home.

Mendeleev developed a passion for the elements, and in 1869 he devised the basis of the periodic table we still use today. He made a card for each element, containing a description of the element and its weight. Because hydrogen was incredibly light and different from the rest, he kept it to one side. Then, starting with *lithium* (*helium* hadn't been discovered by that time), he laid seven cards in a column in order of their weight. Next came *sodium*, which had similarities to lithium, so he placed it next to lithium and started column two (see Figure 14.2). In the third column, so that similar elements were kept alongside one another, he left some gaps, saying, 'I can't know everything, so there will be gaps.' This modesty was the secret of his success: other scientists looking for new elements to fill the gaps now had an idea of what to look for. But they needed a tool to help them.

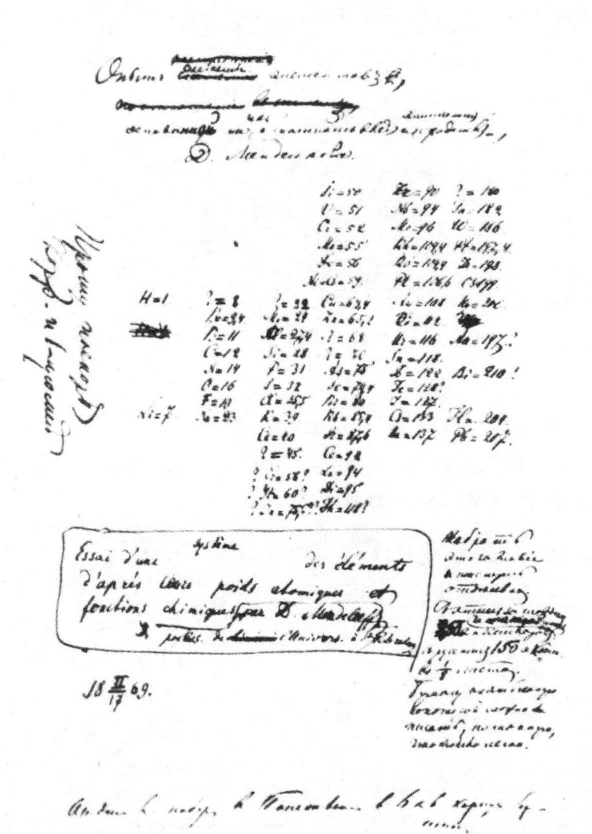

Figure 14.2

Spectacular Spectroscopy

In 1800, William Herschel, still active at 62, discovered invisible *infrared* light, which exists beyond the red end of the visible spectrum. Just a year later, Johann Ritter discovered that light turns silver chloride black, a finding that underpins the principle of negatives in photography. Amazingly he found that this effect was caused by cooler blue/violet light, and that it got even stronger beyond this end of the spectrum. Ritter had discovered *ultraviolet* light, clearly indicating that the true spectrum of light is much wider than the narrow visible band of rainbow colours human eyes can detect.

Isaac Newton, in exploring rainbow colours, had spotted dark lines across the colour spectrum, but he had ignored them. Those lines led to the science of *spectroscopy*, which would prove more revealing to science than even Galileo's telescope. In 1868 French astronomer Jules Janssen observed from these telltale lines that, besides hydrogen, there was another element present in the Sun. Norman Lockyer in England named it *helium* after Helios, the Greek sun god. But it wasn't until 1895 that helium was found on Earth, emanating from uranium ore. After hydrogen, helium is the most plentiful element in the Universe, accounting for about 24 per cent of all matter. Four times heavier than hydrogen, helium didn't fit any known gap in Mendeleev's table. But around 1894, British chemist William Ramsay began finding other similar gases, including *argon* (which had actually been discovered by Cavendish, and which comprises 1 per cent of the Earth's atmosphere), *neon* (meaning 'new'), *krypton* ('hidden') and *xenon* ('stranger'). Together, these elements – none of which mix with others – formed the *noble gases,* and took up an entirely new column in the periodic table.

Slowly, and almost entirely thanks to spectroscopy, all 92 naturally occurring elements were discovered, and the periodic table was complete. But the complex whole is governed by simple mathematics, involving the slightly different structures of each element.

Glowing Discoveries

Michael Faraday (1791–1867) became known as the 'father of electricity': in 1831 he invented a dynamo to generate electricity and a transformer to control it. Today, electricity generated in power stations at, say, 25,000 volts can be transformed up to 400,000 volts for overhead cables, then back down to 240 volts for our homes, and even smaller voltages for our computers. Quite simply, transformers make our modern electrical world possible.

But Faraday was no mathematician, so struggled to explain the electrical phenomena he had discovered. That was left to the brilliant **James Clerk Maxwell** (1831–1879), who mathematically proved Faraday's discoveries using a handful of simple equations. He confirmed, for example, that electricity and magnetism cannot exist in isolation, as one must always accompany the other.

Maxwell saw that electrical charges radiate outwards at about 300,000km (186,000 miles) per second – the speed of light. But these charges can oscillate at very short or very long wavelengths. The entire spectrum of electromagnetic radiation is vast: from ultra-short gamma rays (so small that they pass through anything) via the visible spectrum to long radio waves that can be many kilometres wide. All this was verified by Heinrich Hertz after Maxwell's early death at the age of just 48. We can only imagine the triumphs Maxwell would have achieved had he lived longer. As it was, his work was to have a major influence on the greatest scientist of them all, Albert Einstein.

Michael Faraday, besides finding that electricity can make things 'go', had seen that electric charges in glass tubes can make certain substances (like sodium) 'glow'. **Heinrich Geissler** (1814–1879) produced all kinds of glowing coloured lights in partial vacuum tubes (see plate section). But in the 1870s William Crookes reduced those particles to just 1/750,000th of their number, and this 'absolute vacuum' helped Joseph Swan – and, a year later, Thomas Edison – produce the first electric light bulbs.

Crookes coated the end of a vacuum tube with a reactive substance that glowed when rays of electricity were beamed at it. In 1897 he saw that magnets deflected the rays – in effect stumbling across the concept of television 50 years before it was invented. The negatively charged particles forming these electric rays were named *electrons*, and English physicist J. J. Thomson at the Cavendish Laboratory discovered that they were being stripped from atoms – as he remarked – 'like currants from a bun'. Electrons are incredibly small, and about this time scientists started to wonder what the structure of the atom itself might look like. This would only be explained when scientists could at last observe atoms breaking up.

The Atom Explained

In 1895 **Wilhelm Roentgen** (1845–1923) watched in his laboratory as a luminescent surface some distance from a Crookes tube began glowing, apparently by magic. He discovered that 'invisible rays' were causing the effect, and that these rays could pass through cardboard, a closed door and even a thin metal sheet. He called them *X-rays*, using the algebraic symbol for an unknown quantity. He is credited with starting the second scientific revolution, following Galileo's telescope and support for Copernicus, which had triggered the first.

In Paris the following year, **Antoine Becquerel** (1852–1908) wanted to study the rays being emitted by a uranium salt that glowed in strong sunlight. But after heavy clouds rolled in, he casually popped the uranium salt in a drawer, on top of some photographic plates wrapped in thick black paper. Much later when the plates were unwrapped, it was clear that the film had been exposed – presumably by X-rays…

Becquerel's discovery galvanised one of his students, **Marie Curie** (1867–1934), and she named this phenomenon *radioactivity*. Her husband and former teacher Pierre had discovered *piezoelectricity*, and she used this to measure just how much energy was coming from the uranium salt – it turned out to be much more than could be accounted for by the uranium itself. In 1898 she undertook a long, laborious process in which she single-handedly broke down tons of pitchblende waste (left over in the process of producing uranium) by crushing, washing, sieving and refining it. But eventually she discovered two new elements: *polonium* (named after Poland, her country of birth) and *radium*, so called because it glows green in the dark.

Marie Curie reasoned that as *radiation* seemed to be producing a series of different elements, it was perhaps being caused by atoms breaking apart. Perhaps this would at last explain the internal structure of the atom... Electrons have a negative charge, so presumably the parent atom had to have a positive charge to keep things in balance.

In Manchester in 1911, **Ernest Rutherford** (1871–1937) bombarded a thin film of gold with a stream of electrons, most of which passed straight through. However, many were deflected, and a tiny few bounced right back again. These had to be directly hitting the nuclei of the gold atoms full on, in which case each nucleus of an atom must be tiny, 'like a pea in a cathedral' (as Rutherford exclaimed), or a tennis ball on the centre spot of a football stadium. This meant that at least 99.999 per cent of an atom must be composed of absolutely nothing, and that almost all of its weight must reside in the extremely tiny nucleus.

Rutherford subsequently discovered that atoms contain positively charged particles (which he called *protons*), which are the same as hydrogen atoms, and 1,846 times heavier than electrons. Twenty-one years later, in 1932, James Chadwick saw that for every proton there is a particle of the same size with no charge at all, which he called the *neutron*.

Only hydrogen does not have a neutron, which is why the second element, helium, which has two protons and two neutrons, is not twice as heavy as hydrogen, but four times heavier. Oxygen, the eighth-lightest element, is 16 times heavier that hydrogen – and so on. The structure of the atom, based on very simple maths, was becoming clear.

Marie Curie, along with Rutherford and his colleagues, now saw that radioactivity is the result of unstable elements emitting energy as they break up to form other elements, in a long chain of events. The scientists discovered three types of radiation: *alpha* radiation, in which particles equivalent to helium atoms break away from the atom, *beta* radiation, in which electrons are

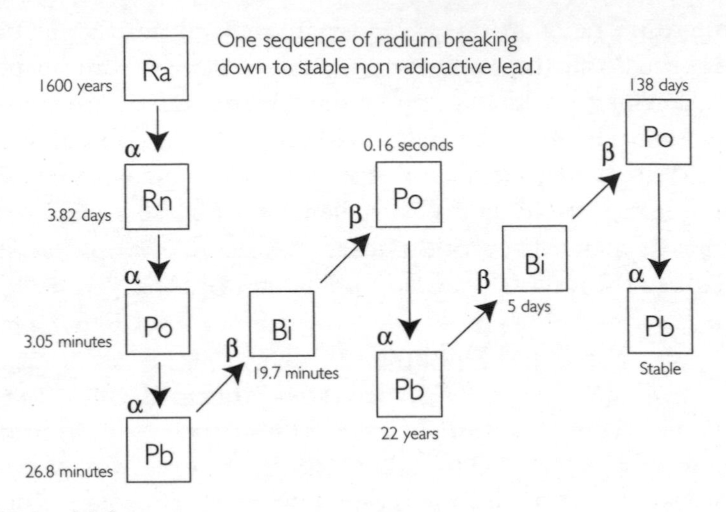

Figure 14.3

stripped from an element, and *gamma* radiation. The first two types are released as a radioactive element breaks down into successively smaller elements and eventually becomes lead, which is non-radioactive and stable (see Figure 14.3). Each stage of the breakdown can vary in time immensely, from a mere fraction of a second to thousands of years.

At last it was possible to produce a model of the atom, with an explanation of how it all worked. But bringing it all together would require one special mind that could attack the problems on several fronts quite independently. This mind belonged to **Albert Einstein** (1879–1955).

Enter – the Greatest

Einstein did not always conform as a student at school in Ulm, Germany, and his teacher predicted he would never make anything of himself. Luckily his uncle Jacob, an engineer, stimulated his curiosity with mathematical puzzles, and he became hooked on maths, reading material beyond what his courses required. A confirmed pacifist, he left Germany for Switzerland to avoid conscription into the army; he eventually graduated from college there, but only by borrowing a friend's lecture notes. In 1901, the same friend's father helped to get him a job in the patent office in Bern where, alone at his desk, he began working on theoretical physics, requiring nothing more than a pencil, some paper and his mind – in addition to the many scientific papers he read and absorbed. After just four years, this unknown office clerk, unconnected to any scientific body, made everyone take notice: in 1905 he published five original papers that were to change the world forever.

Let There be Light

The first paper covered the *photoelectric effect*, by which light falling on certain metals seems to knock energy out of them. Five years earlier, **Max Planck** (1858–1947) had devised his 'quantum theory' (largely ignored by other scientists), which maintained that energy comes in particles that Planck called *quanta* (from the Latin for 'how much?'). According to Planck, the size or energy of a quantum is in direct relation to its frequency: violet light, for example, which has twice the frequency of red light, will have double the energy. This explains why ultraviolet rather than infrared radiation will burn the skin of a sunbather. It's easier to transmit radiation at low frequencies because its energy is lower. Einstein's paper endorsed Planck's ideas, and in 1913 **Niels Bohr** (1885–1962), a Dane who worked with both Thomson and Rutherford, incorporated them into his model of the atom. This explained how atoms form bonds and connect with other atoms to

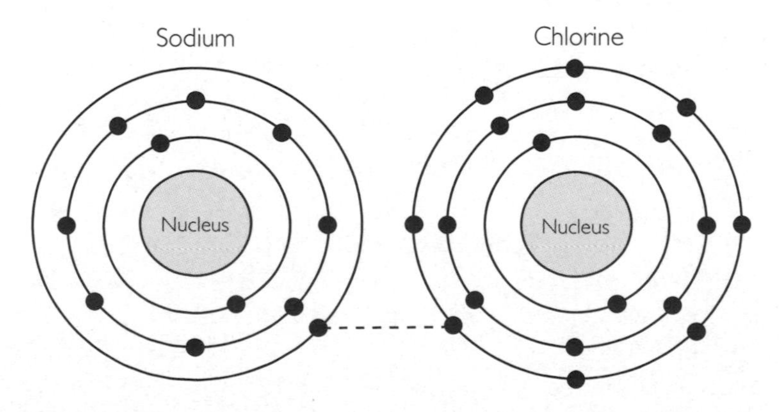

Figure 14.4

produce all known substances, whether gases like CO_2 or solids like steel, glass or concrete (see Figure 14.4).

Bohr's model showed that sodium, at number 11 in the periodic table, has 11 electrons: two (and never more than two) in the inner shell or 'orbit' of the atom; and eight (never more) in the next orbit. But because sodium is number 11, it has a single electron in its third orbit, which gives the element its lively characteristics. Dropped in water, for example, it spits violently as this single electron whizzes off, trying to link with other atoms.

Chlorine, number 17, is a smelly poisonous gas: a little in a swimming pool kills germs, but a lot would kill you. It has two then eight electrons in its first two orbits, and a further seven in its third orbit. There is room for eight in this orbit, so each chlorine atom has a single space. When sodium and chlorine come together, sodium's loose electron fits chlorine's single space perfectly, and they form a stable molecule of sodium chloride (or common salt). Although the model is mathematically simple, it gets very complex with the heavier elements. It also does not explain how light is produced…

In about 1922, **Erwin Schrödinger** (1887–1961), after reading a footnote to one of Einstein's papers, took Bohr's atomic model one giant step further. He showed that when certain elements are bombarded by an energy source, some electrons gain energy and move into a more energetic orbit further from the nucleus. The nucleus, however, insists that these electrons stay part of the atom and pulls them back. In snapping back to their normal orbits, the electrons have to get rid of their extra energy, which is released as light. This is how *all* light is produced. In some elements, such as sodium, just a small bombardment of electrical energy is enough to make the electrons oscillate over and over again to produce light, while not damaging the element in any way. This is why sodium street lamps are so economical and long lasting.

By 1940 scientists were discovering that they could create synthetic elements that were too unstable to hold together for very long. They produced a series of *transuranic* elements (beyond uranium), starting with *neptunium*, and followed by *plutonium*. Today element 115 has been discovered, although these very unstable top-end elements only exist for brief seconds before collapsing.

Many sub-subatomic particles have also been discovered, too numerous to mention in a book on mathematics (protons are made up of *quarks*, for instance). But the basic mathematical explanation of how atoms and elements behave together has been enough to enable us to produce all the materials, drugs and so on covered by the sciences of chemistry, biology and electronics – everything that makes our modern world function. It is all so beautifully mathematical.

Einstein's second paper of 1905 dealt with *Brownian motion*, which had been noticed 75 years earlier, and which describes the movement of extremely light pollen grains in water, which seem to jump about as if they are alive. Einstein explained that water molecules are constantly moving, just as they are in gaseous form, and that their speed is determined by their temperature, as described by Maxwell's maths. According to the law of averages, slightly more particles must move in one particular direction at any one moment, and this imbalance causes the light pollen grains to be moved around.

Special Relativity

But Einstein's 1905 paper that had the biggest impact was so revolutionary it even questioned Newton's laws governing the entire Universe. Newton had imagined travelling like a stone in orbit around the Earth. Einstein imagined what it would be like to travel on a beam of light.

Two Americans, Albert Michelson and Edward Morley, had shown by experiment that if a source of light is travelling away or towards a detector, even at great speed, the light speed of the source will always measure 300,000km per second. But how could this be?

Einstein showed that the speed of light does not change, but the frequency does. As Christian Doppler had discovered using a brass band in an open railway carriage, the same applies to sound, as we know from ambulance and police sirens. Sounds go up in pitch as their sources approach us, but drop in frequency as they speed away. In exactly the same way, light from distant galaxies has exactly the same speed at all times, but the frequency of the light is lower, edging slightly towards the red end of the spectrum. This suggests that the entire Universe is still expanding.

Einstein also maintained that, as speeds increase, even time is changed. If you travel away from a clock at the speed of light, the clock will appear to have stopped, because the only image that can reach you is of the time when you left it. Einstein also suggested that a long, thin space rocket, approaching the speed of light, would foreshorten, and its mass would be focused over a shorter distance. So at the speed of light, the rocket would be squeezed to a single point, with its total mass all focused at that point. This, explained Einstein, is why it is impossible to travel faster than the speed of light (something also confirmed by Maxwell's laws and equations).

The Most Famous Equation of All

Einstein then added an equation of his own, demonstrating just how simple yet overwhelmingly powerful mathematics can be. Both Antoine Lavoisier and John Dalton had shown that when chemical reactions take place, matter is conserved: nothing is lost and nothing is gained – but energy is released. However, physicists like Count Rumford, then Hermann Helmholtz and James Joule, each saw that in all chemical or mechanical reactions, energy is not lost but transferred (so when a cannon is fired its barrel gets incredibly hot). Action or 'work' always produces heat; any kind of mechanical action is caused by energy and results in energy. Just like matter, however, energy is also conserved and can be accounted for, as nothing is lost and nothing is gained. Slowly the concept of *conservation of energy* was formulated.

But it was Einstein who transformed it into a monumental and lasting theory when he presented to the world his famous equation: $E = mc^2$. The mass of an atom (held in the tiny nucleus) is very small, so for the atom to be so powerful it must contain huge amounts of energy. To make the equation work, and to account for and balance this immense energy, Einstein suggested that we must multiply the mass by an absolutely huge number: the speed of light times itself.

This is why atoms are so powerful: when in balance they can hold iron, steel, concrete or bone together. Out of balance, with their electrons able to jump from one atom to another, they produce all our power and energy, whether this is man-made or simply the natural forces of nature that constantly reshape and reform our Universe.

Despite his discovery, Einstein saw that his explanation of special relativity left many questions unanswered. It also confused people, because they tended to forget that it applies, not to our everyday experiences, but only to matter moving close to the speed of light. Special relativity did not seem relatively real in our everyday world. So in 1915 Einstein produced his 'general theory of relativity', which applied to, and explained, everything in our everyday world.

Here is one simple explanation of general relativity. Imagine you are lounging on a beach, and 100yds out to sea a fast yacht is sailing past you from right to left. At the same time an air/sea rescue helicopter is flying along the beach parallel to the sea, also from right to left over your head, and overtaking the yacht. As this is happening, a man on the yacht is busy hoisting a flag up the main mast. What exactly do you, the sailor and the helicopter pilot see happen?

The man on the yacht sees the flag rising 'straight up' above his head. But you saw the flag at deck height to your right, and now it is at the top of the mast, but to your left. So, relative to you, the flag has moved 'diagonally from low right to high left'.

But the helicopter pilot won't have seen it that way. At first he saw the flag ahead and low down. Having passed the yacht, he now sees the flag higher but behind him. So he saw the flag move 'diagonally from low left to high right': the opposite direction to what you saw, and different again from what the sailor saw. In other words, we all see things relative to our own situation: that is general relativity.

A Bent Universe?

When Einstein announced his idea of general relativity, however, he also made the startling announcement that, in space, there is no such thing as a straight line. In essence, he explained, gravity bends space – almost as if each large body in space warps things just by its very weight. Imagine a bowling ball lying on a trampoline and causing it to sag. The path of a golf ball rolled across the surface would bend as it nears the heavy ball because the surface of the trampoline has been warped. This is what happens to light itself: when it passes close to a star, its path is bent by the star's weight or gravity.

According to Einstein, we can prove this by observing the Sun during a full eclipse and noting the apparent position of stars close to the Sun's position in the sky. **Sir Arthur Stanley Eddington** (1882–1944), a Quaker, supported Einstein's humanitarian beliefs in opposing the First World War. They were also in tune scientifically, and in 1919 Eddington travelled to the Caribbean to observe a total eclipse. Sure enough, two known stars appeared to jump further apart as the Sun passed in front of their position – clearly showing that the light from those stars was being bent around the Sun. Because our eyes and brains cannot conceive that the light is bending, the stars appear to us to have jumped further apart in the sky . In fact, in our picture the bend around the Sun looks huge, because of the vast distance to the stars compared with the distance from the Earth to the Sun. The actual bend is only around 1 arc second or 1/3600 of a single degree (see Figure 14.5).

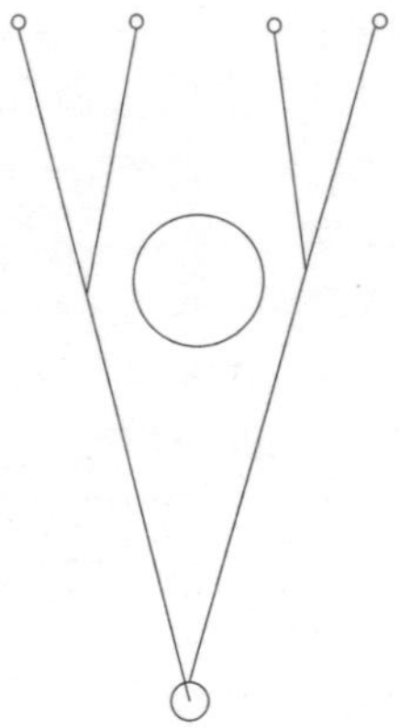

Figure 14.5

Einstein became – rather like Galileo in his day – just about the most famous man on Earth. His fame lay partly in his brilliance, but also in his 'nutty professor' image, thanks to his wild hair and scruffy appearance – he never ever wore socks, for example. This reputation devalued him in the eyes of some scientists, but he cultivated it in the hope that his fame would inspire the young scientists of the future. For that alone, we have a lot to thank him for.

Being Satnav Savvy

Incredibly, it is now more than a hundred years since Einstein proposed his theories of special and general relativity. But we tend not to think of them as important to our everyday lives. Almost every day, however, we rely on them, such as when we operate a satnav in our cars or on our phones: for the system to locate our position it has to consider both special and general relativity – at the same time. Ouch!

Global positioning satellites, all of which are about 12,000 miles above the Earth and spinning in various equatorial, polar or diagonal orbits, take 12 hours to complete an orbit. That's nowhere near the speed of light when time stops, but it is fast in our terms, and because of that speed the clocks on board do run slower: by seven microseconds a day. However, at that height the Earth's gravitational pull is reduced, so the satellite clocks also run faster by 42 microseconds per day. Overall, then, the satellite clock runs 35 microseconds a day faster than the clock in our satnav device back on the ground.

But to accurately locate our position both clocks need to be synchronised. So the frequency of the satellite clock is tweaked by just one digit at its ninth decimal place, enough to locate our position to within about 1ft or 30cm. If the adjustment was one digit at the seventh decimal place however, your satnav would put you 100ft or 30m out. So it could have you driving down the wrong side of a motorway, in which case your life might end relatively quickly…

Today mathematics has enabled everything from more reliable, cleaner and environmentally friendly vehicles, to buildings, energy systems, communications, personal health and longevity, livestock and food produce. All are far more efficient and well controlled than anyone even thought possible just 50 years ago.

Science Grows on You

At the same time as Einstein and others opened up our understanding of the physical sciences, the biological sciences were also becoming clearer. The beauty of plants has attracted the attention of mathematicians for centuries, as we saw with Fibonacci in Chapter 8. Plant growth may appear to be random, but it is always controlled and highly mathematical. Just consider the bilateral shapes of leaves, the rotational symmetry of flowers or the helical shapes of pine cones. Every small leaflet of a fern is a miniature copy of the whole leaf, and their shape is mirrored along each new central stem. The study of such wonders is called *phyllotaxis*.

Alan Turing, in the 1950 paper in which he asked 'Can a machine think?', discussed plant growth in its variety and complexity, and how both ensure that plants prosper through endless seasons and thousands of years. Surely if we could understand the mathematical algorithms that make this possible in plants, and use similar ideas in computers, we could devise machines that could empower us into the future? Generation upon generation, this is slowly happening.

Turing saw that a 'computer' would have to be able to accept a constant stream of information in the simplest of mathematical systems, binary code. Ones and zeroes in groups of set length could tell the computer whether they represented letters, numbers or symbols, or even commands to add, subtract, remember and store or erase information. Today we take all this for granted, but Turing's pure mathematical thinking – together with his experience as an Enigma codebreaker at Bletchley Park in the Second World War – was the major trigger that ensured modern computers happened at all.

Another Eden?

In January 2016, while writing this book, I was asked by BBC TV to explain the Winton Beauty of Mathematics Garden at the world-famous Chelsea Flower Show. It was designed by Nick Bailey, who runs the Chelsea Physic Garden. Established in 1667, the Chelsea Physic Garden contains plants that changed the world, such as rubber, tea and cotton, as well as those used in medicine and industry, plus every poisonous and distinctly smelly plant that will grow in an English climate. Hans Sloane (1660–1753), who created the Physic Garden, made such a profit from its medicinal, edible and floral produce that he bought what is now Chelsea, one of the most valuable pieces of real estate in the whole world.

Nick's mathematical garden (see plate section), set in a huge golden rectangle, has two central motifs. An algorithmic spiral has a round glass pool at its centre, showing the patterns formed by a sunflower head. A huge infinity symbol forms a long, low copper fence with mathematical equations etched along it (including those for photosynthesis and phyllotaxis, most of which are incomprehensible to anyone except modern biologists). The plants – ranging from tiny cactuses to huge and exotic ferns – exhibit a whole range of Fibonacci numbers in their growth patterns.

The TV director generously gave me almost three minutes to explain the garden, perhaps demonstrating a general lack of confidence that maths relating to science can hold an audience for longer. I found that very sad indeed. For the entire Chelsea week, the garden was constantly surrounded by a crowd three or four deep.

CHAPTER 15

The Many Tentacles of Mathematics

When contemplating his career, the aged Isaac Newton remarked that he had been rather like a boy on a beach, sometimes finding a smoother pebble or a more beautiful shell, while a great ocean of truth lay undiscovered before him. This final chapter covers the 300 years since that time – and the discovery, not just of shells or pebbles, but entire beaches of new mathematical ideas.

Inevitably, to cover so many new concepts I will have to be very brief. But the names and events I feature should in themselves be a strong impetus for you to explore further and deeper into the exponential or bottomless well that is mathematics.

Astronomy Was Looking Up

After Edmond Halley's death in 1742, most new advances in astronomy happened in Germany. In 1772, **Johann Daniel Titius** (1729–1796), a maths professor in Wittenberg, spotted – as Kepler had – that the distances between known planets seem to roughly double as they get further from the Sun. Beginning with a simple doubling progression (as shown in the table below), Titius adapted it to give the Earth a distance of 10 units from the Sun (today we call the Earth's distance from the Sun 1 *Astronomical Unit,* or 1 AU).

Doubling progression	0	3	6	12	24	48	96	192	384
To each, add 4	4	7	10	16	28	52	100	196	388
The then known Planets	Mercury	Venus	Earth	Mars		Jupiter	Saturn		
True Distance (Earth = 10)	3.9	7.2	10	15.2		52	95.3		

The numbers fit the actual planetary distances from the sun beautifully, but Titius failed to publish his ideas, so **Johann Elert Bode** (1747–1826) did. He was a self-taught astronomer at the Berlin Observatory and today this is known as 'Bode's law'. Immediately, observers started looking for new planets to fit the gaps.

William Herschel (1738–1822) was born in Hanover in Germany, but when the French army stormed in, his musical parents managed to get him to England, aged 19, along with his brother Alexander. Herschel became a successful organist, and settled in fashionable Bath. He composed 24 choral symphonies, but his music also led him to mathematics and Newton, and then to astronomy.

Too poor to buy a good telescope, he made about 200 attempts to grind his own lenses. Soon his singing sister **Caroline Lucretia Herschel** (1750–1848) joined him. Aged 10 she had survived typhus, and only grew to a height of 4ft 3 in (130cm). But she proved a great lens grinder, and became her brother's devoted assistant, once saying, 'I did nothing for my brother but what a well-trained puppy dog would have done.' However, she also became the first famous female astronomer, specializing in comet hunting – and discovered eight comets on her own.

In 1781 William Herschel discovered a new planet that was 192 Bode units from the Sun, and which fitted the 196 gap in Bode's chart pretty well. He called it *Georgius Siderius* ('George's Star') in honour of King George III. The French, however, refused to accept that name and suggested Herschel. Bode settled the argument by calling it *Uranus*, after the Greek god of the sky. Its orbital time is 84 years; by coincidence William died aged 84 in 1822. Caroline survived to 98.

The Italian astronomer **Giuseppe Piazzi** (1746–1826) visited the Herschels in Slough, close to Windsor Castle, to view their huge 40ft telescope, and the earlier – and slightly better – 20ft version, which had a lens 18.5in (47cm) in diameter (see Figure 15.1). Sadly, he fell off the telescope's ladder and broke his arm. Back in Italy he got a much happier break on the first day of January 1800, when he saw a tiny moving object and logged three positions and times. From these measurements

Figure 15.1

Gauss (who we meet soon) did the maths, confirming that the object was in orbit and its distance from the Sun was 27.6 Bode units, fitting the gap at 28 almost perfectly.

Piazzi named the object *Ceres* after the Roman goddess of agriculture. Today we know that it is slightly egg shaped and very small, at an average of only 930km (580 miles) across. Soon many more lumps of rock were found around this orbit, which had to be the debris from a planet that had either disintegrated or had never actually formed. Astronomers named them *asteroids,* which orbit in the *asteroid belt* – the name we use today. But an asteroid is a small star – really they should have been named *planetoids*.

In 1846 in Berlin yet another planet, *Neptune* (after the Roman god of the sea) was discovered, and logged at 301 Bode units. Finally, in 1930, Pluto was discovered at 396 Bode units, but with an orbit so eccentric it eventually lost its planetary status. But neither really quite fit the Bode sequence, unlike Ceres at 27.6 and Uranus at 192, which fit positions 5 and 8 beautifully.

A Huge Weight Off Our Minds

In 1798, following Newton's revelations about the laws of gravity and attraction at a distance, Henry Cavendish (who we met in Chapter 14) – now aged 66 – decided to try to detect gravity on Earth. He made a box out of hardwood (to eliminate vibrations) in which two lead balls (with a 2in diameter) were suspended 6ft apart by a stiff metal wire that formed a torsion balance. Two large 350lb lead balls, also suspended, were slowly swung towards the smaller balls, which at one point became attracted to the large balls and began twisting the torsion rod under the strain (see Figure 15.2).

After 15 attempts, and by comparing the different results, Cavendish scaled up the figures to give the mass (or rather, the density) of the Earth as 5.48 times denser than water. In effect, he had weighed the Earth. Scientists, later measuring the deflection of pendulums close to the side of a mountain, also

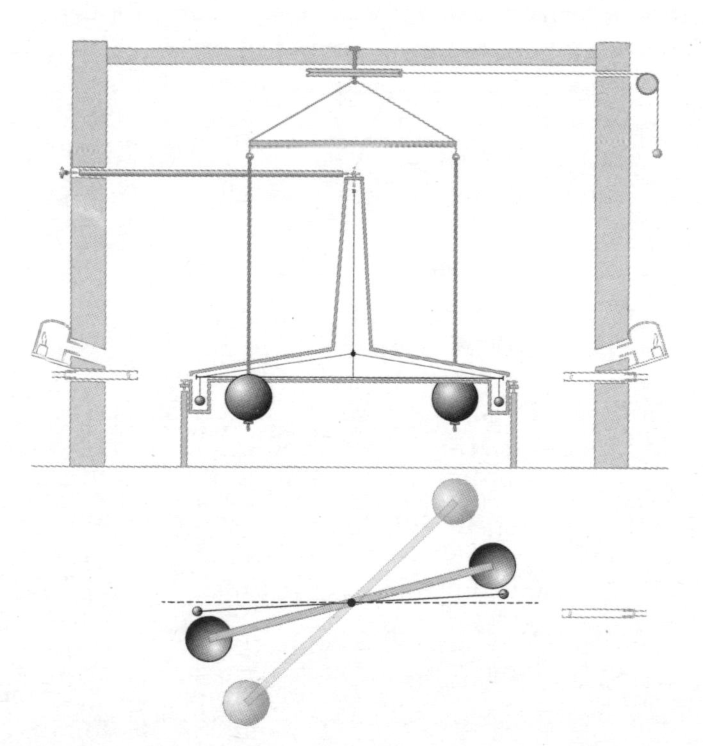

Figure 15.2

declared the Earth's density to be between five and six times that of water. Today's more accurate figure is 5.52 times.

The Rocky Road From Calculating to Computing

Charles Babbage (1791–1871) taught himself mathematics and got into Cambridge, where he befriended the top mathematician of the day, John Herschel (son of William), and another mathematician, George Peacock. The three rebellious lads, led by the more outgoing Babbage, saw that by following Newton and ignoring Leibniz for over a hundred years Britain had fallen far behind the rest of Europe in extending the boundaries of mathematics. They criticised the Royal Society for having more bishops than scientists as members, and in 1815 formed their own Analytical Society.

Babbage was to spend his life pushing for changes for the better. He remarked that it was silly to charge for delivering a letter depending on how far it had to travel, proposing instead just one charge irrespective of where a letter originated or was delivered. This idea was accepted worldwide, and today even ships' containers, whether they are travelling short distances or right across the world, are all charged at the same rate. The ever-inventive Babbage also created skeleton keys, the speedometer and the angled cattle shifter for the fronts of trains, used to clear cows to one side. Today, however, he is very famous for something he never quite achieved.

In 1822, spotting the growing need for accurate mathematical tables, he wondered if the calculations could be better done by a machine. Pascal and Leibniz had both made calculating machines, but Babbage had much greater ambitions. With large amounts of government funding he began work on his *difference engine*, which he hoped would handle polynomial functions and calculate finite differences without the need for multiplication or division – rather as the Greeks had used proportion. Gradually the manufacturer of his machine parts, Clements, began charging more and more for the 250,000 parts he needed to build the 15-ton device. So, after nine years on the project Babbage suddenly scrapped everything and started all over again with an even more ambitious machine.

In the early 1800s a French weaver, Joseph Marie Jacquard, had produced a loom that could weave very intricate patterns by employing punch cards. Descending wires either hit solid card or passed through a hole, which then determined which coloured threads would be uppermost in the weave and the final design. The Jacquard technique is still in use today, but it was the idea of using a binary operating system – a 'hole' or 'no hole' to represent 'yes' or 'no' – that Babbage saw as perfect for computing.

Babbage enlisted the help of Lady Ada Augusta, Countess of Lovelace, the daughter of the poet Lord Byron, as wild and notorious in his day as any pop star of our era. Her mother never allowed her to see her disgraced father, which fostered a wild and tempestuous nature in young Ada. She ran off with her music teacher, after which her mother married her off to a domineering close friend so that they could control her.

But Ada, inspired by Scottish scientist Mary Somerville, became a strong mathematician. After joining Babbage she became the world's very first computer programmer, adapting complex instructions as required by the machine, which was still a long way from being finished.

Costs escalated and funds ran short, so in desperation Ada and Babbage – both being mathematically gifted – devised a system to beat the bookmakers at horse races. But, as you can probably guess, they failed miserably: the art of bookmaking is in never setting fair odds. So the analytical engine was never finished. By the end Babbage must surely have realised that electricity, rather than mechanical devices, would power the future.

Nevertheless, all the parts, drawings and plans for Babbage's machine were preserved, and between 1989 and 1991 his difference engine was completed at the Science Museum in London. Its first calculation produced an answer to 31 decimal places. The analytical engine is also being painstakingly completed, and the hope is that it will be finished by 2021, the 150th anniversary of Babbage's death (see plate section).

As Nice as Pie

Following her success in treating the wounded in the Crimean War in 1854, **Florence Nightingale** (1820–1910) became a leading figure in the design of hospitals. Her ideas for hospital management were meticulous, and she even introduced pie charts to study the mortality of armed forces personnel, whom she saw were in as much danger in a hospital as on the battlefield. She divided the chart into 12 months, but extended or reduced the radius of each section to match the proportion (see plate section). Today pie charts are usually true circles, with the different segments varying in angular size to reflect their proportions.

The Logical Way Forwards

George Boole (1815–1864) was the son of a poor English shoemaker, but made great strides in discovering (also in 1854) a new type of mathematics in which logical statements could be transformed into symbols. Boolean algebra was largely ignored in Boole's lifetime. In 1930, however, the American mathematician and engineer Claude Shannon found it perfect for simplifying electronic circuits in prototype computers. The simple circuit diagram shown in Figure 15.3(1), for example, has three gates. To allow electricity to flow through the circuit we need to close two gates, but closing B and C won't work.

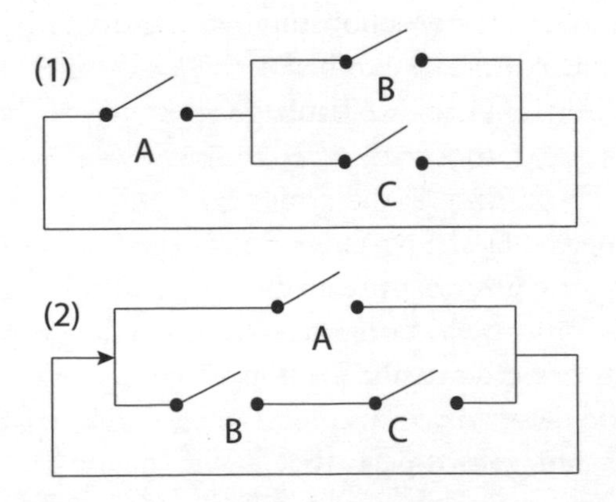

Figure 15.3a

The operation can be written A **and** (B **or** C)

To allow the flow in this second circuit we have two options, to close A or (B and C). Simple rules like this help our modern computers juggle highly complex operations beautifully.

One person who did take notice of Boole's ideas was Englishman **John Venn** (1834–1923), who began devising *Venn diagrams* in the late 1870s, and produced a book on symbolic logic in 1881. Today Venn diagrams feature strongly in modern maths curricula, and introduce students to mathematical logic and set theory.

Whatever the method of statistical analysis you use, it's vital to choose carefully so that you ask the right questions and get relevant results. In Figure 15.3b we have three overlapping circles: A represents Englishmen, B professional snooker players and C people who are left handed. The central area, numbered 1, gives us the total number of left handed English snooker professionals. Fine, but number 2 indicates all the professional players in the world who are neither English nor left-handed. Number 3 gives us no pro snooker players at all, but includes all left-handed Englishmen. Not a great deal of use…

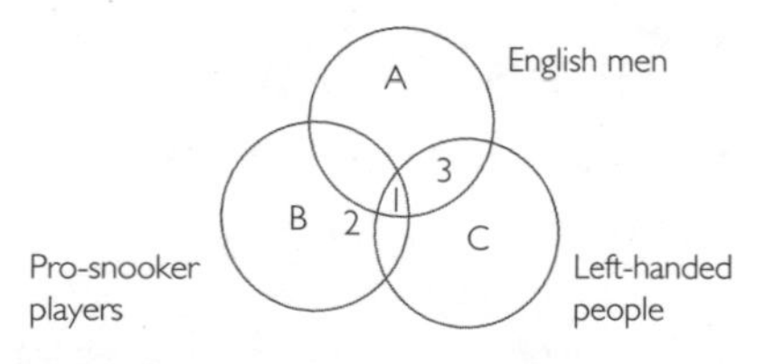

Figure 15.3b

For the system to be powerful, we must use sound mathematical logic when choosing our questions. For instance, the statement 'All crows are black,' is not as easy to prove as you might think. There are three possible ways to proceed:

1. You could find all the crows there are, then check if one isn't black. But how can you be certain you'd recognise a non-black crow?
2. You could collect all black things in the world and count the crows. But if a non-black crow exists, you would have missed it, again. To be absolutely certain, you need a third way.

3. You must collect all the things in the world that are not black, then sift through them to see if there is a crow among them.

As the writer Lord Dunsany once said, 'Logic, like whisky, loses its beneficial effect when taken in too large a quantity.'

Two Very Great Mathematicians

The Wise Old Owl

For a hundred years after Newton, the development of mathematics in Great Britain almost dried up. Most new ideas came from France and Germany, or from Basel's quite amazing Swiss Maths Whizz, **Leonhard Euler** (1707–1783) – say 'Oiler': it means 'owl' – who can be counted among the top half-dozen mathematicians of all time.

He was befriended by the Bernoulli family and recognised as someone clearly very special. He followed Daniel to St Petersburg, where he spent much of his academic life working for the royal family there. Sadly, working day and night in the severely cold Russian climate caused him to go blind in his right eye at the age of just 28. But he and his wife had 13 children while they were there, and he often worked with a child on his knee.

In 1766, aged 58, he lost his left eye, but total blindness hardly seems to have slowed him down. Thanks to his amazing memory he could retain several blackboards-worth of numbers and equations for a very long time. By the time of his death and after 17 years of total blindness, he had produced 500 books and papers, while another 400 appeared after his death. The Swiss edition of his complete works runs to 74 volumes.

Euler became a conduit for new ideas, and established many mathematical terms and symbols we still use today. As we learnt right at the start of this book, the symbol π was suggested by Welshman William Jones in 1706, because it was the first letter of the Greek word for 'periphery'. Euler adopted it in 1737, and it entered common usage soon afterwards. Euler also introduced and clarified the vocabulary of mathematics. To take an example: you might notice that the best football teams draw the largest crowds, in which case the crowd numbers are a function of the team's success. This concept was pounced upon by Euler; to make drawing graphs easier, he introduced f (x) to denote 'function'. So f (x) = 2x + 1 gives a graph for every value of x and its corresponding value for y (see Figure 15.4).

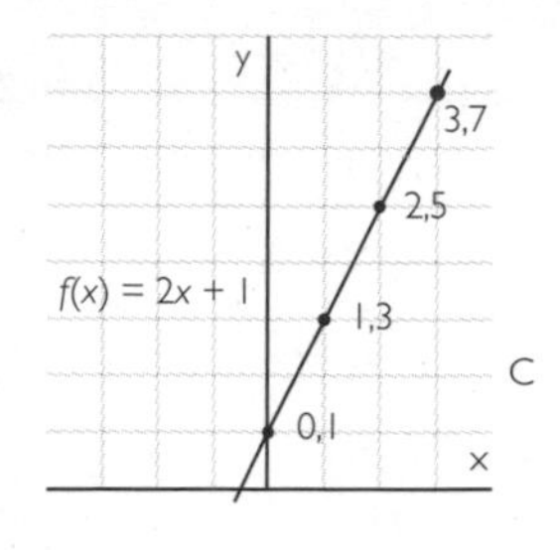

Figure 15.4

Euler formalised maths, and introduced procedures that even today are still accepted as the right way to do things. He also introduced totally new mathematical concepts and produced symbols to explain them. Euler's two most revolutionary ideas were represented by the symbols *e* and *i*.

e by Gum!

Many colloquial expressions are corruptions of terms that were seen as taking the Lord's name in vain. 'Golly gosh!', for example, derives from 'Good God!', and 'Ee by gum!', an expression of surprise or shock used in the north of England, is derived from 'Oh by God!'

In Euler's case, perhaps his most famous creation was the use of and explanation for the number *e,* known as 'Euler's number'. John Napier had first hinted at it with his base 10 logarithms; Euler introduced *natural logarithms* – rather than 10, they use as their base the symbol *e,* which is approximately equal to 2.718 ... Let me explain.

Say you have £100 in a savings account, and you want it to gain interest. The interest could be calculated and added once a year, or divided in half and added every six months, or it could be divided into twelfths and added once a month, or even divided into 365 portions and added each day. Which deal would you go for? Surely, you might say, the more frequent additions will be the best? Well, yes, but not by much, because the increases soon reach an upper limit beyond which they do not progress further.

It's a Matter of Factorials

The value of *e* is calculated using *factorials*: '2 factorial' is written 2! and means 1 x 2 = 2; 3 factorial or 3!, equals 1 x 2 x 3 = 6; and 4 factorial or 4! equals 1 x 2 x 3 x 4 = 24 and so on. To get any factorial you multiply together the integers from 1 to that number.

However, *e* involves 'fractional factorials' as generated by the infinite series, $e = 2 + ½! + ⅓! + ¼!$ ... etc.

2 = 2
2 + ½ = 2.5
2 + ½ + 1/6 = 2.666... 2.66666…
2 + ½ + 1/6 + 1/24 = 2.708333...
2 + ½ + 1/6 + 1/24 + 1/120 = 2.71666...
2 + ½ + 1/6 + 1/24 + 1/120 + 1/720 = 2.7180555...
2 + ½ + 1/6 + 1/24 + 1/120 + 1/720 + 1/5,040 = 2.718253968 ...

So, after just seven progressions, the total value for *e* has practically stopped growing altogether as each successive increase gets smaller and smaller, soon becoming insignificant. In the same way, cutting up interest into smaller pieces applied more frequently, soon becomes so insignificant everyone loses interest.

e is irrational, but the first 15 decimal places, if you break it up, go 2.7 1828 1828 45 90 45 … So now it's quite easy to remember, if that's your idea of a good time.

The '*i*' Has It

The strange number '*i*', also introduced by Leonhard Euler, represents a number that is 'imaginary'.

The square root of 1 is 1, as $1^2 = 1$. But no number squared equals –1, so the square root of –1 is *i*, or $i^2 = -1$.

The mathematical equation known as *Euler's identity* has been compared to a Shakespearean sonnet, and described as 'the most beautiful of all equations'. It is a special case of a foundational equation in complex arithmetic called 'Euler's formula', which the late great physicist Richard Feynman referred to in his lectures as 'our jewel' and 'the most remarkable formula in mathematics'. Euler's identity is written simply as: $e^{i\pi} + 1 = 0$.

Leonhard Euler Spins a Line

Euler loved all branches of mathematics, but enjoyed 'playing' with geometry. He discovered this little gem in 1765, a year before he lost his sight. Take any triangle and then find three centre points. First, find the centre of the circle that circumscribes the triangle – called the *circumcenter.* Now join the halfway point of each side to the opposite angle. These lines cross at the *centroid* point, the 'centre of gravity' of the triangle. Now draw altitude lines from each angle to meet the opposite sides at right angles. Amazingly these three lines also cross at one point, called the *orthocenter.* To his utter delight, Euler found that all three points always fall on a straight line, which today is still called the *Euler line* (see Figure 15.5).

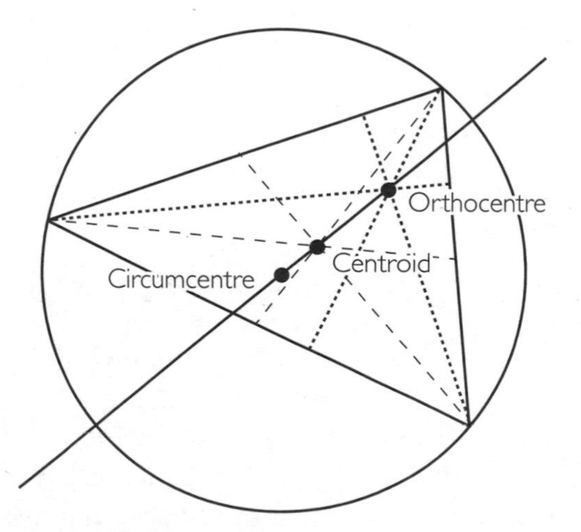

Figure 15.5

Crossing Bridges to a Totally New Mathematical World

Euler was first to explore what was then called *geometria situs* or situation geometry, which led to a new branch of mathematics, now called *topology*. The Prussian town of Königsberg had a river and seven bridges linking both banks with two islands. Visitors were invited to take a walk around the city by crossing each bridge once – and once only. It was actually a prank – as the locals knew, no one was ever able to do it. At long last, Euler explored the problem and explained why.

The land areas A, B, C and D all have an odd number of bridges leading to them, which is precisely why the tour is impossible. However, if you were to add a bridge joining any two areas (giving six options), or remove a bridge (7 options) two areas would then have an even number of bridges. By starting at one odd area and ending at the other odd area, the tour is now simple.

The kite plan shown in Figure 15.6 is a version of the Königsberg problem, with lines for bridges and single points (or 'nodes') for areas. Add or remove any line and two nodes become even. Starting at either odd node, the tour is now possible. So you can tour any similar network, no matter how complex, only if no more than two nodes are odd.

This was the birth of *mapping* or *network theory*, without which it would have been impossible to design modern electronic chips, with their highly complex networks. To

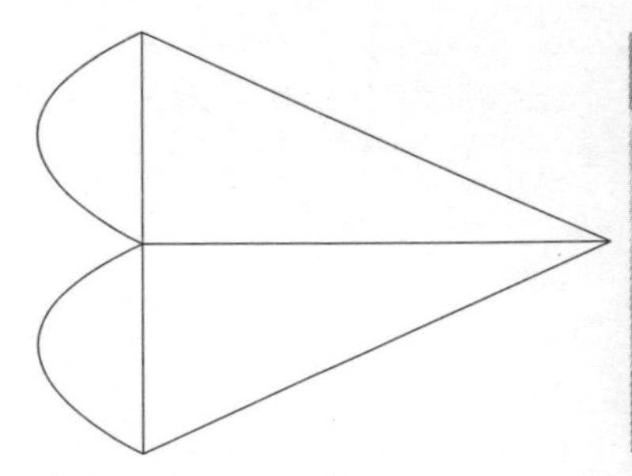

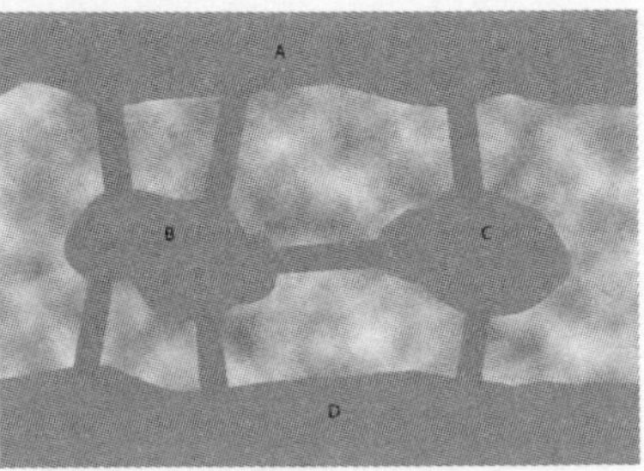

Figure 15.6

make a map of the London Tube, or indeed any of the world's tube networks, actual distances had to be distorted so that distances on the map could be uniform, when in reality they vary from point to point. But topologically the map is still correct, so planning a route from place to place is easy.

The Topsy-turvy World of Topology

At first the basic ideas in topology look obvious, but they soon lead to complex ideas that make us think again about the world around us. A ball, a cube and a pencil are all topologically identical. If you draw a circle anywhere on their surface it will divide the surface into two distinct areas. These items have no holes in them, and are known as *genus 0*.

Items like a mug, a doughnut (with a hole in it) and a person with a blocked-up nose, are included in genus *1*. You can draw a circle on them that would divide the surface into two distinct areas, but if you draw the circle passing through the hole, it doesn't divide the surface into two areas at all, and it is still possible to roam all over the shape without crossing the line of the circle.

Of course, the circle drawn on the person must pass in through his mouth and out at the other end. If his nose isn't blocked then there are four ways in, making it *genus 6*, because any two holes can be linked. But essentially a human – like a worm, or pretty much any other living creature – is a tube that is open at both ends.

In a famous topological problem, providers of three utilities – gas, water and electricity – must be linked to three houses A, B and C so that no service line ever crosses another. If you turn this into a line drawing with six nodes, it's impossible. But if you allow a house to be more than a single point, and if one service is delivered to one wall and a second service to the opposite wall, then the third service could possibly pass under that house, between the ends of the terminals, and on to the previously unreachable house. Problem solved (see Figure 15.7).

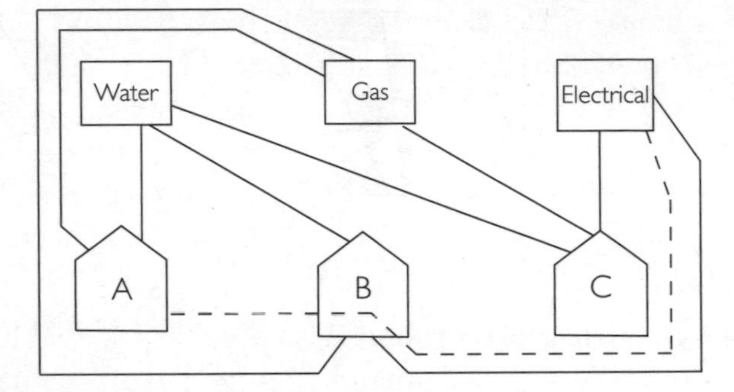

Figure 15.7

The Curse of Masculinity: Hair Today, Gone Tomorrow

I'm not sure that mathematics actually does cause people to tear their hair out ... But the human head of hair does provide a topological enigma. Why does it always have one crown? And why is it topologically impossible not to have a crown, however much you comb it.

A ball covered in hair has twice the problem: combing always produces two crowns and frontiers where hair that is lying in different directions must meet. This explains the Earth's perpetually changing weather systems, because atmospheric temperatures – influenced by the spinning globe and the Sun – vary constantly. There must always be opposing fronts moving towards or away from each other, so there must always be weather, whether we like it or not.

The Four-colour Map Theorem

This is one of the most famous mathematical enigmas. To shade a map produced just using straight borders, so that no two adjacent areas are the same colour, you only need two colours. If you introduce any new straight line, you can keep the situation as it is simply by swapping all the colours on one side of the line (see Figure 15.8).

But according to the 'four-colour map theorem', no matter how complex the map you only ever need a maximum of four colours to shade it so that no two adjacent

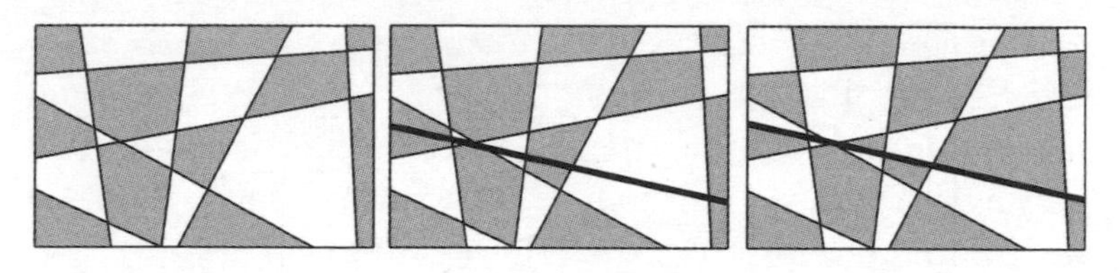

Figure 15.8

areas contain the same colour. This idea was probably first suggested by August Ferdinand Möbius (who we meet in more detail later). But the theorem wasn't proved until 1976, when Kenneth Appel and Wolfgang Haken used a computer to calculate the variables over many hours.

There is a simple yet thought-provoking extension to this idea. If you need only four colours to colour a map, then it must always be possible to shade any map using just three colours for all areas touching the outer edge; you can then use the fourth colour to completely surround the map, and to fill in some internal areas.

Topological Summation

A deep understanding of topology is required in many branches of manufacturing, and topological solutions have revolutionised our ability to make machines that can handle complex packaging or box-making. Weaving machines that make stockings without tying the threads in knots perform small topological miracles. It has always been a simple topological trick to ask someone to tie a knot in a piece of string without letting go of the ends. It's impossible unless you start correctly. First with folded arms, pick up the ends of the string. By just unravelling your arms, the string is now knotted.

Gauss's Work Was Seldom Guesswork

Carl Friedrich Gauss (1777–1855) is arguably the greatest mathematician in history after Archimedes and Newton. As a precocious child, aged three, he could see instantly by looking at several piles of coins which were a coin short, and he was soon correcting his father's wage-calculation errors.

(Mind you, the very young often amaze us with their unschooled abilities. At that age my daughter Zoe could identify from looking at the record the songs on any one of 20 singles records, I can only assume because she recognised the colour and pattern of the printing on each label.)

When Gauss was nine, his severe teacher Herr Buttner set a maths progression question for the class, which young Carl answered in just a few moments, while the other boys laboured for ages. When Buttner checked, Gauss's slate contained just one number – the correct answer. Although it's not known exactly what question was asked, many suggest, as I once explained in a song, that:

Carl Freidrich Gauss, when only nine, he did a sum in record time.
'Please do this sum,' his teacher said: 'Add the numbers from one up to a hundred.'
He made a pair of first and last, that's one hundred and one (1 + 100 = 101)
Then multiplied by fifty pairs, and there the job is done. (101 × 50 = 5,050)

According to legend this was the shortcut Gauss used, but some scholars suggest that the sum was probably more complex than that. Buttner may have been more devious by, for example, setting the following problem:

The series 7, 14, 21, 28 is increasing by what rate? By 7 each time? Okay. Give me the total of the first 100 terms starting with 7 and increasing by 7.

Using the same shortcut – pairing the first and last numbers and multiplying by 50 – we get:

$$7 + 700 = 707 \text{ x } 50 = 35{,}350$$

To make it more complex, the teacher might have started some distance into the sequence. Starting with 28, for example, and increasing by 7 each time, what is the total of the sequence of 100 numbers?

$$28 + 721 = 749 \times 50 = 37{,}450$$

Whatever the problem was, Buttner saw the lad's potential and bought him a better textbook, although before long he was forced to admit, 'I can teach the boy no more!' However, he did have a 17-year-old maths-loving assistant, Johann Martin Bartels, and he put the two together. It worked beautifully.

By the time he was 12, Gauss had exhausted Euclid and was dabbling in non-Euclidean geometry. He saw that Euclid's two-dimensional maths did not work on the surface of a sphere, where a triangle can have angles that add up to more than 180 degrees as Menelaus had discovered (see Chapter 6). But what about the mathematics on a shape like a trumpet horn? (See Figure 15.9) Here a triangle has angles that add to less than 180 degrees, requiring a totally new form of geometry. And in both these types of geometry there are no such things as two parallel lines.

Gauss never published his non-Euclidean findings, because he thought established scholars might laugh at them. However, János Bolyai, the son of Gauss's lifelong friend, spent years contemplating Euclid's parallel postulate, and must have been influenced by Gauss himself. In 1823 he announced his ideas about geometry on curved surfaces. At almost the same time, in Kazan University, an obscure institution in Russia, **Nikolai Ivanovich Lobachevsky** (1792–1856) published remarkably similar ideas. It just so happened that one of the professors at Kazan came from Germany – his name was Johann Martin Bartels, the chap who had been paired with Gauss when he first had these ideas.

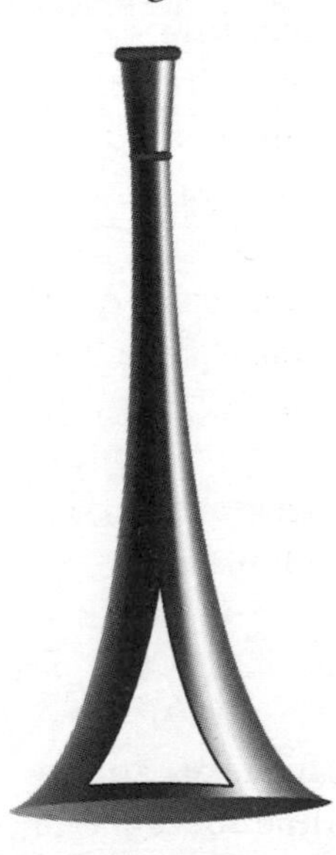

Figure 15.9

Not surprisingly, perhaps, Gauss declined to endorse the work of either man. But neither did he complain: this is essentially how mathematics and science progress, as concepts are passed from person to person until someone declares they've had a new idea and the whole edifice moves forwards. This process is exactly what this book is about.

Forty-three years after Gauss's death, a tiny 20-page book was discovered, containing 146 'original' mathematical discoveries made during his lifetime. The first entry was the *heptadecagon problem*, which he solved when he was only 19 and still planning to become a linguist.

Construct a regular 17-sided figure using just a straight edge and compasses.

This problem had lain unsolved for more than 2,000 years, but Gauss solved it. He then went on to discover a mathematical formula for constructing all regular polygons using only a straight edge and compasses – and actually found 31. Following the 17 sided figure are the 51, 85, 255, 257, … , and 4,294,967,295-sided figures. His first career choice of linguistics now became simply a lifelong hobby.

Another notable entry in this tiny notebook read 'Eureka! Num = Δ + Δ + Δ'. This means that absolutely any integer is the sum of no more than three triangular numbers. The triangular sequence known since the days of Pythagoras is:

1, 3, 6, 10, 15, 21, 28, 36, 45, 55, 66, 78, 91, 105, 120 ...
(see Chapter 2)

So will all numbers comply with the theory? Let's try a few random ones: 150, 151, 152 and 153.

150 = 105 + 45 (just two triangular numbers); 151 = 105 + 45 + 1 (three); 152 = 91 + 55 + 6 (three); 153 = 105 + 45 + 3 (three).

In 1801, aged just 24, Gauss proved the '**fundamental theorem of arithmetic**': that 'every non-prime number is the product of prime numbers in one way only'. Remember that 1 is not classed as a prime number but 2 is; the first few primes are 2, 3, 5, 7, 11, 13, 17, 19 ... Here are some examples that demonstrate the rule:

6 = 2 x 3; 8 = 2 x 2 x 2; 9 = 3 x 3; 10 = 2 x 5; 42 = 2 x 3 x 7; 58 = 3 x 19 and so on.

Today we call this 'factorising' a number, or finding the unique set of factors that produce it. The factors of prime numbers, of course, are simply 1 and the number itself, so 7 = 1 x 7.

In his book *Disquisitiones Arithmeticae*, published in 1801, Gauss introduced modular mathematics. Our everyday clock is a good example of modular maths. If it is 3p.m. now and you add 12 hours, it is not 15p.m., but 3a.m. the next day. After 24 hours, the time is not 27 o'clock, but 3p.m. on the next day. So our clocks tell the time *modulo 12* or *modulo 24*.

As another example, 11, 32 and 53 have no common factors, but if you divide them by 7 they all give the same remainder, 4. So they are all divisible by 7 *modulo* 4. The system proved to be a very powerful part of advanced mathematics, and helped Andrew Wiles solve Fermat's Last Theorem (see Chapter 12).

Gauss spent 20 years working on astronomical calculations, which began on the first day of the nineteenth century, when Giuseppe Piazzi discovered Ceres. Gauss worked out that it was a small planet with an elliptical orbit and calculated its speed and average distance from the Sun. Soon Pallas, Vesta and Juno, smaller sisters to Ceres, were also discovered as Gauss's maths showed astronomers where to look. Some say that Gauss's astronomical years were a waste of his talents, but it's difficult to see how he could have made many more mathematical discoveries in one lifetime.

Gauss had more far-reaching ideas than any other mathematician, all centred on his ability to tackle problems in a whole range of mathematical ways. He developed the surveyors' *heliotrope*, for example, which calculates south using the time and the position of the Sun. A system of mirrors then reflects the sunlight in a completely straight

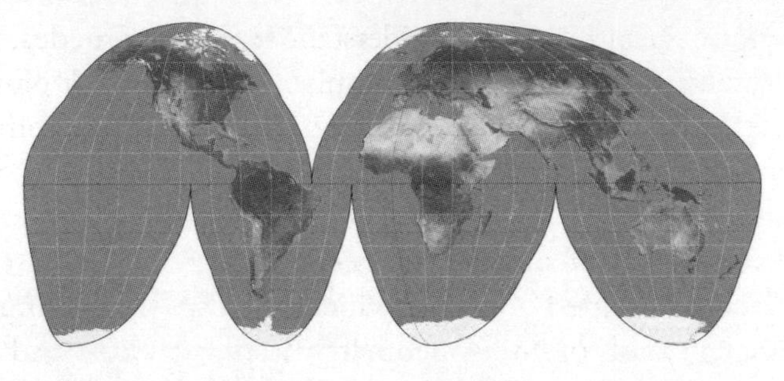

Figure 15.10

line to a distant object, and precise angles can then be read off. It was more accurate than any previous surveying tool (see Figure 15.10).

He also located the accurate position of magnetic north, and (as we saw in Chapter 7) the unit of *magnetic flux* was named after him. When an iron ship is built, the constant hammering of rivets during its construction makes it behave rather like a bar magnet, in that its north and south magnetic poles are always on the line of their original position when it was built. So that the ship's compass can work properly, this magnetic field has to be removed, in a process called *degaussing*.

Gauss additionally studied a new branch of topology, describing knots mathematically according to the number of string crossings involved, making it easy to spot simple, double or more complicated knots. He studied curved surfaces, and explored how many cuts it would take to lay them flat. A cone requires just one cut, for example, while a *torus* or *anchor ring* requires two.

From Archimedes, Gauss knew that the surface of a sphere has exactly the same area as the walls of the cylinder that encloses it. However, a cylinder requires just one cut to lay it flat, while no end of cuts will flatten a sphere. Apparently Gauss tried flattening an orange skin, and preempted Goode's 1923 homolosine map projection often used today.

Gauss idolised Archimedes but grumbled that the great man had stopped short of inventing calculus, despite coming so very close to it. He also couldn't understand why Archimedes, when he'd been looking for a better number system, hadn't plumped for the decimal system – without it much of Gauss's mathematical exploration would have been impossible.

After Gauss's death, aged 78, the King of Hanover erected a statue of him. It stands on a pedestal on which is inscribed his famous 17-sided polygon, which he discovered using only the ancient tools of Greek geometry: a straight edge and a pair of compasses. He will forever be known as the Prince of Mathematicians.

An Infinity of Infinities

After Gauss, many eminent mathematicians arrived on the scene, but it would be impossible for me to feature them all. Here are just two who I feel deserve a mention, but for quite different reasons.

Georg Cantor (1845–1918) was born in St Petersburg to a Danish father. More philosopher than mathematician, Cantor rather put the cat among the pigeons with his views on infinity (although Galileo had preceded him by about 300 years).

First take these two series of numbers:

1	2	3	4	5	6	7	8	9	10	ad infinitum
2	4	6	8	10	12	14	16	18	20	ad infinitum

For every number in the top row, there is always a number that is double it. So there are as many even numbers as there are numbers – which sounds crazy, but if we consider infinity, Cantor reasoned, it has to be accepted as true. Quite soon with arguments like this, you come to realise that defining infinity is infinitely difficult, because the number of situations is also infinite. In fact, performing infinitely more complex cerebral somersaults led Cantor to suffer a breakdown, and he died in a mental hospital.

Cantor's ideas were a source of great worry to other mathematicians, who felt they threatened the purity of mathematical thought. Luckily, another modern mathematician, **Kurt Gödel** (1906–1978), brought a sense of reality to mathematics when he said, 'Whatever mathematical system you use to prove things, there will always be things that those laws cannot prove. If you adjust the system to prove those laws, then other principles will crop up that cannot be proved; and so on.' It is the nature of mathematics.

The Mathematical Explosion

Without doubt, more new mathematical ideas have been discovered and developed in the past 150 years than in the whole of the preceding history of the subject. But it is the development of mathematics through history that I have aimed to cover in this book. If I may, however, I'll explore just a few of the more modern mathematical avenues, especially those that have helped to create a far more colourful and engaging mathematical world.

Come Follow the Band

August Ferdinand Möbius (1790–1868) – who we briefly met earlier – was influenced by Gauss, and became famous in 1865 with a theory that had holes in it from the start. A sheet of paper has two sides and one edge that runs around it. Add a hole, however, and you add an extra edge. In fact, a two-sided sheet can literally have an infinite number of holes and therefore an infinite number of edges.

Möbius then asked, 'Could paper with one edge have just one side?' If you join the ends of a strip of paper to form a loop, it now has two sides, but also two edges. Give the strip a half turn before you join the ends and you form a *Möbius strip*. Pass your finger along either the edge or the face, and after two circuits you come back to where you started. You have a one-edged, one-sided loop of paper. But that's just the start. Now cut along the strip two-thirds of the way across. What you end up with is one loop that has the same length as it did before, but is now one-third the width and, surprisingly, another loop twice as long with two twists – as another surprise, the two shapes are linked (see Figure 15.11).

Symmetry and Curvaceousness!

Snowflakes have six-fold symmetry; it's said that no two snowflakes are ever the same, because there is an infinite number of ways that a collection of water molecules can freeze. People have explored symmetry from the time of Euclid, but in modern times the practice has exploded. In 1904, Swedish mathematician Helge von Koch took an equilateral triangle and placed another triangle a third of its size on the centre of each side; by stages he produced the 'Koch snowflake'. You could repeat the process an infinite number of times but, surprisingly, the area of the final snowflake will be just 1.6 times the area of the original triangle.

In 1915 Wacław Sierpinski produced the 'Sierpinski gasket', although similar patterns have been found in Islamic and even Christian churches from the thirteenth century. Sierpinski began with a black equilateral triangle, then removed an inverted quarter-size triangle from the centre; he then repeated the process.

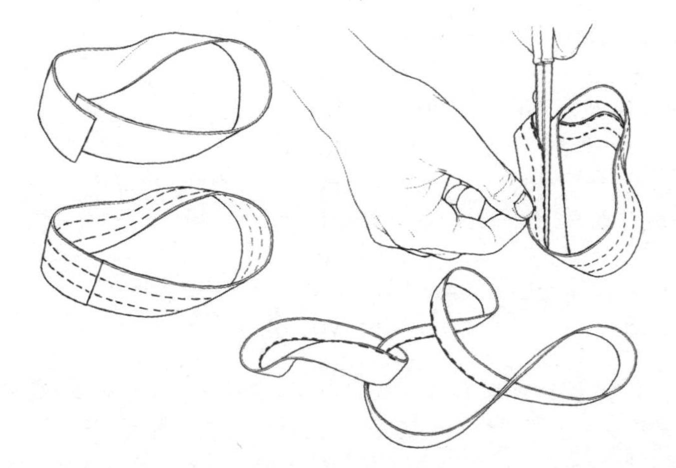

Figure 15.11

He then did the same with squares to produce the 'Sierpinski carpet', removing one-ninth of the square from the centre area, then removing a centre-ninth from each of the surrounding eight squares. He repeated the process, removing centre-ninths from the next eight smaller surrounding squares – and so on.

In 1926 Carl Menger, latching on to the idea of the Sierpinski carpet in three dimensions, produced the Menger sponge. As you progressively remove ever more material, the surface area increases dramatically while the volume of the sponge reduces drastically (see plate section).

The Mathematics of Small Holes, in the Non-mathematical World

An extreme example of these topological ideas is the thin film that protects Wimbledon's main tennis courts in the event of rain. It's made from a material called expanded polytetrafluoroethylene (or ePTFE), which has a fascinating atomic structure. When it's stretched under certain conditions, it suddenly fractures into a practically infinite number of identically sized, hexagonal-shaped holes. These holes are 20,000 times smaller than the smallest water droplet (making the material completely waterproof), yet they are still 700 times larger than a single air molecule. So when it rains and the sliding roof is closed, huge electric fans push air in or out of the arena through this material, balancing the atmosphere inside to match the conditions before the roof was closed. It's a perfect example of how modern mathematics powers technologies that make the many quite features of our 21st world possible.

The Man Who Made Mathematics an Art

'Escher' means 'listen' in Dutch, but **Maurits Cornelis Escher's** (1898–1972) gift was looking and learning. A sickly child, and known as 'Mauk', he attended a special school, where he failed everything except drawing, carpentry and piano – all of which require mathematical thinking. His father, a civil engineer, encouraged him to travel; he fell in love with Italian landscapes, and became addicted to the Islamic tile tessellations of southern Spain (see Chapter 8). This experience inspired topological works of pure genius and the sheer variety of his work is awe-inspiring. He is without doubt the most mathematical artist ever (see plate section).

The Highly Mathematical Architect

As an architect, **Antoni Gaudi** (1852–1926) was a maverick, and his work, mostly in or near Barcelona, is the focus of pilgrimages for architects and art students. His monumental cathedral, the Sagrada Familia, was not finished in his lifetime – in fact it is still not finished today, although young architects from all over the world are constantly trying to interpret Gaudi's intentions so they can complete it. For the last few years of his life, he slept on a mattress on the floor close to the centre of the work in progress. The buildings he did finish include La Pedrera, which was so complex and took so long that for years it was called 'the quarry' (because it looked like one until its sheer finished beauty was revealed); and Casa Batlló, with its balconies of skulls and bones and its coral colouring. All are World Heritage Sites.

Gaudi rarely drew detailed plans of his works, preferring instead to create them first as three-dimensional scale models, so he could add new ideas along the way. He first designed the cathedral in Barcelona as an upside-down model; its arches (hundreds of them) are all based on catenaries (see plate section).

Gaudi had new ideas about every aspect of architecture, and his work is so varied and original that just about every art student in Europe is taken to Barcelona at some point in their course. As far as I know, however, no one takes young engineers or maths students. Yet of all his quotations – and there are many dotted about the building to catch the eyes of tourists – the largest and most prominent, in the nave of the cathedral, is a statement containing just seven words: 'There is no Art. All is Mathematics.'

Everyday Life Is a Mathematical Catastrophe

Topology is often called 'rubber-sheet geometry', and French mathematician René Thom's 'catastrophe theory' (devised in the 1960s) uses a flexible surface to explain how tension and excitement are created in every drama from Shakespeare to Harry Potter (see Figure 15.12).

Thom's warped surface is a graph of events that move forwards from the back edge, with tension moving from low on the right to high on the left. Let's assume a postman calmly approaches a door and a docile dog. Suddenly the dog growls, and the tension increases as the growl gets louder. One of three things can happen. If we follow line A, the dog backs down and the postman delivers his letters and moves on. Along line B, the man shows his anger and the dog backs down. But what if the dog attacks? That takes us along line C and the whole situation jumps onto a totally different plane where for a while chaos reigns and the outcome is totally unpredictable.

Thom's ideas show us that game-changing situations are really what drive the writing of all TV soap operas, with their many plot twists and turns designed to hook us in. But these same pressures and situations mould our real lives too, making our successes and failures just as unpredictable.

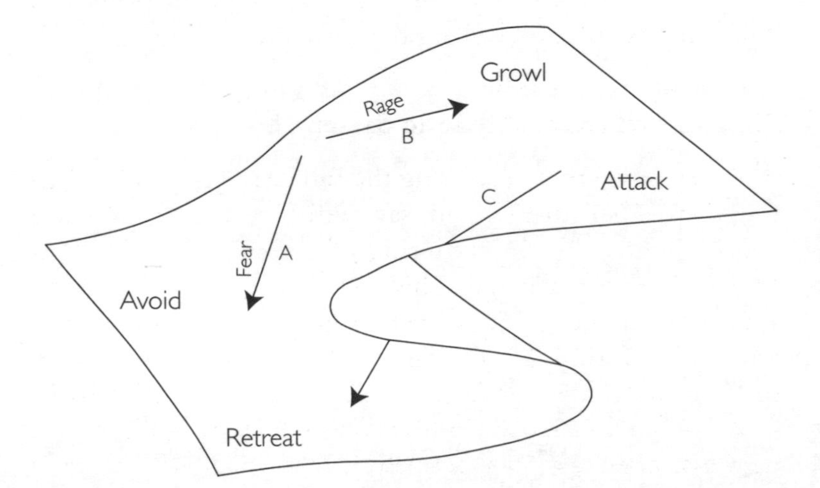

Figure 15.12

Can Mathematics Model Chaos?

The Lever of Mahomet is a puzzle suggested by Richard Courant and Herbert Robbins. Imagine a train with a flatbed trailer on which a serving tray is balanced. If the tray is balanced upright, as soon as the train moves forwards the tray will fall backwards. If the train moves forwards at a constant speed, then it should be possible to balance the tray so that its forwards lean equals the pressure of the air rushing towards it, keeping it in balance. But is it possible to find a starting position for the tray so that it stays balanced on a journey in which both speed and direction might change several times?

Well, it can be done in reverse. You can easily balance a tray on its short edge on your hand, and it will only move out of balance in one of two directions. With practice, you can walk across a room with the tray balanced, and adjust your hand position to allow for changes in the tray's movement when you stop, start, slow down or speed up.

But surely it's impossible to find the starting angle for the tray so it maintains its balance throughout a particular and set journey? Even if the journey involved no movement at all, a balanced tray would become unstable and fall, simply because of the movement of the air. Yet modern mathematicians are being asked to come up with solutions to similar problems. Take, for instance, predicting the weather…

Finding Order in a Chaotic System

It has been said that predicting the future is impossible: even today's weather, in England, say, will have been influenced (albeit slightly) by the fluttering of butterflies in the West Indies a week or two ago. This doesn't stop us ranting at the weather forecasters every time they get things wrong. But by asking for future predictions – for weather, politics or sporting outcomes – we are expecting mathematicians to find ways to calculate the impossible. Everything that happens, when taken collectively, is chaotic and never totally predictable.

Setting up a chaotic system is easy. If you hang a rod or a ruler with a nail through one end, it will swing from side to side and stop in a way that's perfectly easy to explain. But if you hang two desk rulers, pinned together so they can move independently of each other, you only need to set them swinging to observe the most amazing random and totally haphazard movements.

Sorting the Rough From the Smooth

We use mathematics and mathematical models to smooth out problems and make life more understandable. But nature is seldom smooth, and often unpredictably chaotic.

In 1979, while working for IBM, **Benoit Mandelbrot** (1924–2010) wondered how computers might help us study and even measure things so chaotic that most people thought they were immeasurable. According to Galileo, the secrets of the Universe are written in geometric shapes, but Mandelbrot realised that this was too general a statement. In fact, 'clouds are not spheres, mountains are not cones, coastlines are not circles, bark is not smooth and lightning does not travel in straight lines'.

He coined the term 'fractals' for complex shapes, and noticed that although coastlines have no predictable form, whether you view them at your feet on a beach, from the top of a cliff, from a helicopter or even from space, their changeability, though on totally different scales, was always very similar.

Mandelbrot rediscovered the work of French mathematician **Gaston Julia** (1893–1978) who, while fighting in the First World War, had his nose blown off and almost lost his sight. He wore a small mask from then on, but his mind survived intact; he discovered what became known as 'Julia sets'.

A planet is an attractor: in orbit it is attracted to its sun, but it is also attracted to other planets, and even to its own moons. So while it stays in orbit, its orbit changes shape over time. Julia, working with another French mathematician, Pierre Fatou, began to plot the path of attractors that are influenced by changing mathematical equations. The result, his Julia sets, came in many shapes, but were all symmetrical about the centre point on a graph (see plate section). It was laborious work, but everything changed in the late 1970s when computers became more powerful and more sophisticated.

Mandelbrot sets uses complex changing formulae, made from real and imaginary numbers, and plot the results. If a change stays within certain limits it becomes part of the set, and is coloured black. If it moves out of the controlling parameters, it is given a colour that indicates whether it is moving away gradually or quickly. The basic set is viewed through finer and finer scales, which illustrate its increasing complexity. The computer program has to make a calculation for every single pixel in each image, even as the scale increases to about a million pixels in each direction. Above all, however, the results are ever greater examples of the phenomenal and unlimited beauty of mathematics (see plate section).

Summing Up the Wonders Beyond Numbers

Thanks to our modern ability to number crunch, we have been able to produce complex solutions to the four-colour map problem, and unravel the truth about Fermat's Last Theorem. After Mandelbrot, we have discovered more about the maths of cloud formations, or how lightning finds its jagged path to earth. As yet unknown future advances will most certainly come about because of our ever-expanding technological ability.

In this brief and very selective history, I have truly only scratched the surface of what maths is all about. Mathematics will continue to branch out forever, as we encounter new problems and then discover new concepts in every field of science to tackle them, formulating the maths that will clarify them and help us explore further.

Einstein once said to a young audience, hanging on his every word, 'You may think that you have problems with mathematics. I can assure you, mine are much worse!' Any headaches caused by mathematics should never be the result of already established principles. If those before you learnt and understood them, there is no reason why you won't as well: all of these people wanted to pass their understanding on to those who might follow.

There are more mathematicians alive today than ever lived during its long history, and more are being inspired all the time. It's pretty certain that one day, one or two of today's young minds will be judged to be on a par with Archimedes, Newton, Gauss or Einstein. But those new geniuses will most certainly have been inquisitive enough to have trawled through the history of mathematics, taking their lead from the great achievers who went before them.

With today's new empowering technologies in computing and communications, there are very few problems we will not be able to solve. And we are better equipped to find those solutions than any generation that has lived before us.

So please, please, never accept the doubts and worries of those who predict a doomed future for the human race. There will be ups and downs, but just by considering the technological advances in my lifetime, it is as clear as it possibly can be that mankind's ingenuity will make the future brighter than any of us can yet imagine. And the wonders that lie ahead will be explained and enabled by our ever-improving understanding of all things mathematical.

Wow Factor Mathematical Index

Chapter 1: The Most Ancient Mathematical Legend

Egyptian Division and Use of Fractions

The Egyptians found a method for division and also for handling fractions. Both were linked to their multiplication system, and were incredibly sophisticated for their time, seen in both the Rhind and Moscow Papyri.

Division was really an extension of their multiplication system, achieved principally by doubling and halving. To find how many 7s go into 84, for example, they would say, 'How many times must I multiply 7 to get 84?'

As with multiplication, you start with 1 and 7 and double each column.

1	7
2	14
4 /	28 /
8 /	56 /

Stop there – the last two numbers on the right add up to 84. On the left they add to 12. 7 x 12 = 84

Now find the numbers on the right that add to 84 and score them. So 28 + 56 = 84

Score the corresponding numbers on the left and, sure enough, 4 + 8 = 12 and 84/7 = 12.

If a sum did not divide equally, the Egyptians introduced fractions. Let's divide 16 by 3.

1 /	3 /
2	6
4 /	12 / 12 + 3 = 15 but we are looking for 16 so we search for thirds
2/3 of 3 2	
1/3 of 3 /	1 /

So 3 + 12 + 1 = 16 and, sure enough, 1 + 4 + 1/3 = 5 1/3 and 16/3 = 5 1/3

To express a fraction, they placed a bar over the number, so 7 with a bar over it was 1/7.

They only used 1 as the denominator of the fraction, with just one exception: 2/3.

An egg on its side (the sign of the open mouth), placed over a 3, represented 2/3.

The Moscow Papyrus Problem

For me the most amazing revelation in the Moscow Papyrus is the geometric problem number 10, which asks that you calculate the surface area of a semi-spherical basket.

You are given a semicircular basket with a mouth of 4½ units diameter. What is its surface area?

The translation of the solution is given in stages:

Take 1/9 of 9 (as) the basket is half an eggshell. You get 1.
(This seems to refer to the concept that a circle of 9 unit diameter is equal in area to an 8 x 8 unit square.)
Calculate the remainder, which is 8.
Calculate 1/9 of 8. You get 2/3 + 1/6 + 1/18. (There is no explanation as to why…)
Find the remainder of this 8 after subtracting 2/3 + 1/6 + 1/18. You get 7 + 1/9.
Multiply 7 + 1/9 by the diameter 4 + 1/2. You get 32. (64/9 x 9/2. The 9s cancel out, giving 64/2 = 32)
Behold this is its area. You have found it correctly.

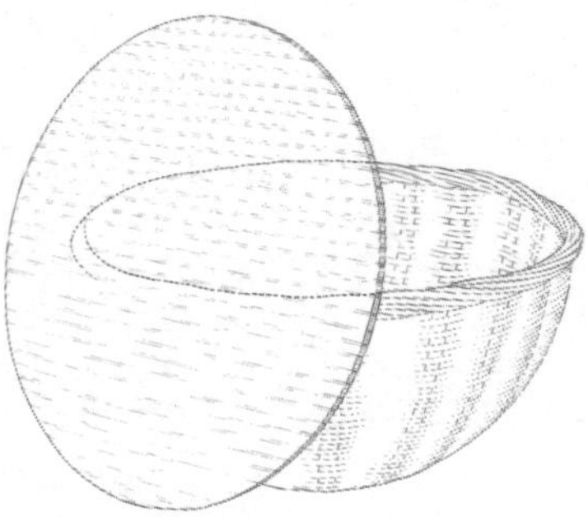
Figure 16.1

It is still wonderfully mysterious as to how they arrived at this solution. But it is correct. Archimedes (see Chapter 4), some 2,000 years later, was to discover this area of the surface of a sphere, but there seems to be no connection to the Egyptian method.

Generating Pythagorean Triples.

The Babylonian tablet Plimpton 322 at the end of Chapter 1 shows that the Babylonians must have known how to generate Pythagorean Triples. The tablet contained 15 triples for right-angled triangles whose smallest angle was between 45 and 30 degrees. It seems that tablet, while still moist, was bent and broken off below the 15th calculation. Why could this have been? Well, it seems that whole numbers with right-angled triangles where the smallest angle is less than 30 degrees, run into the hundreds. Here is a simple way to produce Pythagorean Triples:

Select two numbers 'x' and 'y' and apply this formula:

	$x^2 + y^2$	$x^2 - y^2$	2xy	Triple
Example				
x = 2, y = 1	4 + 1 = 5	4 - 1 = 3	2x2x1 = 4	5, 3, 4
Other examples :				
x = 9, y = 4	81 + 16 = 97	81 - 16 = 65	2 x 9 x 4 = 72	
x = 15, y = 8	225 + 64 = 289	225 - 64 = 161	2 x 15 x 8 = 240	
x = 64, y = 27	4,096 + 729 = 4,825	4,096 - 729 = 3,367	2 x 64 x 27 =3,456	

For the monster at item 4 on the tablet, the starting numbers were 125 and 54.

Start numbers	x	y	Triples	a $(x^2 + y^2)$	b $(x^2 - y^2)$	c (2xy)
	125	54		18,451	12,709	13,500

Sure enough, $18{,}541^2 = 12{,}709^2 + 13{,}500^2$ or 343,768,681 = 161,518,681 + 182,250,000

Lines of these three unit lengths would form a perfect right-angled triangle. Wow.

Chapter 3: The Great Age of Grecian Greeks

Wallace Measures the Diameter of the Earth

A more modern scientist who was also influenced by ideas from the ancient Greeks was Alfred Russell Wallace who, with Charles Darwin, developed the theory of evolution. Wallace first demonstrated his knowledge of Greek maths around 1870, when he read a letter in *The Times* written by a chap who had visited the Old Bedford Level, a perfectly straight canal more than 20 miles long, used to drain the Fens in East Anglia to protect some of the finest farmland in the UK.

As the correspondent explained, 'I placed my telescope to my eye and gazed down the Old Bedford Level. I could see no discernible curve. Therefore "The Earth is assuredly Flat and I will give a £500 prize to anyone who can "prove" otherwise.' Not only was he an idiot, he was a wealthy one at that.

Fired up by this challenge, Wallace set off for the Old Bedford Level, stopping off at Cambridge to borrow a couple of shallow punts. Arriving at the canal, on a beautifully calm day, he placed a theodolite (a telescope used to measure levels) set at 1 yard above the waterline, in each punt along with an observer. (see Figure 16.2).

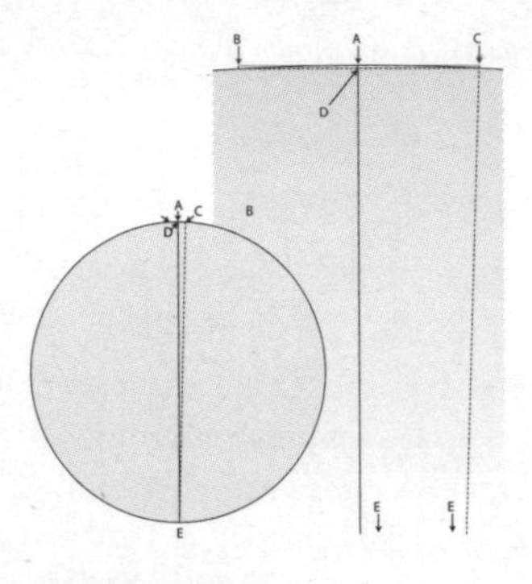

Figure 16.2

Each of the punts was pushed away from a chosen point on the canal. The men on board each boat had flags to indicate whether, having checked the line of sight on their respective theodolites, they wished to be moved further apart or closer together. Eventually both men indicated that the line of sight between the two telescopes coincided exactly with the surface of the water between them.

In modern measurements the distance between the two was 7.12km. Taking half that distance in Figure 16.2, and squaring it, Wallace calculated the diameter of the Earth: 3,560m^2 is 12,673.60km. Nowadays we know that the average diameter of the Earth is 12,742km, so Wallace's simple calculation was only half of one percentage point out.

Mesolabe Compass – Dividing any line equally

We featured the mesolabe compass in Chapter 3. Using this device, it's possible to divide any line into an equal number of pieces, without even knowing the length of the line.

Starting from the left end, choose any point along the base line of the compass as your line. Let's assume you want to divide that line exactly into five equal pieces. Draw a line through the point and the 5-unit mark on the upper line. Now draw parallel lines through the 4, 3, 2 and 1 points. You have now divided your selected base line into five pieces of perfectly equal length.

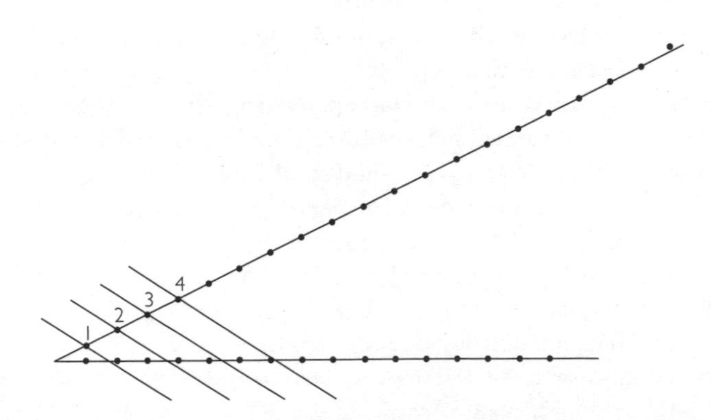

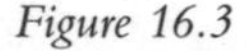

Figure 16.3

Chapter 4: Archimedes – the Greatest Greek of Them All

Archimedes' Lever to Move the Earth

Archimedes said, 'Give me a lever long enough and I could move the Earth.'

How long would Archimedes' lever need to be?

If the Earth was a perfectly round sphere, resting at the very end of a lever, the fulcrum could be *any* distance from that contact point – even 1mm along. Then if the point of leverage was 2mm further along, the weight of the Earth would be effectively halved.

The Earth weighs about 6×10^{24}kg, and if Archimedes weighed 100kg, or 10^2kg, then he would need a lever to give him a 6×10^{22} advantage: a lever 60,000,000,000,000,000,000,000mm or 6×10^{16}km long.

A lever reaching the Sun would be 148,800,000km, or 14.88×10^7km. So a lever a billion (10^9) times as long as the distance from the Earth to the Sun, or 1 billion AU, should do it. There are 63,240 AU in a light year, so 1 billion/63,240 = 15,812 light years away. Once there he would need somewhere firm to stand, and he would have to hope that the lever does not bend. Archimedes was right in principle, and the principle is the thing.

Archimedes' Extended Number System

Archimedes suggested how we might count every grain of sand in the entire Universe. We might fill a box 1cm cubed with sand grains and count them… Now there are 100^3 or 1,000,000 cubic centimetres in a cubic metre. There are $1{,}000^3$ or 1,000,000,000 cubic metres in a cubic kilometre, making 1,000,000,000,000,000 cubic centimetres in a cubic kilometre. We need to multiply that by the number of grains in the original cubic centimetre – and we are already into very large numbers just for one cubic kilometre.

In Archimedes' day, the largest number was a myriad, or 10,000, and a myriad myriads was 100,000,000.

But Archimedes set up a second order of myriad myriads as $100{,}000{,}000^{100{,}000{,}000}$, then suggested a third order that set that number to the power of 100,000,000 again – and so on.

Today we count huge numbers, using scientific notation, which works in a similar way to the system suggested by Archimedes. So let's ask, 'How many atoms in my body?'

Well, I'm around 90kg, so will have about 9.10^{27} atoms in my body – about 9,000,000,000,000,000,000,000,000,000.

99 per cent of them will be hydrogen, oxygen and carbon, with perhaps 1 per cent of trace elements like calcium and iron.

But 99 per cent of me will be made of these three elements in the approximate ratios:

$\frac{2}{3}$ hydrogen, or 6,000,000,000,000,000,000,000,000,000 atoms;
$\frac{1}{4}$ oxygen, or 2,250,000,000,000,000,000,000,000,000 atoms; and
$\frac{1}{10}$ carbon, or 900,000,000,000,000,000,000,000,000 atoms.

But as each oxygen atom is 16 times heavier than a hydrogen atom, you can see that we are mostly oxygen all locked up with hydrogen and carbon (12 times heavier than hydrogen), which together form almost everything in our body. You can go further and show that, in a similar way, by far the greater part of a mountain is oxygen.

The Cattle of the Sun

This was the puzzle set by Archimedes and sent to Eratosthenes in Alexandria. But be warned. It may have been a joke… It was only solved in 1880. It goes:

> *Compute the number of Cattle of the Sun, oh stranger and if you are wise, apply your wisdom and tell me how many once grazed upon the plains of the island of Sicilian Thrinacia.*

The cattle form four herds: milk white, sleek dark-skinned, tawny and dappled.

Each herd had a great multitude of bulls in these ratios.

White bulls = half plus a third ($\frac{5}{6}$) of the dark-skinned bulls + all the tawny bulls.

Dark-skinned bulls = a quarter plus a fifth ($\frac{9}{20}$) of the dappled bulls + all the tawny bulls.

Dappled bulls = one-sixth plus one-seventh ($\frac{13}{42}$) of the white bulls + all the tawny bulls.

Of the cows, the white cows = $\frac{1}{3}$ + $\frac{1}{4}$ ($\frac{7}{12}$) of the dark-skinned herd.

The dark-skinned cows = $\frac{1}{4}$ + $\frac{1}{5}$ ($\frac{9}{20}$) of the whole dappled herd.

The dappled cows = $\frac{1}{5}$ + $\frac{1}{6}$ ($\frac{11}{30}$) of the whole tawny herd.

The tawny cows = $\frac{1}{6}$ + $\frac{1}{7}$ of the whole white herd.

The white bulls + the dark-skinned bulls form a perfect square.

The tawny bulls + the dappled bulls form a perfect triangular number.

Solving the bull and cow equations produces a total of 50,389,082 cattle. But in that solution the white bulls and the dark skinned bulls, when added together, do not produce a perfect square. It was only in 1880 that the general solution was found (by A. Amthor), who gave the answer of 7.76×10^{206544} cattle. This is far more than could possibly fit in the observable Universe, let alone Sicily. What solution Archimedes had, if any, has not survived, but some suggest that the whole thing might well have been a huge Archimedean leg-pull.

The Semi-Platonic Solids

Archimedes worked out that there can only be 13 semi-Platonic solids. Remember the formula for all regular solids: faces + vertices = edges + 2. To make models, draw and enlarge the nets on paper, then cut and fold them. Here are the 13.

1. Truncated Tetrahedron 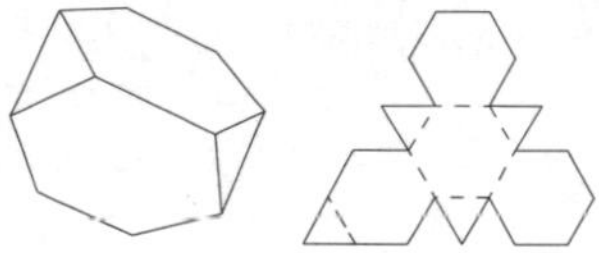	To *truncate* means to cut off one's head – or in this case each vertex. It has to be done so that every final edge is exactly equal in length. The results are four triangular faces and four hexagonal faces. Each edge is one-third the original length, and each vertex joins two hexagons to a triangle. Its formula is 4F(3 edges) + 4F(6 edges) + 12V = 18E + 2.
2. Cuboctahedron 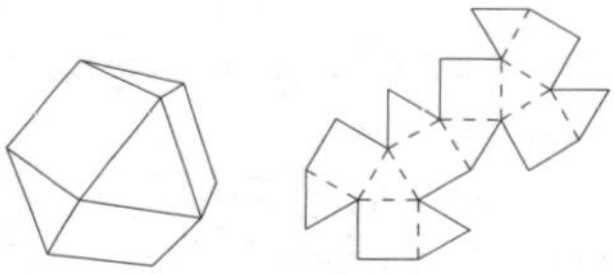	The corners are cut off a cube; each cut is made at the centre of each original edge. Amazingly it still has six square sides, but the original corners have been replaced by triangles. Its formula is 6F(4) + 8F(3) + 12V = 24E + 2.
3. Truncated Cube 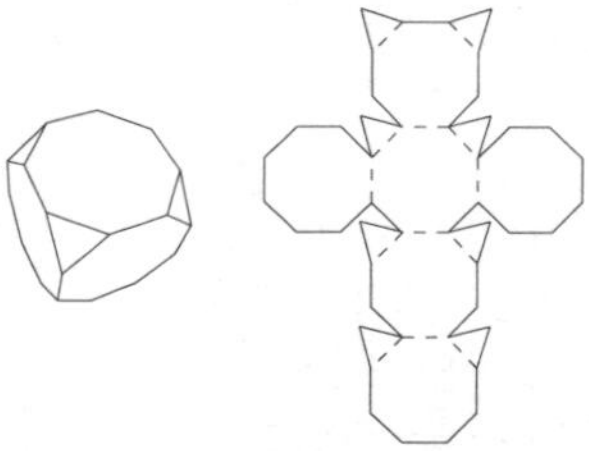	Once again, the corners are cut off an original cube, but the measurements are more subtle, making each original face into a regular octagon. Its formula is 8F(3) + 6F(8) + 24V = 36E + 2.
4. Truncated Octahedron 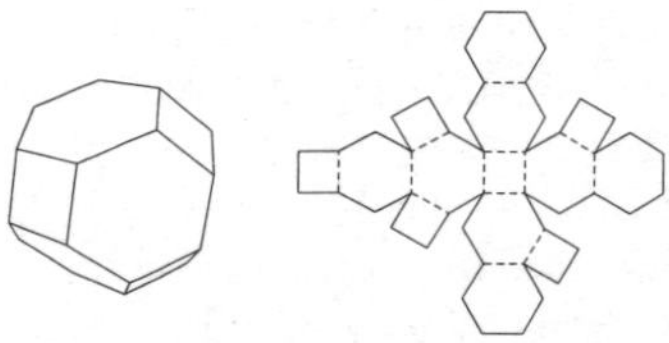	First find points one-third of the way along each edge, then lop off each corner to produce regular squares and hexagons. Its formula is 8F(6) + 6F(4) + 24V = 36E + 2.

5. Small Rhombicuboctahedron 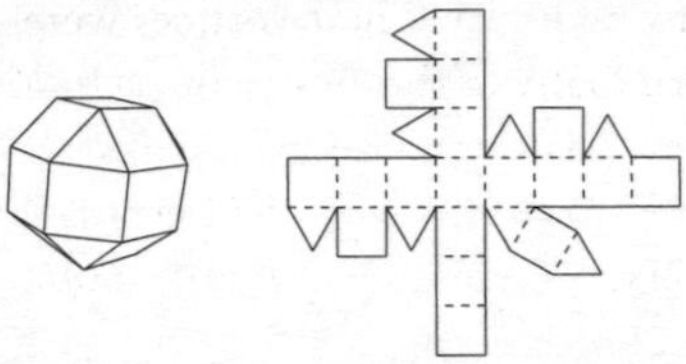	A rhombus is a quadrilateral or four-sided figure. Made from an original cube, it requires two sets of cuts. It's the easiest to understand and make from the net. Its formula is $8F(3) + 18F(4) + 24V = 48E + 2$.
6. Great Rhomicuboctahedron 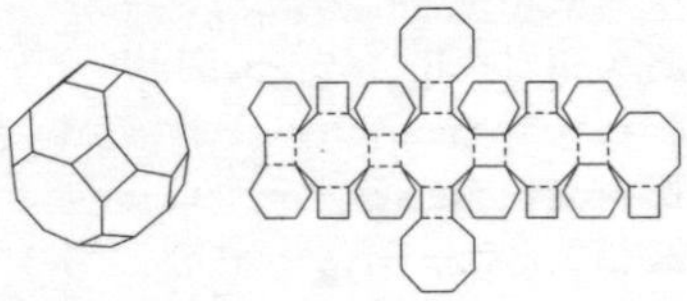	This semi-regular solid is made from three separate shapes, as the net shows: 6 octagons, 8 hexagons and 12 squares. Its formula is $6F(8) + 8F(6) + 12F(4) + 48V = 72E + 2$.
7. Snub Cube 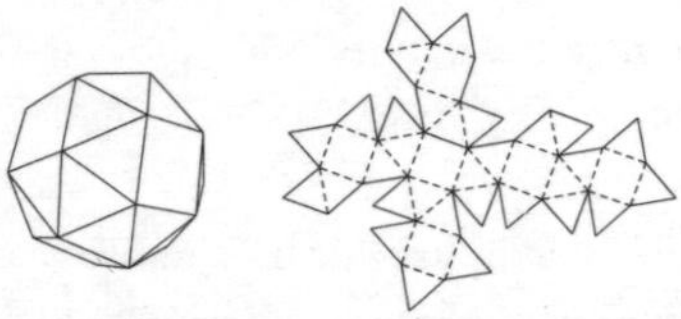	Its six square faces are set at an angle to the other square faces. As the net shows, each pair of squares is linked by two triangles. Its formula is $6F(4) + 32F(3) + 24V = 60E + 2$.
8. Icosadodecahedron 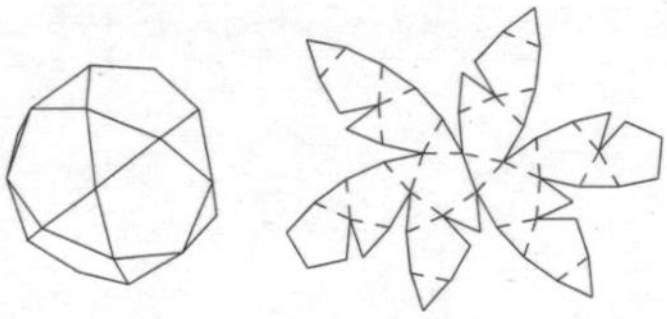	A dodecahedron is truncated so that each pentagon is smaller, with a corner for each previous edge and a triangle for each corner. Its formula is $12F(5) + 20F(3) + 30V = 60E + 2$.
9. Truncated Dodecahedron 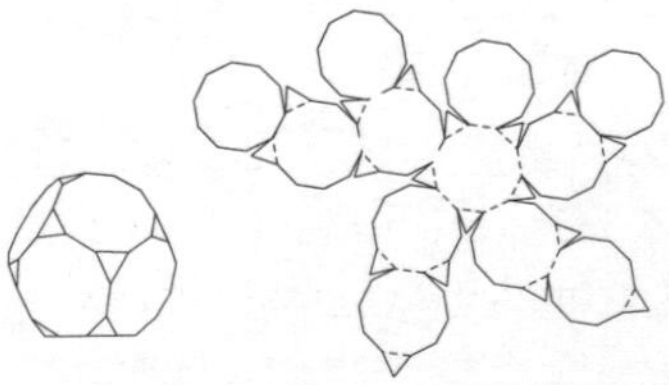	Each pentagon has been turned into a decagon and each vertex is now a small triangle. Its formula is $12 F(10) + 20F(3) + 60V = 90E + 2$.
10. Truncated Icosahedron 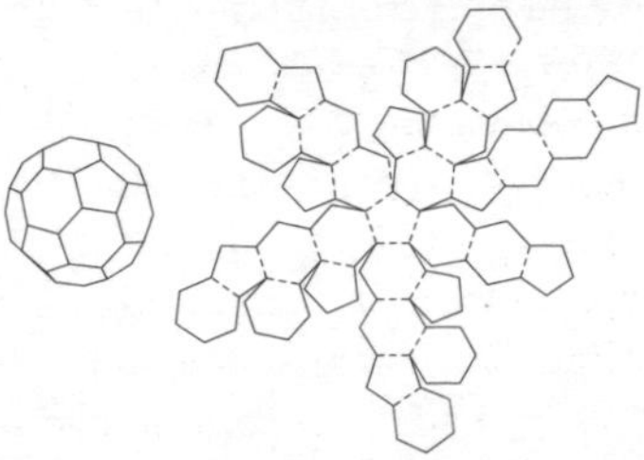	The icosahedron's 12 vertices have been cut off to make 12 pentagons and now 20 hexagons. Its formula is $20F(6) + 12F(5) + 60V = 90E + 2$.

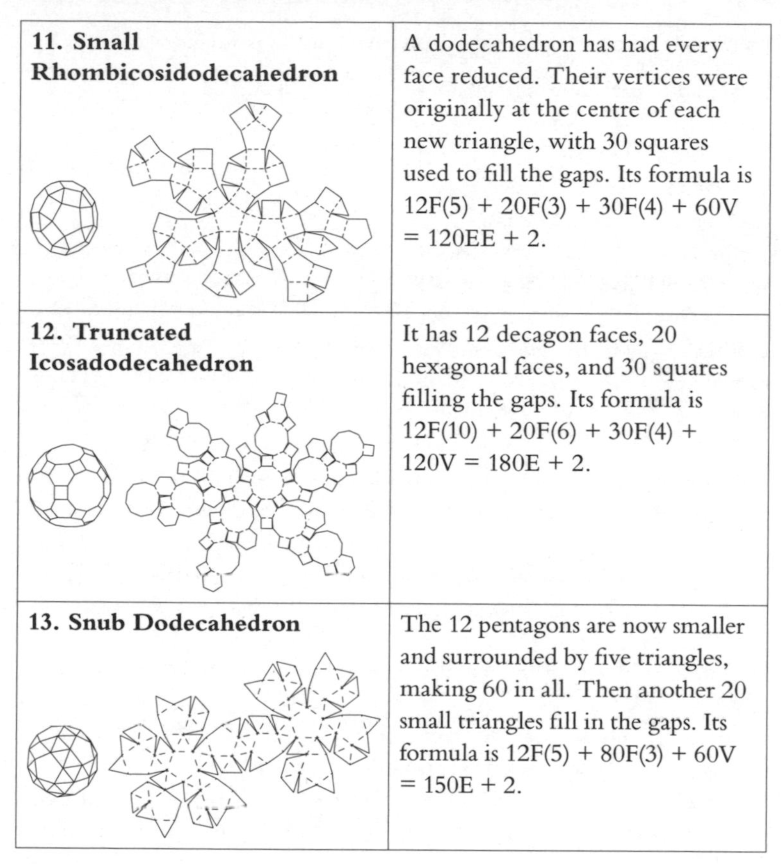

11. Small Rhombicosidodecahedron	A dodecahedron has had every face reduced. Their vertices were originally at the centre of each new triangle, with 30 squares used to fill the gaps. Its formula is 12F(5) + 20F(3) + 30F(4) + 60V = 120EE + 2.
12. Truncated Icosadodecahedron	It has 12 decagon faces, 20 hexagonal faces, and 30 squares filling the gaps. Its formula is 12F(10) + 20F(6) + 30F(4) + 120V = 180E + 2.
13. Snub Dodecahedron	The 12 pentagons are now smaller and surrounded by five triangles, making 60 in all. Then another 20 small triangles fill in the gaps. Its formula is 12F(5) + 80F(3) + 60V = 150E + 2.

Figure 16.4

The Method

Archimedes had to find a method for solving problems that went beyond all 'then known' geometry. So it seems he wondered whether he could solve geometric problems by imagining that triangles and parabolic sections actually had weight.

He had laid down the principles of levers and how two items of different weight could be balanced on a seesaw, provided that the weight times the distance from the fulcrum was the same on both sides. But this means you would have to know the centre of gravity for each item.

Euclid showed how to find the centre of gravity for a triangle. Draw a line from each angle to the centre of the opposite side. The three lines will always converge at one spot, which is the triangle's centre of gravity. This point is always one-third along the line from the side and two-thirds from the angle (see Figure 16.5). Archimedes was to find this fact very useful.

Archimedes set out to find the area under a parabolic segment and produced this diagram.

Strangely, the original diagram was set at an angle, which could have been his way of saying, 'The method works for any shaped segment of a parabola, not just one symmetrical about one axis,' but this is just my guess. I've included two diagrams, as the one with a horizontal base is easier to explain.

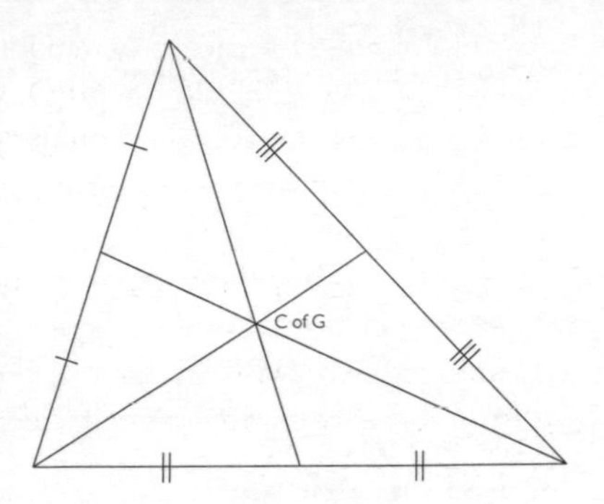

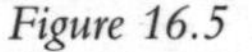

Figure 16.5

First, Archimedes took the segment of a parabola ABC with the centre point of the base at D. He then drew the triangle ABC. A line from D rises vertically through B and an equal distance to E. Another vertical line rises from A.

He now drew a line from C through B and the same distance again to F. Next a line from C passes through E and the same distance again to G. This line is a tangent to the parabola. AFG is parallel to DBE.

The area of triangle AGC must be four times the area of triangle ABC as both have the same base AC, but AG is four times the height of DB. But what is the relationship between these areas and that of the parabolic segment ABC? That's what Archimedes wanted to know…

Archimedes' sheer genius shines brilliantly now: using the third diagram, he solves this problem using very thin slices, balanced weights and centres of gravity.

First, CF is extended to a point H where CF equals FH. This line will become the seesaw balance with the fulcrum at F.

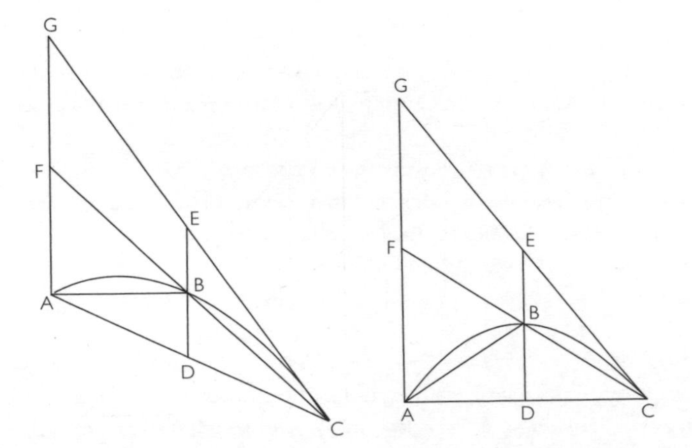

Figure 16.6

Now Archimedes took any random vertical line dropping through triangle AGC. We can call it MNPO, where N lies on CF so that MN = NO. Also on this line is point P on the parabolic curve.

Then he began the most amazing manipulations – rather like a grocer measuring peanuts on a balance.

He copies the line PO to pass through H at the far end of the lever arm HFC and explains, 'I want to compare the large triangle AGC with the parabolic segment ABC by comparing proportions for the random line MO and that part passing through the parabola PO. I say that MO: PO is in the same ratio as AC: AO.'

Look at the diagram and see if you agree? But Archimedes then said, 'This will always be the case, no matter where we drop the line MNPO.'

'What if I dropped every conceivable random vertical line MO through triangle AGC and placed every section PO at H, at the other end of the balance beam? Eventually I would have a copy of the whole parabolic section resting at point H...'

'Now I say that the parabolic section at H will balance the triangle AGC where its centre of gravity rests on the beam HFC. But where is that centre of gravity point? It is one-third of the way along FC from F at a point X.'

As the distance HF is three times FX, for the two to balance the triangle AGC must be three times the area of the parabolic section ABC. But we started by showing that the triangle AGC is four times triangle ABC.

So the parabolic segment must be two-thirds of the rectangle it fits into, and four thirds of the triangle inscribed within it. Success – and for Archimedes that must have truly been a Eureka moment!

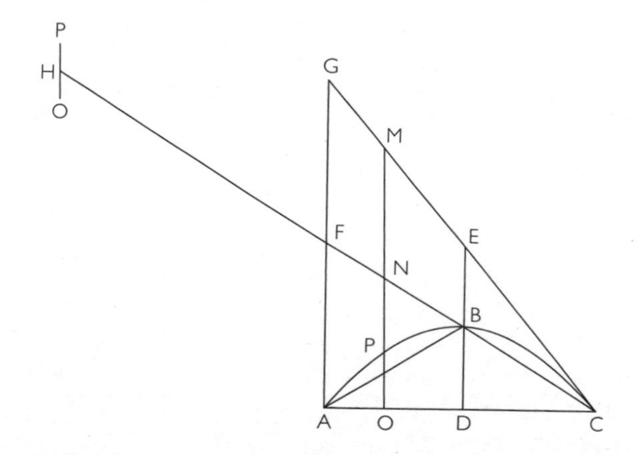

Figure 16.7

The Method in Three Dimensions

However, this was only the start. Archimedes realized that calculating the area under a plane parabolic shape by slicing it up would also work in three dimensions.

If you take a parabola and revolve it around its central vertical axis, you produce a *conoid* or three-dimensional parabola. Archimedes reasoned that if a parabola is two-thirds of the rectangle it fits into, then surely a parabolic conoid must be two-thirds of the cylinder it fits neatly into.

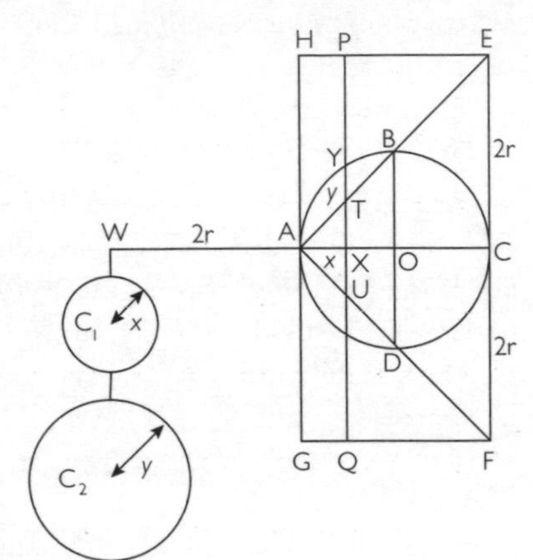

Figure 16.8

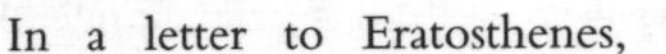

In a letter to Eratosthenes, Archimedes explained the Method in Proposition 1, and using the same method in three dimensions, the volume of a conoid in Proposition 4. In fact, we have evidence of three separate proofs for the area of a parabola, and each would extend to prove the volume of a conoid.

Now that he had these proofs to lean on, he may have simplified it even more by comparing a cylinder, sphere and double-napped cone, as we show in the text (see Chapter 4).

Chapter 5: The Glory That Was Alexandria

Apollonius' Nesting Circles

Given three circles, can you produce a fourth circle that will touch all three?

There are eight solutions: one in which the new circle is outside all the other three; one in which the circle is inside all three; three with two circles inside and one outside; and three with two circles outside and one inside (see Figure 16.9).

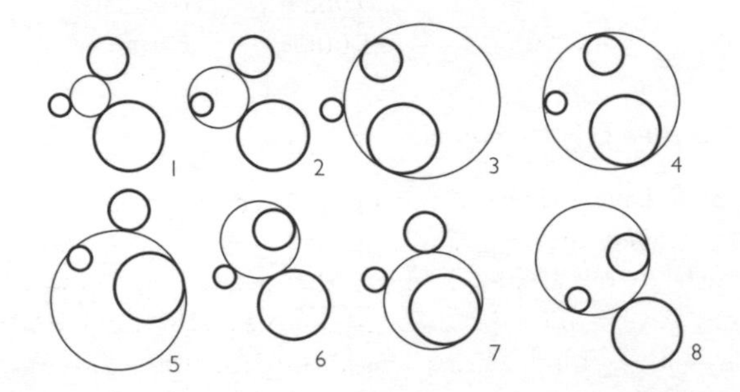

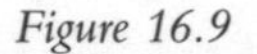

Figure 16.9

Amazingly, they can all be found using just a straight edge and a pair of compasses.

Hipparchus' Trigonometric Ratios

Using Hipparchus' system, with any right-angled triangle, knowing just one side and one other angle enables you to calculate all the sides and angles (see Figure 16.10).

Here we've used a 3, 4, 5 triangle.

Sine θ is always the *opposite* over the *hypotenuse* – or, for this triangle, 4/5, or 0.80.

You can see that as θ gets larger, so the sine value gets closer to 1.

As θ gets smaller the fraction value becomes ever smaller, right down to 0.000.

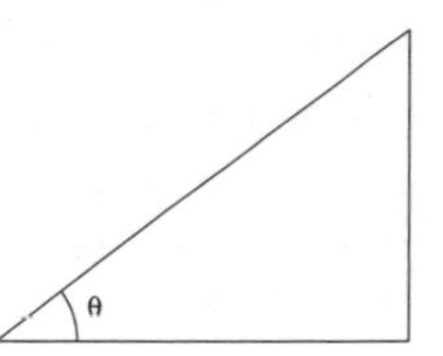

Figure 16.10

Cosine θ is the *adjacent* over the *hypotenuse* – or, for this triangle, 3/5, or 0.60.

As θ gets larger, the value gets smaller, down to 0.000.

As θ gets smaller, the cosine value gets closer to 1.

Tangent θ is the *opposite* over the *adjacent* – or, for this triangle, 4/3 or 1.3333.

Here, when θ equals 45 degrees, the tangent value is 1.

When θ is less than 45 degrees, the tangent value is less than 1.

When θ is greater than 45 degrees, the tangent value is greater than 1.

Hipparchus worked out the complete table of trigonometric numbers. Here are just a few:

Angle (θ)	Sine	Cos	Tan
90°	1.000	0.000	Infinity
75°	0.9639	0.2588	3.7321
60°	0.8660	0.5000	1.7321
45°	0.7071	0.7071	1.0000
30°	0.5000	0.8660	0.5774
15°	0.2588	0.9659	0.2679
0°	0.0000	1.0000	0.0000

Appolonius and Hyperbolas

All hyperbolas have *asymptotes*, a pair of crossed straight lines that in effect never quite touch the hyperbolic curves, with one exception: if the napped cones are cut absolutely down the middle, the cut produces two triangular slices bounded by their asymptotes. Any other cut produces hyperbolas with asymptotes that would only touch the curves at infinity (see Figure 16.11).

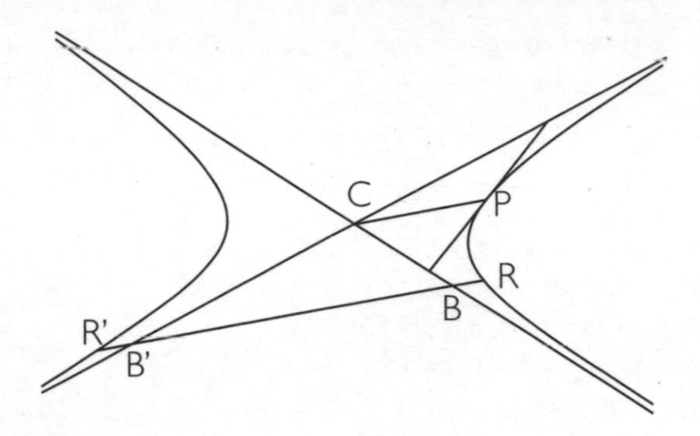

Figure 16.11

For any tangent line on either curve that connects the two asymptotes, the point of contact with the curve will always be halfway along that tangent.

Take the line from C where the two asymptotes cross to the spot where the tangent touches a curve P, and draw any parallel line RR^1. This line reaches both curves and cuts through both asymptotes at BB^1. The lengths RB and R^1B^1 will always be the same, and when they're multiplied together will equal CP^2 in every case – wow!

Chapter 6: Total Eclipse of the Greeks

Heron's Formula: calculating the area of any triangle – just from knowing the lengths of its three sides

First of all, add the three side lengths and divide by two: $(a + b + c)/2$

Next take each side from this number to get three more values.

Now multiply the four values together.

The square root of this total will be the area of the triangle.

So here are two examples:

The area of a triangle with sides of 7, 8 and 9 units.

$7 + 8 + 9 = 24/2 = 12$

$12 - 7 = 5$

$12 - 8 = 4$

$12 - 9 = 3$

So $\sqrt{(12 \times 5 \times 4 \times 3)} = \sqrt{720} = 26.83$

Let's test this with a triangle with three sides of 5 units (see Figure 16.12).

To calculate the area you would use base x height/2

From Pythagoras' Theorem the height would be $5^2 - 2.5^2 = 25 - 6.25 = \sqrt{18.75} = 4.33$

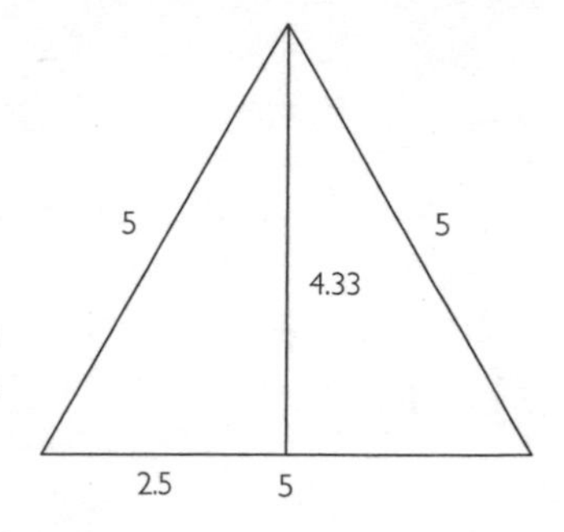

So the area is 4.33 x 2.5 = 10.825.

Using Heron's Formula:

5 + 5 + 5 = 15/2 = 7.5

7.5 – 5 = 2.5 (3 times)

So we need – √(7.5 x 2.5 x 2.5 x 2.5) = √117.1875 = 10.825. Heron's Formula works every single time.

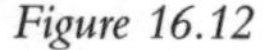
Figure 16.12

The Theorem of Menelaus

Take any triangle and draw a straight line through it. The line will cut two of the sides, and if you extend the third side far enough, the line will cut that as well.

The theorem says that if a straight line crosses the three sides of a triangle (and one side is extended beyond the vertices of the triangle), then the product of the three non-adjacent line segments is equal to the product of the three remaining segments.

So we can quickly check if the theorem works, I've drawn a triangle, ABC, with sides of 5, 6 and 7 units (see Figure 16.13). I've also drawn a straight line that passes through one point (D) on AB that is 2cm from B, and another point (E) on CA, that's 1cm from C. The line continues on to cross an extended BC at point F. CF measures 3cm.

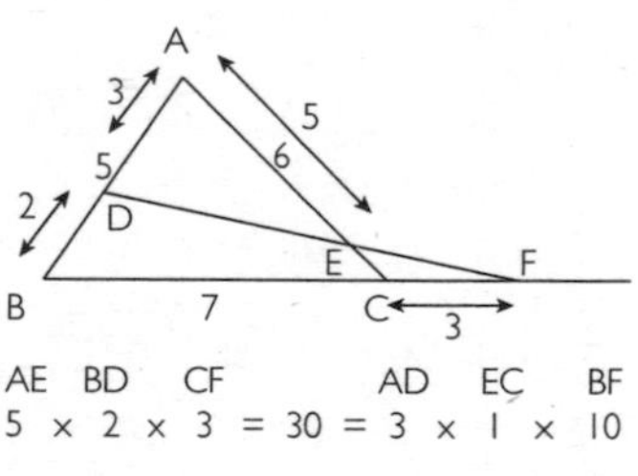

Figure 16.13

From now on, we can ignore the new dividing line. The theorem states that:

The unconnected segments, AE.BD.CF / AD.EC.BF = -1 (because the length CF is outside the side that originates it and is measured backwards, so the result is negative).

So AE x BD x CF will always equal AD x EC x BF.

And 5 x 2 x 3 = 30 is equal to 3 x 1 x 10 = 30.

Amazingly, this works for absolutely any triangle, which makes it such a beautiful mathematical discovery. Euclid missed it, but the theorem was probably known around the time of Hippocrates and his mesolabe compass (see Chapter 3).

To prove the theorem, you have to draw a line from each of the angles of the triangle that meets the dividing line at right angles (see Figure 16.14).

These lines form three sets of similar triangles where all the angles in each triangle are equal. This means that the lengths of the lines in those

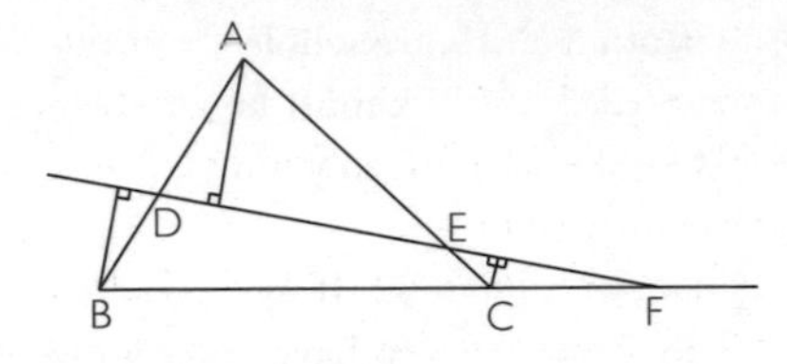

Figure 16.14

triangles are always in proportion to one another, and comparing similar proportions is what made Greek maths so powerful.

You can even generalise the theorem for any polygons, not just triangles. Even more amazingly, Menelaus found that the theorem still works for triangles drawn on the surface of a sphere – if, but only if, all the lines are parts of great circles of the sphere.

The Birth of Algebra, via Diophantus

Diophantus began with quite straightforward problems that require a little thought. In many of his problems, he took a solution and then set an algebraic problem to find it. He then progressed to more complex and powerful examples. Here we go.

Find two numbers that add up to 10 and whose combined cubes are 370.

We can assume the two numbers are not the same. If they add up to 10 then we can write them as (5 + x) and (5 – x). So what is x? If it's a whole number, it must be 1, 2, 3 or 4.

By trial and error we find that if x = 2, then (5 + 2) = 7 and (5 – 2) = 3, and 7 + 3 = 10.

Their cubes are 7 x 7 x 7 = 343, and 3 x 3 x 3 = 27. Sure enough, 343 + 27 = 370.

The problem is solved.

Here are some more quite simple examples. The first one has two unknowns:

1) 11x + 12y = 58. What are x and y, given that they are whole numbers?

Let's explore. If y = 1, then 11x + 12(1) = 58, so 11x = 58 – 12 = 46, making x a little over 4.

So if x = 4, what then? Then 12y would be 58 – 44 [11 x 4] = 14, and y cannot be a whole number.

Let's try x = 3. Now 58 – 33 [11 x 3] = 25. Once again, at 25/12, y cannot be a whole number.

Let's try x = 2. Now 58 – 22 [11 x 2] = 36, which is 12y. So y = 3 and x = 2. QED.

2) $3x - 4y = 29$. Both x and y are whole numbers. What are they?

3x must be greater than 29, so x must be greater than 10. Let's try 11.

If $x = 11$, then $3x = 33 - 29 = 4y$, so y must equal 1. So $x = 11$ and $y = 1$.

But that's not the only solution.

If $x = 27$ then $3x = 81 - 29 = 52$. If $4y = 52$ then $y = 13$.

Diophantus hardly ever seems to have considered that there might be more than one solution, at least in the early problems he set. He usually went for the simplest result.

3) From the same 'required' number (the number that we're trying to find), subtract two given numbers so that the ratio of one to the other is a given ratio. Let the two numbers be 100 and 20, and let the large remainder be 3 times the smaller.

Then, as Hippocrates did with the mesolabe compass, Diophantus started with unity (1).

Let the required number be 1x.

So $1x - 100$ gives the smaller remainder, and $1x - 20$ gives the larger one, which must be three times as big.

So $1x - 20$ must equal $3x - 300$:

$1x - 20 = 3x - 300$

$(3x - 300) - x - 20 = 2x - 280$

So $x = 140$.

Let's check that works with our original numbers: $140 - 100 = 40$, and $140 - 20 = 120$, which is three times 40.

4) Find two numbers so that when they are added they make 20, and when they're multiplied they make 96. And if you square half of their sum (20), the result must be greater than 96 by a square number. Well:

Half their sum is 10 and $10 \times 10 = 100 - 96 = x^2 = 4$ (a square number). So $x = 2$.

Two numbers that add up to 20 could be $10 - 2$ and $10 + 2$, or 8 and 12.

So $8 + 12 = 20$, and $8 \times 12 = 96$.

Now had the original question been 'Find two different numbers which add up to 20 and which when multiplied together and taken from 100, leave a square number,' then the answer would have been 'all of them' or 'every possibility in whole numbers'.

So:

$10 + 10 = 20$ $10 \times 10 = 100$

Now take any pair that adds up to 20 $9 + 11 = 20$ $9 \times 11 = 99$ from 100 leaves $1 = 1^2$

$8 + 12 = 20$, $8 \times 12 = 96$ from 100 leaves $4 = 2^2$

$7 + 13 = 20$ $7 \times 13 = 91$ from 100 leaves $9 = 3^2$

$6 + 14 = 20$ $6 \times 14 = 84$ from 100 leaves $16 = 4^2$

$5 + 15 = 20$ $5 \times 15 = 75$ from 100 leaves $25 = 5^2$

$4 + 16 = 20$ $4 \times 16 = 64$ from 100 leaves $36 = 6^2$
$3 + 17 = 20$ $3 \times 17 = 51$ from 100 leaves $49 = 7^2$
$2 + 18 = 20$ $2 \times 18 = 36$ from 100 leaves $64 = 8^2$
$1 + 19 = 20$ $1 \times 19 = 19$ frpm 100 leaves $81 = 9^2$
And even $0 + 20 = 20$ $0 \times 20 = 0$ from 100 leaves $100 = 10^2$

Diophantus was fascinated by square numbers, and found several instances where two square numbers equal a third, such as 9 + 16 = 25, 36 + 64 = 100, 25 + 144 = 169, 81 + 144 = 225 and 64 + 225 = 289.

When later scholars discovered Diophantus' maths, part of the attraction must have been the deceptive simplicity of his chosen problems that actually led to more complex mathematical ideas.

Chapter 7: Maths Origins, Far and Wide

Liu Hui's Calculations for a 3,072-agon

Liu Hui surpassed Archimedes' 96-agon, going a further five doubling stages to create a polygon with 192, 384, 768, 1,536 and finally 3,072 (6×2^9) sides and a value for π of 3.14159.

Of course, he didn't actually have to draw a shape with that many sides. Instead, he derived the measurements using a formula involving Pythagoras' Theorem, twice.

Figure 16.15 shows part of a hexagon in a circle with radius *r*, a chord *c*, a line *b* bisecting the chord, and *d*, the line that continues from the chord to the circle. Lastly we have a new chord *e*.

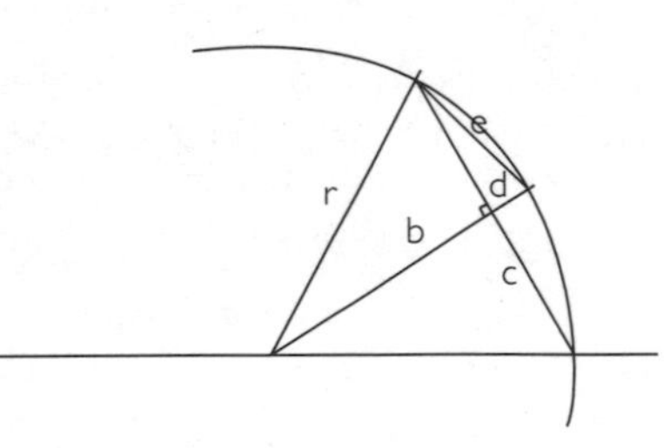

Figure 16.15

From Pythagoras we can work out the length of *b*, then *d*, then *e*– $r^2 - \frac{1}{2} c^2 = b^2$. Now find the square root of b^2, then $r - b = d$.

Next find $d^2 + \frac{1}{2} c^2 = e^2$ and the square root of that is *e*.

Now with *e* as your new chord *c*, because the radius is the same you can calculate the new bisector, *b*, then calculate a new *d* and onto the next new *e*. Each stage produces smaller and smaller numbers, of course, but you don't have to construct the circle – or could you?

Imagine, for instance, that you start with a radius of, say, 1km. After 10 stages your new chord would be a little over 1m long, and 3,072 of them would be fairly close to the distance around a circle with a 1km radius.

Chui Chang and Chinese Maths Problems Pre-first Century BC

Problem 1. A novice weaver improves every day, and each day she doubles the output of the day before. In five days she produces five *chih* (units) of cloth. How much did she produce on each of the five days?

The Chinese solved this problem using the principle of 'false position', also used by the Egyptians.

These days, we might say that the first day's output was x, so over the five days the weaver produces $x + 2x + 4x + 8x + 16x = 31x$ chih.

Here's their solution. The weaver produces 31 units in five days, and only 1 on the first day, which equals 1/31 of 5 chih = 5/31 chih. In the following days she produces 10/31, 20/31, 40/31 and 80/31 chih, giving a total of 155/31 = 5 chih.

Problem 2. Picture a tree 20 chih tall and 3 chih around. An arrowroot vine grows from the base and winds itself around the tree seven times to the top. How long is the vine?

We can find the answer using Pythagoras' Theorem again, by making the vine the hypotenuse of a right-angled triangle. The upright side is 20 chih, and the base is 7 x 3 chih = 21 chih.

So, $20^2 + 21^2 = 400 + 441 = 841$. The square root of 841 = 29 chih: the length of the vine.

The Chinese seem to have used Pythagoras' Theorem in more ways than any other civilisation. A 1977 book by Swetz and Kao examining the Chinese preoccupation with this theorem even had the amusing title *Was Pythagoras Chinese?*

How to Transform a Rectangle into a Square (From the Baudhayana Sulbasutra)

Among their other achievements, the Vedics also found a geometric way to turn a rectangle into a square. Take a rectangle ABCD and draw two lines parallel to AB, the first forming a square on AB and the second dividing the rest of the rectangle exactly in two. Now turn the top rectangle through 90 degrees and drop it down the side of the original rectangle so that C lies on an extension of the line AB. Now with AC as radius complete the square ACQR. This forms a new square equal in area to the original rectangle plus the small square at top right.

To get rid of that area, set your compasses at radius CQ and draw an arc from Q to meet the side of the original rectangle at S. Now take the length BS to form the square CTUV. This square is equal in area to the original rectangle ABCD. The Vedics almost certainly used Pythagoras' Theorem to prove this.

$$CV^2 = CS^2 - SV^2 = AC^2 - BC^2 = (AC + BC)(AC - BC)$$

For those of you who would like to see it in figures, if AB is 40mm and AD 90mm, then the area of the rectangle would be 3,600mm^2 and the square would have a side of 60mm.

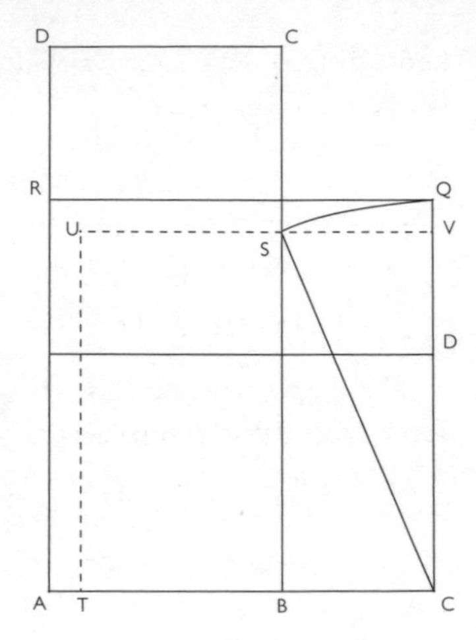

Figure 16.16

Vedic Maths Examples

Method for multiplying numbers that are just over 100 or 1,000:

Let's multiply 1,034 x 1,008.

Multiply the 'extra' numbers that are over 1,000.

34 x 8 = 272. That's the back end of the answer.

Now add 1,034 + 1,008 = 2,042, and subtract 1,000 = 1,042. That's the front end.

So 1,034 x 1,008 = 1,042,272.

Method for multiplying two numbers, one slightly more than 100 and one slightly less:

This method is easier with a little practice.

Let's multiply 104 x 92.

4 x 92 = 368

100 x 92 = 9,200

Total = 9,568

The Peruvian Quipu Knot System

The quipu (see plate section) had three main types of knot: a short overhand knot; a 'long knot', consisting of an overhand knot with one or more additional turns; and a figure-of-eight knot – although to end a number or sequence the figure-of-eight knot would be given an extra twist.

Numbers were represented as a sequence of knot clusters in base 10.

In the tens position, and higher power positions, a digit would be a close row of simple knots. So 40 would be four simple knots. On successive strings, hundreds, tens and units would all be in parallel.

Digits in the 'units' position were represented by long knots. So 4 was a simple knot with four turns, but the digit 1 would be a figure-of-eight knot. Because unit digits were tied in a special way, they indicated clearly where a number ended.

For example, if 's' represents a simple knot, 'L' represents a long knot, 'E' represents a figure-of-eight knot, and 'X' represents a space, then:

541 would be represented by 5s, 4s, E.

603 would be represented by 6s, X, 3L.

206 followed by the number 71 would be represented by 2s, X, 6L, 7s, E.

Chapter 8: Mathematics Was Never a Religion

Al-Khwarizmi Basic Algebra

Question: 'The root squared plus 10 of the root equals 39. What is the root?'

In modern notation we would write this as: $x^2 + 10x = 39$

Then convert this into: $x^2 + 10x - 39 = 0$

We, like Al-Khwarizmi, might see that 39 has factors 3 and 13, with a difference of 10.

So $(x + 13)(x - 3) = 0$

So the root $x = 3$ and $x^2 + 10x = 39$ becomes $9 + 30 = 39$.

But Al-Khwarizmi's language and terminology were far more complex.

His arithmetical solution went like this:

1. Half the simple root $10/2 = 5$
2. Square the result $5 \times 5 = 25$
3. Add this (25) to 39 $25 + 39 = 64$
4. Find the square root $\sqrt{64} = 8$
5. From 8 take (½ root) $5 = 3$. So $3^2 + 10 \times 3 = 39$.

By way of confirmation, he also solved the problem graphically (see Figure 16.17.

He drew a square of side x and added side rectangles 2.5 units long, thus adding 10x.

The area of the cross created is now $x^2 + 10x$, which we are told equals 39.

If you add the areas of the four corner squares you get 25 square units.

So the area of the large square is $39 + 25 = 64$ square units, of which the square root is 8 units, which is the side of the large square. So the centre square must have a side of length $8 - 2.5 - 2.5$ for each of the side rectangles. So $x = 3$.

Thabit's Explanation of How Archimedes Drew A Regular Heptagon

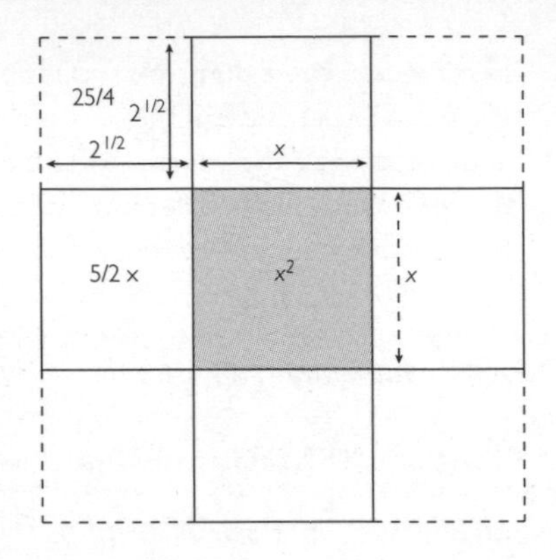

Figure 16.17

This amazing solution, discovered by Archimedes, was found among Thabit's writings just under a hundred years ago. It is another example of Archimedean maths that makes its own rules.

So, how do you construct a regular heptagon? First, take a line AB and form a square ABCD on it, with diagonal AC (see Figure 16.18). Extend the line AB to the right. Now perform a *neusis* – a new way of doing something.

Place a ruler through points D and B. Then, keeping it on point D, swing it right of B to a point X that forms the two (shaded) triangles BXH and CDG, which are equal in area. It's not explained how Archimedes knew when he had found the right position for X that meant the triangles were equal, so we can only assume he found it by trial and error. There is no easier way, because the two triangles are not similar and do not have sides of the same length, although they do have one angle (AXD and XDC) that is the same.

Now draw a perpendicular line FGY that runs through G. With a compass point at Y and with a radius equal to AY, draw an arc. Then, with the compass point at B and the radius set to BX, draw another arc, so the

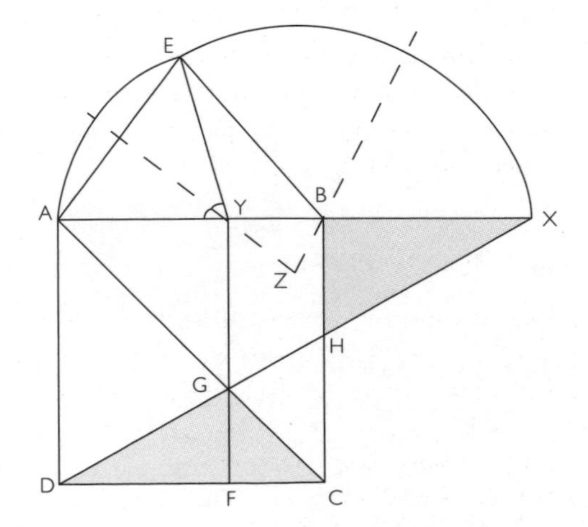

Figure 16.18

two arcs cross at a point we'll call E. Now construct bisectors of the angle AYE and EBX, and call the point where they cross Z. With Z as the centre and AZ as the radius, form the circle AEX, in which a heptagon with side AE will fit perfectly. It's quite brilliant, but no explanation is given as to how Archimedes discovered it.

The Mathematics of Islamic Design

Students of Islamic Art are taught to start at the very beginning, with a single point – their point of departure. It has no mass or dimension – it is the smallest, slightest dot. Moving on from that point we cover the shortest distance to any other point along a totally straight line. If we now swing that line around the original point, we produce a boundary that, as it arrives at its starting point, becomes a perfect circular domain. It is Euclid made poetic.

Islamic circles are said to have 'sixness': they naturally cluster in sixes around a central circle of the same size. In a plane, this resembles the natural geometric form of froth or bubbles. By staggering the method of overlapping, any number of fascinating, and essentially circular, patterns can be formed using a very small number of distance variations (see Figure 16.19a and b).

But circles can also be given 'fourness' by adding square and diagonal lines or 'threeness' by adding triangles, and by overlapping or varying sizes. A vast number of new patterns emerges, all based simply on unit-sized circles and straight lines (see Figure 16.19c).

Hexagons can also be used as central motifs from which new patterns spring (see Figure 16.19e and f).

These base patterns are used in a wide variety of Islamic tile designs. And if straight lines are curved uniformly, there is no end to the variety of designs that can result, as our colour pictures from the Alhambra in Granada show (see plate section).

Tessellation and Tiling Geometry

In geometric terms, covering a plane surface with repeating images is surprisingly easy and can take so many forms. It seems so sad and unadventurous for us to use almost exclusively square or rectangular tiles in our bathrooms.

In essence a four-sided figure of any shape can tile a plane, if the orientation of alternate tiles is reversed. This Roman mosaic floor design uses the same rhombus or diamond shape in three shades and orientations to stunning effect. Please see more examples in the plate section.

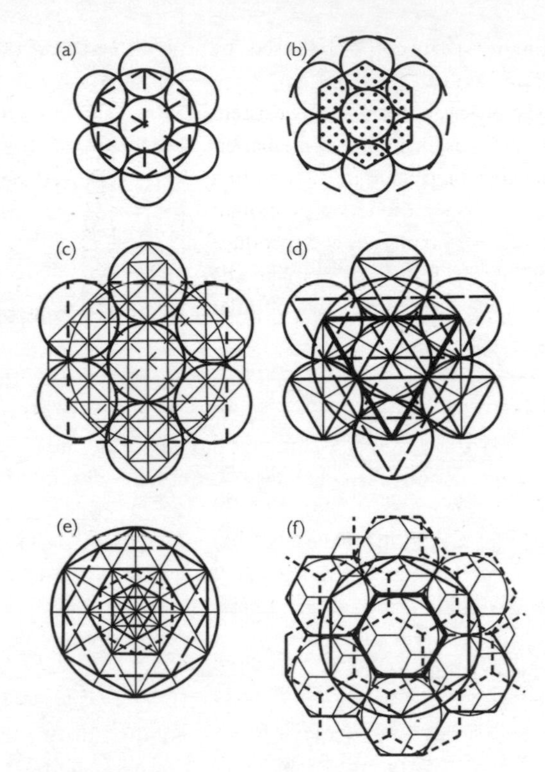

Figure 16.19

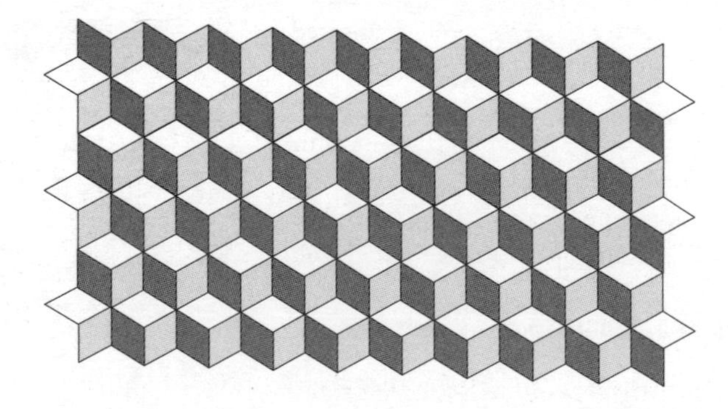

Figure 16.20

Chapter 10: The Huge Awakening and a New Age of Learning

Pacioli Fonts

Pacioli saw that the printed word was the most important component in the spread of learning. So he set about designing a complete set of letters of the alphabet that could be used by printers far and wide.

To produce some form of consistent style he applied his skills in mathematics. Here is his capital letter 'E', which he first set in a square, then used two diagonals and eight circles of varying sizes to produce the elegant curves required to make the letter distinctive. This font is still popular today (see Figure 16.21).

Figure 16.21

Pacioli also made an effort to explain the sheer variety of possibilities in geometric shapes. For this work, he employed a far better artist than he himself was. He gave that job to Leonardo da Vince. Here are a couple of Leonardo's simple three-dimensional impressions of the same geometric shape (see Figure 16.22).

First (Figure 16.22a), the solid elevated dodecahedron, where each of the 12 faces supports a shallow pentagonal pyramid. Each face is subtly shaded differently to give a better impression of its three-dimensional quality.

Next (Figure 16.22b) is a Truncated Icosahedron with 12 pentagonal and 20 hexagonal sides. But it is made far more complex to draw by being vacuous or hollow. Leonardo's draughtsmanship and use of shading here is quite beautiful. We can only assume he made a model and then drew it.

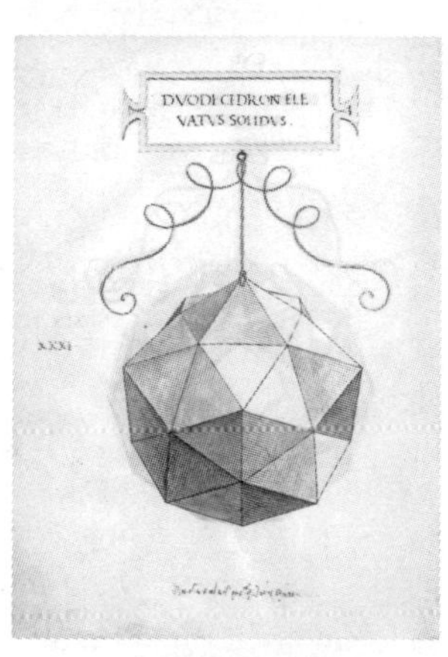

Figure 16.22

The Three Brides and a Boat Puzzle

Solution: chap A takes his wife across, leaves her and comes back. The other two women then cross and wife A comes back. The other two men now cross and either couple comes back. The two men cross and the third wife comes back. Now two wives cross, then the odd man comes back and picks up his wife for the final crossing.

Chapter 11: The New Age of Mathematical Discovery

Early Slide Rules – Straight and Circular

Slide rules are totally different from ordinary rulers, which have measurements in either inches or millimetres that are perfectly regular. Slide-rule measurements are logarithmic and multiplication is reduced to simple cross referencing.

In using a slide rule, rather than solve one multiplication problem, you achieve a whole 'times table' of solutions with just one setting. At the same time other scales on the same device can give you other times tables or even a series of square roots.

For example, to multiply any number by 2, locate the 2 on one ruler next to the 1 on the other. Now keeping the two rulers in that position and moving along the rule, you find 2 opposite 4, 3 opposite 6, 4 opposite 8 and 5 opposite 1 which you now read off as 10 (see Figure 16.23)

If the desired solution has not appeared by the time you reach the right-hand end of the scale, you slide the top ruler to align your start number to the number 1 at the other end of the scale. Now you can continue with the 2 times table for 5 is above a 1, which you now read as 10. Then 6, 7, 8, 9 and 10 all show their results as 12, 14, 16, 18 and 20. Slide rules usually came with a cursor, so you could accurately compare slides and read off the required result.

In our example (see Figure 16.24), as well as scales C and D giving the 2 times table and A and B the 4 times table, you can see that scale A gives the squares of the numbers on slide D, while slide D gives the square root of the numbers on slide A.

Slide rules also came in circular form and even as cylinders. When the scientific calculator came along in the late 1960s, it could pack even more power into a small and even more convenient device, with a huge number of alternative functions, and the days of slide rules were sadly numbered.

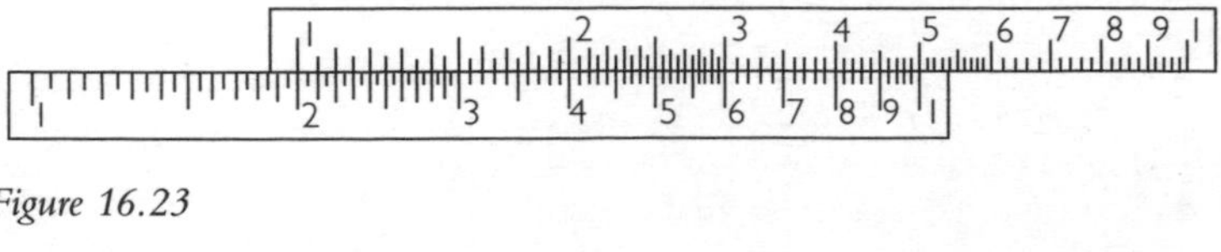

Figure 16.23

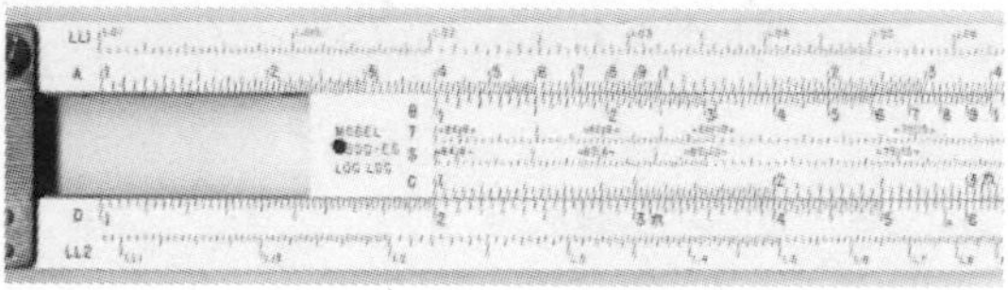

Figure 16.24

Kepler's Nesting Planets

With no calculators in those days, Kepler settled down to some very complex mathematical calculations regarding the distance between the planets. He eventually came up with a startling mathematical pattern.

He set the Sun at the centre and placed an octahedron around it, just large enough so that the orbit of Mercury fitted neatly inside its faces. Amazingly, the orbit of Venus then seemed to fit onto the octahedron's outer points.

Kepler then placed an icosahedron snugly around Venus's orbit and saw that the Earth's orbit fitted reasonably around the vertices of that figure.

Next he placed a dodecahedron around the Earth's orbit and found that Mars's orbit matched the outer points of that.

Then he placed a tetrahedron around Mars's orbit, with Jupiter's orbit snugly around the outside.

Lastly the cube that encompassed Jupiter's orbit allowed Saturn's orbit to fit around its extremities (see Figure 11.8, p. 309). Success!

Kepler's Third Law of Planetary Motion

Kepler, after much trial and error, finally found that when he multiplied the orbit time of any planet by itself, amazingly it was equal in ratio to its distance from the Sun, cubed. If we call the Earth's orbit time 1 unit of time and the Earth's distance from the Sun (150,000,000km) 1 AU (Astronomical Unit), then by making comparisons we find that the law works pretty well for all the planets. Amazing.

So, compared to the Earth's orbit time of 1, Mercury's is 0.241, which squared gives 0.05808. Mercury's distance from the Sun is 0.387 AU which, when cubed, gives 0.05796.

Alternatively, Mars's orbit time is 1.88 Earth years which, squared, gives 3.5344. Mars's distance from the Sun is 1.524 AU which, cubed, gives 3.5396.

Here is a full list, including newer planets:

Planet	Period	Dist. from Sun (AU)	Period^2	Distance^3
Mercury	0.241	0.387	0.05808	0.05796
Venus	0.616	0.723	0.37946	0.37793
Earth	1 earth year	1 AU	1	1
Mars	1.88	1.524	3.5344	3.5396
Jupiter	11.9	5.203	141.61	140.85
Saturn	29.5	9.539	870.25	867.98
Those not discovered at that time:				
Uranus	84.0	19.191	7056	7068
Neptune	165.0	30.071	27225	27192
Pluto	248.0	39.457	61504	61429

Galileo's Pendulum Experiments

Galileo saw that a pendulum swings until it gradually comes to a stop. So a downwards force affects it, and the height it reaches is never increased, but gradually lessens until all the pendulum's energy has gone, or a new force makes it swing again.

As Galileo's pendulum swung he tried putting a pencil in the way of the string. The weight still swung up to the same height as when it was unimpeded. If the pencil was below that height, the weight looped over the pencil as it tried to attain the same height. Try it (see Figure 16.25).

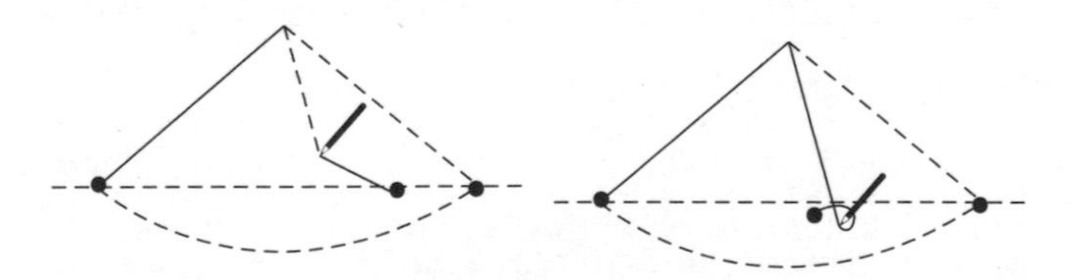

Figure 16.25

Chapter 12: How to Calculate Anything and Everything

Basic Graph Shapes and Their Formulae

The advantage of Descartes's discovery of Cartesian Coordinates was that it turned maths into pictures. To produce simple or complex geometric shapes now simply required a formula set out in terms of x and y. See Figure 12.3 for a series of simple graphs.

With Cartesian Coordinates, all the major mathematicians of the time began looking to create new and exotic shapes. Here are just three:

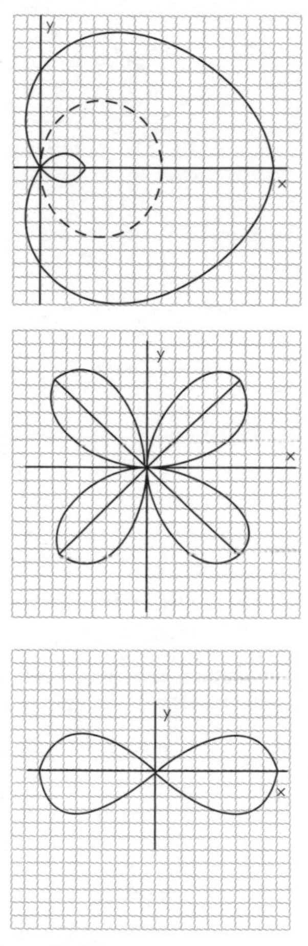

a) Etienne Pascal (Blaise's dad) produced the lemicon, which is a kidney shape generated by a circle, so that any point on the circle is always equidistant between two points on the lemicon. Its formula is $(x^2 + y^2 + ax)^2 = b^2 (x^2 + y^2)$

b) Guido Grandi (1671–1742), an Italian monk, produced his Rose of Grandi or four-leaved clover shape with the formula $(x^2 + y^2)^3 = 4a^2x^2y^2$.

c) In 1694, Jacob Bernoulli produced his lemniscate which is Latin for ribbon. He constructed it while exploring hyperbolas. Its formula is $(x^2 + y^2)^2 = a^2 (x^2 - y^2)$.

Figure 16.26

It can easily be drawn with a simple three-arm linkage. Take two rods and make a third rod 1.414 or √2 times that length. Use this longer rod to mark two fixed points. Now connect the shorter rods to these two points with the longer rod between them to complete the linkage. With a pencil through a small hole in the very centre of the longer rod, you can trace out Bernoulli's lemniscate. It comes as a surprise that John Wallis had suggested this shape as the symbol for Infinity some 40 years earlier. Bernouilli must have seen what an apt choice it was, as this ribbon-like swirl appears to go round and round forever.

An Idea by Fermat as an Extension of Logarithms

Fermat explored the basic idea of logarithms and came up with three beautiful propositions. Here is a double sequence: first the natural numbers called *exponents*, then below that the natural doubling numbers all reduced by 1 (or unity), which Fermat called *radicals*. In other words, the numbers in the second row are powers of 2 minus 1. So below 2 we have $2^2 - 1 = 3$, and below 3 we have $2^3 - 1 = 7$.

1	2	3	4	5	6	7	8	9	10	11	12	13
1	3	7	15	31	63	127	255	511	1,023	2,047	4,095	8,191

1. In all cases, if the upper number is a compound number (not prime), the one below it will also be compound.
2. If the upper number is a prime number, its radical minus 1 is a multiple of its double.
3. If the upper number is a prime, then the lower number is divisible only by multiples of that number plus 1. So for 11: 2,047 = 23 x 89, which is (2 x 11 + 1) x (8 x 11 + 1).

Projecting Maths ideas into Three Dimensions

With encouragement from 16-year-old Gerard Desargues, Blaise Pascal imagined a cone with its peak as a light source beaming downwards. He found that if he sliced the cone anywhere to form a circle or an ellipse and drew within it a hexagon with no two sides parallel, then the extended opposite sides of the hexagon would all meet at three points that were co-linear or exactly in line (see Figure 16.27).

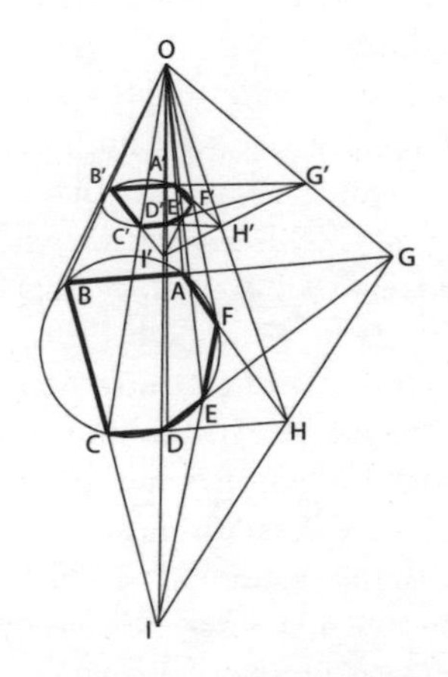

Figure 16.27

He also saw that any other cross section of the cone, at any angle, with a projection of the same hexagon inscribed within it, would have exactly the same properties, and the co-linear points and the lines they formed would also always all be in line with the apex of the cone.

Though the ellipse and the circle are at different angles within the cone, every point and line extension is also in line with the cones apex at O.

Today we can imagine that though distorted, the images of a projected film would have the same picture detail, no matter where or at what angle you placed the cinema screen to receive them. But in Pascal's day film projection was still 300 years into the future.

In fact, as the most exciting new revelations in mathematics of those times were the co-ordinate geometry of Descartes and Fermat, after Pascal the development of projective geometry fell out of fashion for a long time.

Chapter 13: A Mathematician With Gravitas

Billiard Ball Collisions

The art of playing snooker or billiards is purely mathematical. Newton's Cradle demonstrates that all the impetus of a swinging ball when it strikes another ball (or balls) directly will transfer to the ball that's been struck. If that ball is touching other balls in a line, then all the energy will pass through the sandwiched balls to be absorbed by the ball at the free end of the line.

If a ball strikes another at an angle of 45 degrees, then half its impetus will be transferred to the object ball, and the two balls will continue forwards 90 degrees apart. Both are deflected by 45 degrees, and with equally halved momentum.

However, billiards and snooker are played on a stretched baize cloth, and a skilled player can impart various spinning actions on the ball struck. Striking the ball near its top edge imparts forwards spin or 'run'. When it hits another ball, only part of its motive force is transferred to that ball, while it runs forwards as well instead of stopping (as it would in the Cradle).

Striking the ball lower than its centre point gives it backspin. On striking another ball, the original ball continues spinning backwards, and as soon as it has re-established its grip on the cloth after the collision, it moves backwards. Skilled players can make the ball spin back the entire length of a 12ft table. But backspin can be applied more subtly, so that its force is reduced by the time it reaches the object ball some distance away, and on the point of collision, the cue ball stops dead. The skills professional players can display continue to make snooker one of the most watchable and calming games seen on television today – and it is all applied mathematics.

The Calculus

There are two sorts of calculus and it was Wallis (see Chapter 12) who first saw that they were really two sides of the same coin.

First comes the problem of finding areas under a curved line. This is done by dividing the area into rectangles of regular width, calculating the integral area of each and adding them all to get the area of the whole.

So it is known as **integral calculus**. If we imagine a fence made of equal-width, square-ended planks, the area of the planks is the height of all of them combined, times the width of just one of them. The thinner the planks, the closer will their combined area be to the area under the curve. In fact, at the point when the planks becomes a straight line of zero width, the combined length of all of the lines will give you the entire area under the curve. So calculus is all about achieving solutions as close as you can get to the real answer.

Now if we intend the top of our fence to be a curve, then the total area of the planks will not include the almost triangular shapes at the top of each plank. It was Pierre de Fermat who first showed how most of this extra area could be calculated.

If you draw a tangent line through the point where a plank meets the curve, and create a triangle then using the angle ABD and trigonometry, or by simply multiplying AD x BD/2 you can calculate this triangular area, and once again by using thinner planks and smaller triangles, a much closer figure for the total area of the fence can be calculated. So calculus is all about achieving a closer and closer approximation.

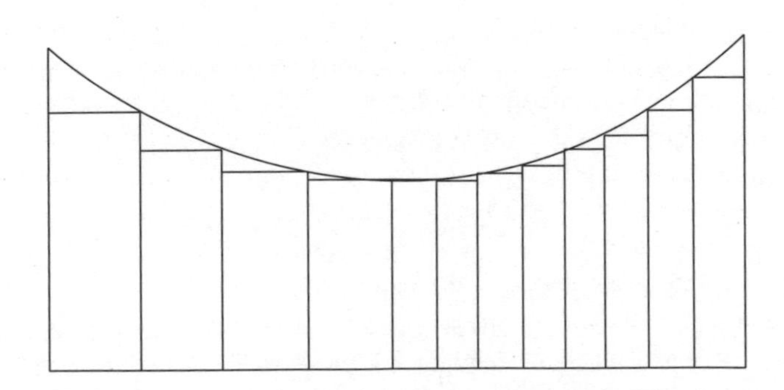

Figure 16.28

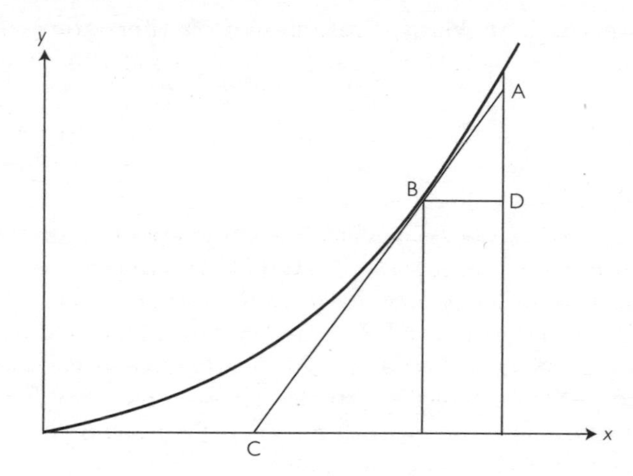

Figure 16.29

Newton had seen that Kepler had solved some problems in integral calculus already with his exploration of the paths of planets sweeping out equal areas in equal times as they orbit the Sun.

But he also had Galileo's explanations of a cannonballs flight, and that was to help him solve problems in **differential calculus**, or the way the curved path of the cannonball or the curved orbit of the Moon around the Earth changes over time.

Calculus required complex calculation, but when at last early computers arrived on the scene they were immediately used to help gunners find their targets more accurately. But soon they were number crunching vast calculations to plot the curved path of rocket to first orbit the Earth, then reach and orbit the Moon and finally to deliver two men to its surface on 20 June 1969, and bring them back safe and sound four days later.

This mathematics was extremely complex because of the three-body problem in that the rocket is affected firstly by the Earth's gravity, which it must overcome to escape into space, and then the Moon's smaller but significant gravity as the rocket gets closer. But on top of that there is the Sun's gravitational pull acting on the rocket, the Earth and the Moon at all times.

Calculus is used today to solve thousands of problems in architecture, like calculating the stresses and strains on a bridge from traffic and the natural elements, and in engineering by designing the complex body shapes of modern cars, and aircraft wings that are reliable and perform perfectly as expected under all conditions, and have flaps that change their geometry for complex take-off and landing situations.

Chapter 15: The Many Tentacles of Mathematics

Newton Explains His Discovery of Universal Gravitation

Now, finally, just over 20 years ago, I set up a Historic Character Agency, with Isaac Newton as our first educational lecturer. David Hall joined us shortly afterwards, and still presents Newton today. At a photoshoot at the Royal Society Library, my first 'Newton' posed with a copy of the *Principia*. There was some alarm when I asked if we could have a quill pen and ink, as we had spotted an error and wished to correct it.

In trying to comprehend Newton's true feelings after he discovered gravity, and having seen his simple drawing of the Earth and planetary orbits (Figure 13.1), I wondered how he might have described it poetically in the language of his day. Over time, I came up with this poem:

Ahha, my deep felicitation, I am Isaac Newton, man of celebration,
Dubbed a Knight of this fair nation, due to my great cerebrillation.
Oft times while seeking clarification of problems cloaked in obfuscation
I would climb a hill or upthrustation, unless prevented by precipitation.
One day I aimed my concentration at the problem of flight or trajectation,
A stone thus, in latteration, would fall to Earth without hesitation,
A curve its immediate deviation ... Whereas a stone thrown with elation,
With the utmost acceleration, with super-human ejaculation, might?
Fly the Earth in parallelation, o'er the boarders of this nation,
Across the oceans aquallation, around the Earth's great globulation,
And like the Moon in its gyration, describe a perfect circulation,
arriving with un-expectation, at my rear-side elevation. Ouch.
Eureka, what a revelation, me thinks I've discovered Gravitation,
For stone and Moon in their gyration, desire a true delineation,
To go straight on without deviation, so why go round in circulation?
It must be the Earth's weight and Gravitation.
The Earth's great weight is the persuasion, that holds us down to our frustration,
And the Moon's acceleration equals the pull of gravitation,
causing it without cessation, to turn in constant revolation.
Further more, every particle in creation is attracted to every other's station,
And that attraction is in relation to their duel mass combination,
and also in inverse relation to the squares of their separation.
These laws of universal Gravitation, the results of my amalgamation,
Are due cause for celebration, and not with sombre meditation,
But with elation, affectation and highly mirthful levitation! Haha!

(Sadly Newton was only known to have laughed once in his entire life...)

Bibliography

Books that have helped me to build my knowledge of all things mathematical over the years.

Aaboe, A., 1998. *Episodes from the Early History of Mathematics*. American Association of Maths.

Asimov, I., 1982. *Asimov's Biographical Encyclopaedia of Science and Technology*. Doubleday.

Bergamini, D., 1960. *Life Science Library: Mathematics*. Time Life.

Bernice, D. D., 1978. *Mathematics, Ideas and Applications*. Academic Press.

Bernstein, W., *Masters of the Word*, 2013. Grove Press.

Bronowski, J., 1973. *The Ascent of Man*. BBC.

Crane, N., 2002. *Mercator: The Man who Mapped the Planet*. Phoenix.

Cohen, I. B., 2005. *The Triumph of Numbers*. W. W. Norton & Co.

Courant R., and Robbins. H., 1941. *What is Mathematics*? Oxford University Press.

Covo Torres, J., 1999, *Maya Calendar*. Dante.

Critchlow, K., 1976. *Islamic Patterns*. Thames and Hudson.

Daintith, J., and Nelson, R. D., 1989. *The Penguin Dictionary of Mathematics*. Penguin.

Davis, P. J., and Hersh, R., 1980. The *Mathematical Experience*. Bïrkhauser.

Fanelli, G., and Fanelli, M., 2004. *Brunelleschi's Cupola*. Mandragora.

Fauvel, J., and Gray, J., 1988. *The History of Mathematics: A Reader*. Macmillan.

Ferguson, K., 1988. *Tycho and Kepler*. Headline.

Gaukroger, S., 1995. *Descartes: An Intellectual Biography*. Oxford University Press.

Gayling, A. C., 2005. *The Life of Descartes*. Simon & Schuster.

Gilbert, J., 1971. *Charting the Vast Pacific*. Aldus.

Gillings, R. J., 1972. *Mathematics in the Time of the Pharaohs*. The MIT.

Hogben, L., 1968. *Mathematics for the Million*. W. W. Norton & Co.

Hollingdale S., 1989. *Makers of Mathematics*. Penguin.

Jacobs, H. R., 1970. *Elementary Algebra*. Freeman.

Jacobs, H. R., 1970. *Mathematics, a Human Endeavour*. W.H.Freeman & Co Ltd.

Joseph, G. G, 1991. *The Crest of the Peacock*. I. B. Tauris.

Jost, E., and Maor, E., 2014. *Beautiful Geometry*. Princeton/Oxford University Press.

Lindmayer, A., 1990. *Algorithmic Beauty of Plants*. Springer-Verlag.

Mankiewicz, R., 2000. *The Story of Mathematics.* Casselle and Co.
McWhirter, N. (ed), 1985. *Guinness Book of Answers.* Guinness.
Mitchell, J. (ed), 1980. *The Joy of Knowledge Encyclopaedia.* Guild Publishing.
Needham, J.,1981. *Shorter Science and Civilisation in China,* vols 1 and 2. Cambridge University Press.
Netz, R., and Noel, W., 2007. *The Archimedes Codex.* Weidenfield and Nicolson.
Newman, J. R., 1956. *The World of Mathematics.* George Allen and Unwin.
Northrop, E. P, 1944. *Riddles in Mathematics.* Pelican.
Pacioli, L., and da Vinci, L., c.1509. *De Divina Proportione [On the Divine Proportion].* Unknown.
Pedoe, D., 1958. *The Gentle Art of Mathematics.* Pelican.
Reston Jnr, J., 1994. *Galileo, A Life.* Cassell.
Rouse Ball, W. W., 1908. *A Short Account of the History of Mathematics.* Dover.
Sawyer, W. W., 1955. *A Prelude to Mathematics.* Pelican.
Sawyer, W. W., 1943. *Mathematician's Delight.* Pelican.
Strandh, S., 1979. *Machines: An Illustrated History.* Mitchell Beazley.
Struik, D. J., 1948. *A Concise History of Mathematics.* Dover.
Thomas, I. (translator),1993. *Greek Mathematical Works, Volume II: Aristarchus to Pappus.* Loeb Classics.
Thompson, D. W., 1942. *On Growth and Form.* Dover.
Vincent, R., 1999. *Geometry of the Golden Section.* Chalagam.
Weinberg, S., 2015. *To Explain the World: The Discovery of Modern Science.* Penguin Random House.
Wells, D., 1991. *Curious and Interesting Geometry.* Penguin.
Wells, D., 1986. *Curious and Interesting Numbers.* Penguin.

And of course, the works of Martin Gardner (Pelican), who started it all for me while I was still a teenager.

Image Credits

Figures 1.5, 1.8, 1.9, 2.2, 2.4, 2.10a, 3.1, 3.2, 3.4, 3.10, 4.1, 4.4, 4.5, 4.6, 4.9, 5.1, 5.2, 5.3, 5.6, 5.7, 5.13, 6.3, 6.4, 6.9, 6.10, 6.11, 6.12, 6.13, 7.2, 7.12, 7.14, 8.1, 9.1, 9.4, 9.5, 9.6, 9.7, 10.12, 12.7, 15.2, 15.11, 16.1, 16.2 © Marc Dando.

Fig 1.1: Leonello Calvetti/Science Photo Library/Getty Images. Fig 1.3: Marcello Bertinetti/National Geographic Magazines/Getty Images. Fig 2.13: William A. Casselman (http://www.math.ubc.ca/~cass/Euclid/ybc/ybc.html). Fig 2.14: Creative Commons Attribution-Share Alike 3.0/pbroks13. Fig 6.7 Fouad A. Saad/Shutterstock. Fig 6.14: Morphart Creation/Shutterstock. Fig 6.17: Hein Nouwes/Shutterstock. Fig 7.3: Science History Images/Alamy. Fig 8.3: James Gordon/Creative Commons Attribution 2.0 Generic. Fig 9.2: Kean Collection/Getty Images. Fig 10.1: Alinari Archives/Getty Images. Fig 10.3: Universal History Archive/Getty Images. Fig 10.4: Art Collection 2/Alamy. Fig 10.5: Rob Atkins/Getty Images. Fig 10.6: Universal Images Group/Getty Images. Fig 11.8: Hulton Archive/Getty Images. Fig 11.11: Print Collector/Hulton Archive/Getty Images. Fig 11.17: Alinari Archives/Getty Images. Fig 11.18: De Agostini Picture Library/Getty Images. Fig 12.2: Royal Astronomical Society/Science Photo Library. Fig 12.6 Universal Images Group/Getty Images. Fig 12.14: Alexander V Evstafyev/Shutterstock. Fig 12.15: Print Collector/Hulton Archive/Getty Images. Fig 13.1: Science History Images/Alamy. Fig 13.4: Science & Society Picture Library/Getty Images. Fig 13.5: Dotted Yeti/ Shutterstock. Fig 14.1: Science and Society Picture Library/Getty Images. Fig 14.2: Paul Fearn/Alamy. Fig 15.1: Universal Images Group/Getty Images. Fig 15.6: Johnny Ball. Fig 16.21: Art Collection 2/Alamy. Fig 16.22a: DEA/Veneranda Biblioteca Ambrosiana/Getty images. Fig 16.22b: Science History Images/Alamy Stock Photo

Colour section photo credits (t = top, b = bottom, l = left, r = right, c = centre)

Part 1: P. 1: DEA Picture Library/Getty Images (t); Murat Hajdarhodzic/Shutterstock (c); Print Collector/Hulton Archive/Getty Images (b). P. 2: Print Collector/Hulton Archive/Getty Images (tl); Hulton Archive/Getty Images (tr); Old Paper Studios/ Alamy (b). P. 3: Bertl123/Shutterstock (tl); Heritage Images/Hulton Archive/Getty Images (tr); Jodie Coston/Getty Images (cl); ccarlstead (cr); Marzolino/Shutterstock (b). P. 4: VW Pics/Universal Images Group/Getty Images (t); INTERFOTO/Alamy (c); DEA/G. DAGLI ORTI/De Agostini Editorial/Getty Images (bl); Public domain (br). P. 5: Yang Yin/Getty Images (t); Manfred Gottschalk/Getty Images (ct); Angelo Hornak/Corbis Historical/Getty Images (cb); Gregory James Van Raalte/Shutterstock (b). P. 6: Art Collection 3/Alamy (t); Religious Images/Universal Images Group/Getty Images (c); Monty Fresco/Stringer/Hulton Archive/Getty Images (bl); Anton Balazh/ Shutterstock (br). P. 7: Radicalrobbo/Creative Commons Attribution 3.0 Unported (tl); Art Collection 2/Alamy (tr); Fratelli Alinari IDEA S.P.A./Corbis Historical/Getty Images (c); Culture Club/Hulton Archive/Getty Images (b). P. 8: Public domain (t); Eric Vandeville/Gamma-Rapho/Getty Images (bl); Historical Picture Archive/Corbis Historical/Getty Images (br).

Part 2: P. 1:Historical Picture Archive/Corbis Historical/Getty Images (t, b). P. 2: Science & Society Picture Library/Getty Images (tl); Photo Scala, Florence/Scala Archives (tr); De Agostini Picture Library/Getty Images (b). P. 3: Bettmann/Getty Images (t); Print Collector/Hulton Archive/Getty Images (bl); DAVID PARKER/ SCIENCE PHOTO LIBRARY/Getty Images (br). P. 4: Colour plan: Will Tubby/ Architectural drawings: Stephanie Wilcox/Water bowls sketches: Giles Raynor (t); Science History Images/Alamy (c); Science & Society Picture Library/Getty Images (b). P. 5: Elena Korchenko/Alamy (t); Niabot/Creative Commons Attribution 3.0 Unported (c); Public domain (b). P. 6: Alfred Pasieka/Getty Images (t, c, bl, br). P. 7: Tracy Packer Photography/Getty Images (t); Ken Welsh/Getty Images (c); Tuul & Bruno Morandi/Corbis Documentary/Getty Images (b). P. 8: Lluís Vinagre – World Photography/Getty Images (t); White Images, Scala, Florence/Scala Archives (c); Paolo Gallo/Shutterstock (b).

Index

A'h-mose 14–15
abacus (abaci) 191–2
Abu al Wafa 221
acoustics 150–1
Adelard of Bath 89, 226–7
aeolipiles 160–1
Agrippa, Marcus 147–8, 167, 354
air compression 162
Al-Karaji 221, 225
Al-Khwarizmi 216–18, 221, 227, 232, 284, 300, 455–6
Alberti, Leon Battista 271–2, 278
Alcmaeon 55
Alcuin of York 215
Alexander the Great 68, 69, 81–2, 118
Alexander VI, Pope 250
Alexandria 68, 81–3, 91, 118–20
  Apollonius 133–6
  Aristarchus 122–3
  Claudius Ptolemy 170–1
  Cleopatra 143–6, 147
  Ctesibius 124–8
  Diophantus 171–4
  Eratosthenes 128–33
  Hipparchus 136–41
  Hypatia 176–7
  Menelaus 168–70
  Pappus 174–7
  Poseidonius 142
  Seleucus 123
  Strato 121–2
  Theon 176
Alexis 191
Alfonso of Castile 304
algebra 74, 171, 173–4
  Al-Khwarizmi 217–18, 227, 455–6
  Boole, George 410–11
  Descartes, René 334, 335–7
  Diophantus 450–2
  Viète, François 299–300
Alhazen 223–4, 237, 238, 265
amicable numbers 219
Amthor, A. 102
Analytical Society 408
Anaximander 38–9
Antikythera Device 127
Apollonius 81, 106, 133–6, 141, 219, 280, 299, 446–8
Appel, Kenneth 419
aqueducts 152–4
Aquinas, Thomas 237
Arabia 89, 170, 171, 201, 202, 213
  Abu al Wafa 221
  Al-Karaji 221, 225
  Al-Khwarizmi 216–18
  Arabic numerals 202, 265
  Europe 226–32
  ibn Al Haythen 223–4, 227
  Ibn-Sina (Avicenna) 221–3
  Islam 211–13, 215–16, 225–6, 457
  Khayyam, Omar 224–5
  Thabit ibn Qurra 219–20
Archimedes 11, 28, 92, 122, 124, 128, 153, 158, 195, 200, 219–20, 298, 300, 343, 349, 364, 372, 419, 425, 434
  Archimedes' Claw 104
  area under a parabola 111–12
  Cattle of the Sun 101–2, 440
  conics 106–17
  Eureka 92–4, 148
  extended number system 439–40
  lever to move the Earth 439
  Method 112, 443–5
  Method in three dimensions 446
  Mighty Five 94–8, 159
  military inventions 102–5
  Ostomachion 100–1
  Sand Reckoner 101
  search for Pi 105–6
  Semi-Platonic Solids 102, 441–3
  Thabit ibn Qurra 456
Archytas 59–60, 79, 97
Aristarchus 122–3, 142, 171, 291–3
Aristocles *see* Plato
Aristotle 68–9, 70, 81, 110, 122, 197–8, 236–8, 317–18, 323, 326, 328, 343, 346
arithmetic 56, 217, 196, 453
armillary spheres 138, 305
Aryabhata 201–2
astrolabes 136–7, 219, 243
astronomy 56, 404–6
  ancient China 182–4
  ancient India 197
  Aristarchus 122–3
  Brahe, Tycho 303–7
  Copernicus, Nicolaus 290–4
  Eudoxus 79–80
  Galileo Galilei 315–31
  Gauss, Karl Friedrich 423–4
  Hipparchus 123, 136–41, 142, 170–1, 290, 292
  Huygens, Christiaan 356
  Kepler, Johannes 306–15
  Ptolemaic system 170–1, 288, 290, 291, 293, 307, 323, 325
  Regiomontanus 289–90
  Sumerians 30–2
  Tychonian system 306
Atlantis 67–8
atoms 61–4, 392–9
Augustus 147, 186, 192
Avicenna 221–3

Avogadro, Amedeo 387–8
Aztecs 206, 209, 255

Babbage, Charles 408–9
Babylonians 33–5, 36, 38, 42, 53, 136, 171, 180, 183, 192, 198
Bacon, Roger 237–9, 240, 262, 290, 317
Bailey, Nick 402–3
Balboa, Vasco Núñez de 252, 255
Ball, W. W. Rouse 100–1
barometers 343–4, 344–5
Barrow, Isaac 89, 370
Baudhayana 197, 453–4
Becquerel, Antoine 392
Bede, the Venerable 213–15
Bellini, Giovanni 278
Bernoulli family 379
  Bernoulli, Daniel 379, 412
  Bernoulli, Jakob 379, 380–1, 382, 464
  Bernoulli, Johann 379, 380
Big Bang 237
Big Dipper 30
Bilancetta, La 317
billiards 465–6
binary number system 17–18
binomials 221
Black Death 240–2
Bligh, William 263
Bode, Johann Elert 44, 405, 406
Bohr, Niels 395–6
Bolyai, János 422
Bombelli, Rafael 289
Bonaparte, Napoleon 19
Boole, George 69, 410–11
Borelli, Giovanni 358
Boyle, Robert 358–9, 360, 373, 379
brachistochrone problem 380
Brahe, Tycho 303–7, 310, 314, 357
Brahmagupta 202, 218
Brand, Hennig 359
bricks 197
Briggs, Henry 301–2
Brownian motion 397
Brunelleschi, Filippo 266–70, 271, 354
Bruno, Giordano 314–15, 327
Buddhism 69, 200
Byron, Ada Augusta 89–90, 409

Cabot, John and Sebastian 250–1
Caesar, Julius 143–5, 148
Caesarion 144–5, 146
calculus 308, 334, 352, 378, 425, 466–8
Caligula 153
Callinicus 212
Cantor, Georg 425–6
Cardan, Jerome 281, 286–8
carpenter's squares 77–8
Cartesian coordinates 333, 349, 463
Cassini, Giovanni 356
catapults 103–4
catastrophe theory 430–1
catenaries 354, 430
Cavalieri, Bonaventura 302, 342–3, 367, 378
Cavendish, Henry 384–5, 386, 390, 401, 407–8
Cavendish, William 384
Caxton, William 271
cells 362
Central America 203–6, 209, 255
  Mayans 31, 203, 205–10, 254
Chadwick, James 393
Chang Heng 195
chaos, modelling 431–2
Charlemagne 179, 214–15
Charles Martel 212, 214
chemistry 359, 373
chessboard puzzle 220
Chhin Shih Huang Ti 181–2
China 270–1
China, ancient 32, 48, 180–7
  abacus 191–2
  counting rods 187–91
  magic squares 192–4
  Pi 195–6
Chiu Chang 188–9, 196, 453
chords 87–9
Chou Pei Suan Ching 184
Christianity 165, 174, 177–8, 211, 226
Chuquet, Nicolas 272–3, 281–4, 301
church design 150
Cicero 117, 142
circles 32–3, 87–9, 107
  great circles 169
  squaring the circle 79
clay tablets 28–30, 33–5
Clement IV, Pope 238
Cleopatra 143–6, 147
clepsydra 126–7
clocks 263, 270, 290–1, 324
  pendulums 317, 340, 356–7, 360
Columbus, Christopher 246, 248–50, 251, 254–5, 289
comets 289, 304
communications technology 17–18, 401, 434
compasses 184–5, 322–3
  mesolabe compass 71–4, 109, 128, 438
computers 402, 408–9
cones 106–7, 112–14
  ellipses 107–8, 280
  parabolas 108–11
  surface areas 115–17
Confucius 69, 180–1
Constantine the Great 177
Constantinople 177–9, 212
Cook, James 155, 263
Copernicus 123, 262, 290–4, 307–8, 323, 325, 327, 328, 339, 392
Cortes, Hernando 255
cosines 138, 303
counting rods 187–91
Courant, Richard 431
cow-catchers 408

Crookes, William 391–2
Crusades 232–4
crystallography 308
Ctesibius 124–8, 148, 151, 160
Curcavi, Pierre 349
Curie, Marie and Pierre 392–3
Cuzco, Peru 203–4
cycloids 329–30, 349, 353, 379–80
cylinders 112–17

Da Gama, Vasco 247
Da Vinci, Leonardo 49, 148, 270, 273–8, 300, 319
  Vitruvian Man 274–7
Dalton, John 360, 386–7, 398
Dark Ages 213, 216, 232, 242, 243, 265
Darwin, Charles 61, 437
Davy, Humphry 154
day length 137, 207
De La Roche, Estienne 282
decimals 52–4, 72, 77, 187, 190, 202, 217, 298, 425
degaussing 185, 424
Del Monte, Francesco 320–1
Delamain, Richard 303
Delian problem 76–7
della Francesca, Piero *Flagellation* 267
Delle Colombe, Ludovico 326
Democritus 61–2, 63, 66, 147, 359
Desargues, Gerard 345, 464
Descartes, René 219, 332–3, 339–42, 345, 355, 356, 357, 377, 381, 383, 465
  Cartesian coordinates 333, 349, 463
  Geometry, The 74, 333–5, 353, 367
  rainbows 337–8, 369
  turning maths into pictures 335–7
Diaz, Bartholomew 247
differential calculus 467
differential gears 185–6
Diocles 110
Diocletian 177
Diogenes 191
Dionysius Exiquus 214
Diophantus 171–4, 188, 217, 289, 349–50, 450–2
dioptra 163–4
dispensers 161–2
displacement 92–4
dodecahedrons 48
Doppler, Christian 397
double-entry bookkeeping 190, 272
Drake, Francis 262–3
Dudeney, Ernest Henry 8, 90–1
Dürer, Albrecht 278–81

Earth 57–9, 122–3, 136, 137
  circumference 357
  curvature of the Earth 39
  density 407–8
  diameter 437–8
  distance around 130–1, 171, 254
  distance from the Moon 139–41
  distance from the Sun 132, 141, 404
  Ptolemaic system 170–1, 288, 290, 291, 293, 307, 323, 325
  surface area 115–16
  Tychonian system 306
earthquakes 360
Eddington, Sir Arthur Stanley 400
Edison, Thomas 391
egg timers 112–13
Egypt, ancient 14–15, 31, 42, 48, 57, 79, 105, 118, 126, 158, 184, 195
  Cleopatra 143–6, 147
  Moscow Papyrus 25–8, 115, 435, 436
  Pyramids 19, 19–25
  Rhind Mathematical Papyrus 14–17, 25, 27, 79, 435
Eilmer of Malmesbury 239
Einstein, Albert 10, 89, 92, 378, 379, 391, 394–5, 434
  Brownian motion 397
  E = mc2 398–9
  general relativity 399–400, 401
  photoelectric effect 395–7
  special relativity 397–8, 399, 401
electricity 391–2
  electronic circuits 410–11
electrons 63, 392, 393–4, 396, 399
elements 384, 388–9, 390, 396–7
ellipses 107–8, 280, 374
Empedocles 55, 60–1
energy, conservation of 398
equals sign 288
equations 284–8
Equator 58, 126, 132, 251, 261
equinoxes 169, 183, 214, 292, 366
Eratosthenes 101, 128–33, 139, 140–1, 142, 144, 148, 155, 171, 254, 292, 357, 440, 445
Escher, Maurits Cornelis 429
Euclid 83, 169–70, 177, 344, 421–2, 427, 443, 449
  chords 87–9
  Elements 83–4, 219, 227
  examples of Euclid's maths 84–6
  influence 89–90
  parallel postulate 90–1
  proof of Pythagoras' Theorem 86–7
Eudoxus 79–81, 84, 107, 112
Euler, Leonhard 19, 23, 219, 299, 303, 412–13
  Euler line 415
  Euler's identity 414–15
  Euler's number 413
  factorials 414
  topology 415–17
exploration 243–4, 254–6
  Columbus, Christopher 248–50
  mapping the known world 256–64
  New World 250–4
  Zheng He 244–7

factorials 414
falling objects 297–8, 317–20
Fantis, Segismundus 281
Faraday, Michael 154, 386, 391
Fatou, Pierre 433
Feng Hsiang Shih 183–4
Fermat, Pierre 346–7, 349–52, 367, 378, 464–5, 467
  Fermat Points 350–1
  Fermat's Last Theorem 349–50, 423, 433
Feynman, Richard 414–15
Fibonacci series 229–30, 349, 403
  Lucas numbers 231–2
finger-counting 31–3
Fiore, Antonio Maria 285–6
fire engines 126, 160
fireworks 237–8
Fitzgerald, Edward 224–5
Flamsteed, John 358
Fletcher, Christian 263–4
Florence cathedral 267–9
fortifications 154–5
Four Elements 61–2, 66
fractals 432–3
fractions 158–9

Galileo Galilei 10, 46, 69, 89, 103, 122, 238, 262, 278, 281, 294, 298, 312, 315, 332, 339, 343, 357, 358, 359, 400, 467
  cannonball trajectories 321–2
  Church 326–9
  cycloids 329–30, 379–80
  falling objects 317–20
  fortress design 320–1
  infinity 330–1, 425
  La Bilancetta 317
  military compass 322–3
  pendulums 315–17, 462
  pulse measurement 317
  telescope 323–6, 361, 390, 392
  thermometer 323
gases 359, 385–7, 390
Gassendi, Pierre 298
Gaudi, Antoni 355, 429–30
Gauss, Karl Friedrich 185, 364, 419–25, 434
Gay-Lussac, Joseph Louis 387
Geissler, Heinrich 391
Gelon II of Sicily 101, 103
general relativity 399–400, 401
Genghis Khan 186
geometry 56, 74, 90
  Apollonius 133–6
  Dürer, Albrecht 280–1
  Euler, Leonhard 415
  Islamic art 225–6
  projective geometry 464–5
  spherical geometry 169–70
  tessellation and tiling 457–8
Gerard of Cremona 227
Gilbert, William 239, 249, 261–3, 312, 317
Gillings, Richard 24
Giotto 265–6
Giovanni de Dondi 290
Gödel, Kurt 426
Golden Ratio (Golden Mean) 51–3, 69, 150, 229–30, 231–2
  human measurements 276–7
Golenishchev, Vladimir 25
Gothic architecture 234–6
GPS (global positioning satellites) 401
Grandi, Guido 463
graphs 335–7, 463–4
gravity 220, 278, 317, 364–7, 376, 377, 399–400, 468
Great Bear 30
Greece, ancient 192, 195, 196, 201
  Alexandria 81–3, 118–46
  Archimedes 92–117
  Aristotle 68–9
  Democritus 61–2
  Euclid 83–91
  Eudoxus 79–81
  Hippocrates of Chios 69–79
  oared ships 120–1
  philosophers 59–61
  Plato 65–8
  Pythagoras 41–56
  shape of Earth 57–9
  Socrates 64–5
  Thales 36–41
  Zeno 62–4
Greek fire 212
Greenwich Meridian 138
Gregorian calendar 209, 213, 225, 305, 353
Gregory, James 299
Grosseteste, Robert 236–8
Gunter, Edmund 302
Gutenberg, Johann 271, 289

Hadrian 166
Hagia Sophia, Constantinople 178–9
Haken, Wolfgang 419
Halley, Edmond 358, 373–4, 377, 404
Halley's Comet 289
Hannibal 102–3
Hanno 58–9, 243–4, 247
Harrison, John 263, 324
Heath, Thomas Little 89
height, measuring 37–8
Heisenberg, Werner 63–4
heliotropes 423–4
Helmholtz, Hermann 398
Henry the Navigator 243–4, 247
Herodotus 24, 57–8, 373
Heron 152, 157–65, 224, 269, 296
  fractions 158–9
  Heron's formula 157–8, 448–9
  inventions 160–4
Herschel, Caroline Lucretia 405

Herschel, John 408
Herschel, Sir William 338, 390, 405, 408
Hertz, Heinrich 391
hexagons 174–5
hexahedrons 47
Hiero II of Syracuse 92–3, 98
Hinduism 52, 180, 202, 216, 217, 232, 265, 281, 282
Hipparchus 123, 142, 170–1, 290, 292
  distance from Earth to the Moon 139–41
  inventions 136–8
  trigonometry 138–9, 447
Hippasus 53
Hippocrates of Chios 69–71, 83, 84, 110, 128, 333
  duplicating the cube 76–7
  lunes 74–6
  mesolabe compass 71–2
  square roots 73–4
  squaring the circle 79
  trisecting an angle 77–8
Ho Thu 193–4
Homer *Iliad* 99, 169
Hooke, Robert 354–5, 358–9, 359–61, 378, 379, 383
  Micrographia 361–2
  Newton, Isaac 371–3, 377, 383
Horace 192
horizon, calculating distance to 39–40
Hunayn ibn Ishaq 219
Huygens, Christiaan 317, 340, 349, 354, 355–7, 358, 360, 377, 378, 380
hydrostatics 93–4, 114, 298
Hypatia 176–7
hyperbolas 107, 135–6, 447–8

ibn Al Haythen (Alhazen) 223–4, 227
Ibn-Sina 221–3, 224
icosahedrons 48
Incas 255
India, ancient 190, 196–7
  origin of zero in India 200–2
  Pythagorean Triples 197–9
  Vedic maths 199–200, 454
Industrial Revolution 74, 334
infinity 330–1, 425–6
infrared 338, 390, 395
integral calculus 466
invariance theorems 40
inverse square law 310, 373
irrationals 50–1, 52–4
Islam 211–13, 215–16
  direction of Mecca 219
  Islamic art 225–6, 457

Jacquard, Joseph Marie 409
Jainism 200
Janssen, Jules 390
Jerusalem 165–6, 232
Jesus Christ 165, 214, 287
Johnny Ball Reveals All 108
joint, universal 360
Jones, William 19, 23, 412
Joule, James 398
Julia, Gaston 433
Julian calendar 144
Justinian 66, 178

Karaka 218
Kekulé, Friedrich 388
Kepler, Johannes 84, 89, 107, 224, 281, 294, 301, 306–15, 323, 327, 343, 361, 383, 461–2, 467
  Newton, Isaac 364, 372, 373, 375–6, 382–3
Khayyam, Omar 224–5
knots 425
  quipus 204–5, 454–5
Kublai Khan 186

Lagrange, Joseph-Louis 383
latitude and longitude 79, 131, 139, 171, 218
  mean speed theorem 240, 333
Lavoisier, Antoine 398
Leibniz, Gottfried Wilhelm 74, 174, 308, 311, 334, 353, 377–80, 382, 408
Leonardo of Pisa 218, 221, 281, 402
  Fibonacci series 229–30, 349, 403
  Liber Abaci 227–8
Let the Force be with You 160
Lever of Mahomet 431
levers 95–6, 438–9
Leyden Jar 384–5
L'Hopital, Guillaume François 379, 380
libraries 178
  Alexandria 68, 82–3, 128, 144, 176
  Bait al-Hikma 216
light 310
  general relativity 400
  Snell's law 338–9
  special relativity 397–8, 399
  speed of light 357–8
Lincoln, Abraham 89
Lincoln Cathedral, England 23
Lippershey, Hans 323
Liu Hui 195, 452–3
Lo Shu 193
Lobachevsky, Nikolai Ivanovich 421–2
Lockyer, Norman 390
logarithmic spirals 380–1
logarithms 283, 300–2, 342, 412, 464
longitude 263, 324, 357
  latitude and longitude 79, 131, 139, 171, 218
Lucas numbers 231–2
Lucas, Edouard 231
Lucretius 147
lunar eclipses 122, 157, 209
lunes 74–6
Luther, Martin 259, 283, 293

Magellan, Ferdinand 253–4
magnetic north 424
magnetism 239, 261–2
Malpighi, Marcello 361
Mandelbrot, Benoit 381, 432–3
maps 131, 139, 171
  four-colour map theorem 418–9
  globes 259
  Mercator, Gerardus 256–61, 263
  network theory 417
Marc Antony 144, 145, 147
Marcellus 103, 105, 117
Masefield, John 119
mathematics 7–10, 384, 433–4
matter 384–8
Maxwell, James Clerk 384, 391, 397, 398
Mayans 31, 203, 205–10, 254
mean speed theorem 240–2
measurements 274–7
mechanics 91, 239, 343
Mela, Pomponius 155–6
Menaechmus 81, 107, 112, 333
Mendeleev, Dmitri 388–9, 390
Menelaus 168–70, 421, 449–50
Menger, Carl 428
Menzies, Gavin 246–7
Mercator, Gerardus 256–61, 263
Mersenne, Martin 340, 349
mesolabe compass 71–4, 109, 128, 438
method of exhaustion 80
Michelangelo 315
Michelson, Albert 397
microscopes 237, 328, 361–2
military compasses 322–3
minus sign 288
mirrors 104, 108, 110, 158
Möbius, August Ferdinand 419, 426–7
modular mathematics 423
Mohammad 211
Moon 31, 91, 122–3, 127, 137, 324, 376
  distance from Earth 139–41
  tides 214, 327, 366–7
Morley, Edward 397
Morley, Frank 78
Moscow Papyrus 25–8, 115, 435, 436
multiplication 13–14, 282–3
  multiplication sign 303
  Rhind Mathematical Papyrus 14–17, 25, 27, 79, 435
  Vedic maths 199–200, 454
music 42–4, 56, 59–60, 315
  Music of the Spheres 44, 66, 310

Napier, John 283, 300–2, 314, 342, 413
Nazca lines, Peru 204
Necho 57
Needham, Joseph 180, 246
negative numbers 188–9, 190, 283
Nero 153
network theory 417
neutrons 393
New World 250–4
Newton, Isaac 9, 46, 74, 89, 92, 123, 174, 214, 224, 308, 310, 327, 331, 334, 358, 362, 363, 380, 382, 384, 399, 404, 408, 412, 419, 434, 467
  fluid motion 376
  fluxions 353, 367–8, 377–8
  gravity 220, 278, 311, 364–5, 376, 377, 468
  laws of motion 374–5
  measurements 367–8
  Newton's Cradle 375, 465–6
  Principia 374–8
  rainbows 338, 368–9, 390
  religion 372, 373
  solitary years 363–5
  speed of sound 376–7
  telescopes 369–72
  tides 366–7
Nightingale, Florence 410
Nollet, Jean Antoine 385
North Pole 20, 116, 257, 262
number systems 17–18, 206–10, 217
  Archimedes 439–40
  very large numbers 283–4
number theory 44–7, 173

Ockham's razor 239–40
octahedrons 48
Octavian 145–6, 147
odometers 151–2
Oldenburg, Henry 371–2
Olmecs 205
optics 91, 224, 237, 238
Oresme, Nicole 240, 333
organs 125–6, 160
Osiander, Andreas 293, 294, 307–8
Oughtred, William 302–3, 352, 353

Pacioli, Luca 190, 272–3, 276–7, 278, 281, 285, 334, 459–60
  Pacioli fonts 459
Pantheon, Rome 167–8, 267
paper 186, 270–1
Pappus 174–7, 343, 349
parabolas 107, 108–12, 320–2, 352
  parabolic flight 103–4
parallax 139–40
parallel postulate 90–1
parallelogram of forces 296–7
Parmenides 55
Pascal, Blaise 146, 344–9
  air pressure 345–6
  calculating device 345, 378, 409
  Pascal's theorem 344–5
  Pascal's triangle 221, 225, 287, 348–9
  pistons 346
  probability theory 346–7
  projective geometry 464–5
Pascal, Etienne 463
Peacock, George 408

pendulums 315–17, 356–7, 462
pentagons 48
Peregrinus 239
perfect numbers 340
periodic table 388–9, 390
perpetual motion 269, 297
perspective 167, 265–7, 271–8, 278–81
Peru 203–5
Pharos Lighthouse 82, 118–19
Phi 51, 52, 229–30, 231–2
Phidias 51
Philip II of Macedon 68, 81
Philolaus 55, 59, 66
Phoenicians 57–9
phosphorus 359
phyllotaxis 402, 403
Pi 19, 23, 24–5, 180, 412
 ancient China 195–6
 Archimedes 105–6
Piazzi, Giuseppe 405–6, 423
Picard, Jean 357, 358, 374
pie charts 410
piezoelectricity 392
pistons 346
Pizarro, Francisco 255
Planck, Max 395
planes 47–8
planets 44, 56, 66, 137, 139, 405–6
 Copernicus, Nicolaus 290–4
 Galileo Galilei 324–5
 Kepler, Johannes 308–10, 314, 461–2
 planetary orbits 80, 122, 136, 140–1, 310–12, 365, 366, 433
 Ptolemaic system 170–1, 288, 290, 291, 293, 307, 323, 325
 Rudolphine Tables 307, 314
 Tychonian system 306
plant growth 401–2
Plateau, Joseph 351
Plato 48, 65–8, 79–80, 84, 148–9, 328
Plimpton, G. A. 33
Pliny the Elder 156–7, 213
Plough 30
plumb bobs 196
plus sign 288
Plutarch 98
Pole Star 30, 38, 136, 183
Polo, Marco 186, 249
polynomials 221
Pompey 142, 143
Pont-du-Gard aqueduct, Nimes 154
Poseidonius 131–2, 142, 171, 248, 254
Priestley, Joseph 385
prime numbers 128, 423
 Mersenne primes 340
 sieve of Eratosthenes 129
printing presses 270–1
prisms 368–9
probability theory 346–7
protons 393, 397
Ptolemy I of Egypt 82–3, 118, 122
Ptolemy II of Egypt 82, 118
Ptolemy III of Egypt 99, 119
Ptolemy XIII of Egypt 143
Ptolemy XIV of Egypt 144
Ptolemy XV of Egypt (Caesarion) 146
Ptolemy, Claudius 39, 169, 170–1, 218, 219, 224, 239, 248, 254, 260, 289
 Ptolemaic system 170–1, 288, 290, 291, 293, 307, 323, 325
pulleys 97–8
pulse measurement 317
Pyramids, Egypt 15, 19–28, 82
Pythagoras 34, 35, 41–4, 65, 69–70, 84, 171, 197–8, 315
 from planes to solid shapes 47–8
 irrationals 50–1, 52–4
 number theory 44–7
 Order of Pythagoreans 44
 Pythagoras' legacy 55–6
Pythagoras' Theorem 48–50, 74–5, 83–4, 115, 123, 157
 ancient China 180, 184
 ancient India 197
 Euclid's proof 86–7
 proving 48–50
 Viète, François 299
Pythagorean Triples 33–5, 197–9, 349–50, 436–7

QED 27, 49, 87, 114
quadrature problem 79
quadrilaterals 176
quipus 204–5, 454–5

radioactivity 392–4
rainbows 337–8, 368–9, 390
ramps 94–5
Ramsay, William 390
Recorde, Robert 288
Regiomontanus 289–90
relativity 397–8, 399–400, 401
Renaissance 148, 155, 168, 170, 174, 213, 218, 232, 234, 237, 242, 265
 astronomy 289–94
 Brunelleschi's cupola 266–70
 Dürer, Albrecht 278–81
 equations 284–8
 mathematical art 265–6
 mathematics 281–4, 288–9
 Oriental influence 270–1
 perspective 271–8
Rheticus (Von Lauchen, Georg) 292–3
Rhind Mathematical Papyrus 14–16, 25, 27, 79, 435
rhumb lines 258–9, 263
Ritter, Johann Wilhelm 386, 390
Robbins, Herbert 431
Robert of Chester 227
Roemer, Ole 357–8, 374
Roentgen, Wilhelm 392
Rome, ancient 102–5, 141–2, 201, 213
 Agrippa 147–8
 architecture 166–8

Cleopatra 143–6, 147
Heron 157–65
Pliny the Elder 156–7
religious turmoil 165–6
Vitruvius 148–55
rowing boats 120–1
Royal Society 353, 354, 357, 358–62, 371–2, 373, 377, 378, 383, 408
Rudolphine Tables 307, 314
rule of three 71
Rumford, Benjamin 398
Rutherford, Ernest 393, 395

Sagredo, Gianfrancesco 324, 328
Salviati, Filippo 328
sash windows 360–1
satellites 401
Schrödinger, Erwin 396
science 384
atoms 392–9
Beauty of Mathematics Garden 402–3
Einstein, Albert 394–401
electricity 391–2
elements 388–9
global positioning satellites 401
matter 384–8
plant growth 401–2
spectroscopy 390
Scipione del Ferro 285, 287
screws 97
sea navigation 57–9
Seleucus 123
Shakespeare, William 315, 414, 430
Shannon, Claude 410
shapes 47–8
ships 94
oared ships 120–1
Syracosia 98–9, 119–20
Shu Shu Chin Chang 190
sidereal day 137
Sierpinski, Waclaw 428
Siliotti, Alberto *The Pyramids* 23–4
sines 138, 221, 303
Singmaster, David 113
skeleton keys 408
slide rules 302–3, 460–1
Sloane, Hans 403
slot machines 161
Snell, Willebrord 338–9
snooker 7, 411, 465–6
snowflakes 308, 427–8
Socrates 64–7, 69, 70, 315
solar eclipses 31, 36, 209, 400
solids 47–8
Five Platonic Solids 48, 66, 84
Semi-Platonic Solids 102, 441–3
stellated solids 313–14
Somerville, Mary 409
Sosigenes 144, 213
Sostratus 118
special relativity 397–8, 399
spectroscopy 390
speedometers 408
spheres 112–17, 425
spherical geometry 169–70
Spinoza, Baruch 356
springs 360
square numbers 45–7
square roots 53, 54, 73–4, 200
squares 149–50
magic squares 192–4, 222
staircases 149
statistical analysis 411–12
steam engines 160–1
Stevin, Simon 295, 297–8
Stifel, Michael 283, 301
Strabo 144, 155
Strato 121–2
Su Sung 186
Sulla 68
Sumerians 28, 36, 97, 136, 191, 196
astronomy 30–2
clay tablets 28–30
finger-counting 31–3
Sun 30–1, 39, 58, 80, 122–3, 126, 132, 136, 137, 237, 262, 264, 312, 400
distance from Earth 132, 141, 404
mirrors 104, 110
Ptolemaic system 170–1, 288, 290, 291, 293, 307, 323, 325
sunspots 326
tides 214, 327, 366–7
Tychonian system 306
Sun Tsu 189–90
sundials 38, 113
Sushruta 200
Swan, Joseph 391
Swanee Whistle 125
Swift, Jonathan 361
symmetry 427–8

Taccola, Mariano di Jacopo 269–7
tangents 138, 303, 352
Taoism 181
Tartaglia 281, 285–7, 321, 460
tautochrone problem 380
telescopes 136, 140, 209, 237, 307, 312, 356, 361
Galileo Galilei 323–6
Newton, Isaac 369–72
tessellation 457–8
tetrahedral sequences 47
tetrahedrons 47
Thabit ibn Qurra 194, 219–20, 227, 456
Thales 36–40, 42, 57, 65, 69–70, 92, 119, 195, 213, 262
Thales' Theorem 40–1, 73, 87–8
Theaetetus 77, 84
theodolites 164
Theodoric of Frieberg 238
Theodorus 53–4, 65
Theodosius 176
Theon 176

Theon of Smyrna 54
thermometers 323
Think of a Number 8–9
Thom, René 430–1
Thomson, J. J. 392, 395
tides 214, 327, 366–7
tiling 457–8
Tirthaji, Bharati Krishna 199–200
Titius, Johann Daniel 404–5
Toltecs 206
topology 415–17, 424–5
  four-colour map theorem 418–9
  mathematics of small holes 428–9
  rubber-sheet geometry 430–1
  topological problems 417–18
  topological summation 419
Torricelli, Evangelista 323, 30, 343–4, 350, 381
Toscanelli 248, 249
Trajan 166–7
trajectories 103–4, 321–2
triangles 149–50, 157–8
  magic triangle 222–3
  Pascal's triangle 221, 225, 287, 348–9
  right-angled triangles 33–5, 38, 41, 48–50
  spherical triangles 221
triangular numbers 45, 47
triangulation 256, 260
trigonometry 35, 38, 138–9, 219
  trigonometric ratios 447
Tsu Chhung Chih 195
Tsu Keng Chih 195, 298
Tunstall, Cuthbert 288
Turing, Alan 402

Uccello, Paolo 271

Van Leeuwenhoek, Anton 361
Van Musschenbroek, Pieter 384–5
Van Roomen, Adriaan 299
Vedics 197–200, 454
Venn, John 411–12
Vespucci, Amerigo 251–2, 253
Viète, François 298–300, 352
Vitruvius 66, 148–55, 269, 272, 321
  acoustics 150–1
  aqueducts 152–4
  city walls 154–5
  De Architectura 148
  odometers 151–2
  staircases 149
  triangles 149–50
Viviani, Vincenzo 358
Voltaire, François 364
Von Guericke, Otto 344, 359
Von Koch, Helge 427
Von Lauchen, Georg (Rheticus) 292–3
Von Lindemann, Carl Louis 79

Wallace, Alfred Russell 437–8
Wallis, John 352–3, 367, 375, 464, 466
watches 360
water clocks 126–7, 166, 186–7
Watt, James 161
wedges 95
wheels 96–7
Whittington, Dick 241
Wickins, John 368
Wiles, Andrew 350, 423
William of Ockham 239–40
William of Rubruck 238
Wimbledon tennis courts 428
Winton Beauty of Mathematics Garden, Chelsea Flower Show 402–3
Wren, Christopher 303, 353–5, 358, 362, 373, 375, 376, 377

X-rays 392
Xenophon 64

Yang Hui 193, 194
year length 30–1, 132, 183, 207–8, 305
Yehlu Chhuh-Tsai 186
Yu 192–3

Zapotecs 206
Zeno 55, 62–4, 81
zero 80–1, 190, 200–2, 207, 217, 227, 232
Zheng He 244–7